Bernhard Hofko

Hot Mix Asphalt under Cyclic Compressive Loading

Bernhard Hofko

Hot Mix Asphalt under Cyclic Compressive Loading

Towards an Enhanced Characterization of Hot Mix Asphalt under Cyclic Compressive Loading

Südwestdeutscher Verlag für Hochschulschriften

Impressum/Imprint (nur für Deutschland/only for Germany)
Bibliografische Information der Deutschen Nationalbibliothek: Die Deutsche Nationalbibliothek verzeichnet diese Publikation in der Deutschen Nationalbibliografie; detaillierte bibliografische Daten sind im Internet über http://dnb.d-nb.de abrufbar.
Alle in diesem Buch genannten Marken und Produktnamen unterliegen warenzeichen-, marken- oder patentrechtlichem Schutz bzw. sind Warenzeichen oder eingetragene Warenzeichen der jeweiligen Inhaber. Die Wiedergabe von Marken, Produktnamen, Gebrauchsnamen, Handelsnamen, Warenbezeichnungen u.s.w. in diesem Werk berechtigt auch ohne besondere Kennzeichnung nicht zu der Annahme, dass solche Namen im Sinne der Warenzeichen- und Markenschutzgesetzgebung als frei zu betrachten wären und daher von jedermann benutzt werden dürften.

Coverbild: www.ingimage.com

Verlag: Südwestdeutscher Verlag für Hochschulschriften GmbH & Co. KG
Heinrich-Böcking-Str. 6-8, 66121 Saarbrücken, Deutschland
Telefon +49 681 37 20 271-1, Telefax +49 681 37 20 271-0
Email: info@svh-verlag.de

Approved by: Wien, TU, Diss., 2011

Herstellung in Deutschland (siehe letzte Seite)
ISBN: 978-3-8381-3298-3

Imprint (only for USA, GB)
Bibliographic information published by the Deutsche Nationalbibliothek: The Deutsche Nationalbibliothek lists this publication in the Deutsche Nationalbibliografie; detailed bibliographic data are available in the Internet at http://dnb.d-nb.de.
Any brand names and product names mentioned in this book are subject to trademark, brand or patent protection and are trademarks or registered trademarks of their respective holders. The use of brand names, product names, common names, trade names, product descriptions etc. even without a particular marking in this works is in no way to be construed to mean that such names may be regarded as unrestricted in respect of trademark and brand protection legislation and could thus be used by anyone.

Cover image: www.ingimage.com

Publisher: Südwestdeutscher Verlag für Hochschulschriften GmbH & Co. KG
Heinrich-Böcking-Str. 6-8, 66121 Saarbrücken, Germany
Phone +49 681 37 20 271-1, Fax +49 681 37 20 271-0
Email: info@svh-verlag.de

Printed in the U.S.A.
Printed in the U.K. by (see last page)
ISBN: 978-3-8381-3298-3

Copyright © 2012 by the author and Südwestdeutscher Verlag für Hochschulschriften GmbH & Co. KG and licensors
All rights reserved. Saarbrücken 2012

Not that the incredulous person doesn't believe in anything. It's just that he doesn't believe in everything. Or he believes in one thing at a time. He believes a second thing only if it somehow follows from the first thing. He is nearsighted and methodical, avoiding wide horizons.

If two things don't fit, but you believe both of them, thinking that somewhere, hidden, there must be a third thing that connects them, that's credulity.

U. Eco: Foucault's Pendulum

Foreword

This publication summarizes scientific work that aims for a better understanding of the material behavior of hot mix asphalt under cyclic dynamic compressive loading. The results and findings were derived during my work as a research assistant at the Research Center of Road Engineering, Institute of Transportation at the Vienna University of Technology. It represents a personal milestone of a four year research journey through the field of material science of bituminous bound materials. The journey was exciting and full of new experience and findings; it was at times exhausting, even long-winded with detours and bumps along the road. But most importantly, it was lined by people standing by to accompany me, discussing critical intersections or doubting seemingly obvious and all too easy paths to prevent me from getting lost along the way. They cheered me up when it was needed most, loved and hated test equipment, the materials, data and numbers all together with me.

So, it is not my obligation but a privilege to thank these supporters and companions along my way. First and foremost, I would like to express my deep gratitude to *Prof. Ronald Blab*, head of the research center, who did not only support me to put this doctoral thesis on a sound basis by numerous discussions and suggestions; he is the one who encouraged me always to go the extra mile. Not only did he support me scientifically, he never loses sight of the fact that satisfying results can only be created in an environment that takes care of each individual, not only as a scientist but also as a human being. I would also like to especially thank my second reviewer *Prof. Manfred Partl* for his inspiring comments to this study and our in-depth discussions, which were an inestimable contribution in rounding off this thesis.

Each member of our institute's laboratory, which has been my scientific home for most of my time at the institute, has been a great support over the years. To mention them in alphabetical order, thank you so much to *Georgi Chankov, Markus Gmeiner, Lisi Hauser, Lukas Kirchmaier, Thomas Riedmayer, David Valentin* and *Michael Wagner.* You all contributed to my thesis in one way or the other – thank you for that. Special thanks go to *Matthias Mader* and *Charly Kappl,* who supported me greatly in fighting off obstacles with test equipment and running parts of the test program. This saved me valuable time to concentrate on finding clues and connections from a vast volume of test data.

Let's not forget the rest of our institute, *Ingrid Amesberger-Redl* and *Karin Beck* – you saved (maybe not my life but certainly) my sanity when bureaucracy and administrative duties took me by surprise, which happened more than once. Thank you for that! And who had to listen to my doubts and troubles in my scientific work for uncountable hours – special thanks for that to you, *Josef Füssl;* also for finishing your doctoral thesis a little sooner than I did and thus showing me that it can indeed be done. I enjoy our after-work meetings very much.

Last but not least – family and friends. Thank you for being my tightly woven net where I could relax, let off steam, feel at home knowing that you will be there no matter what. My parents, *Herta & Otto Hofko*, who supported me throughout my education with unlimited trust and who gave me support whenever I needed it, as well as the freedom to become who I am today. The same is true for my sister *Silvia* and my brother *Michael*. Thank you so much for being there!

Why are the people I care most about last in this listing? – probably because I feel that words cannot express my gratitude in an adequate way. Thank you, *Angelika*, for your loving support, your never ending patience and your truly unique view of things, thank you for your music. Life is wonderful with you! Thank you, *Olivia,* for making our life even more vivid and bright. Just looking at you is the most wonderful gift one could possibly receive in life.

Therefore, and for uncountable other reasons, I would like to dedicate this doctoral thesis to my wife, *Angelika Treml-Hofko*, and my daughter, *Olivia Smilla Hofko*.

ABSTRACT

This doctoral thesis is aimed towards an advanced characterization of the material behavior of hot mix asphalt (HMA) under cyclic dynamic compressive loading. The triaxial cyclic compression test (TCCT), which today is mainly employed according to (EN 12697-25, 2005) for the assessment of the resistance to permanent deformation, is thoroughly reviewed. Four main objectives were developed to enhance the output of this test type.

The first part of the thesis introduces an alternative assessment method for the characterization of the resistance to permanent deformation for standard TCCTs carried out according to (EN 12697-25, 2005). Presently, the creep curve is approximated by a linear function within its quasi-linear part. The slope of the linear, the linear creep rate f_c, is the benchmark parameter. However, the standard does not define the starting point and range of the quasi-linear part of the creep curve. To overcome this shortcoming, it is shown that during the TCCT each specimen reaches a state of constant viscoelastic parameters after a certain number of load cycles. An unambiguous definition for this point in the test is given. Within this state of constant viscoelastic parameters, the creep curve is linear in the log/lin scale and can thus be approximated by a logarithmic function with high fit quality. An excellent correlation was found between the linear creep rate f_c and the logarithmic creep rate b. Furthermore, a significant benefit can be achieved when not only the axial but also the radial deformation of the specimen is recorded in the TCCT: In this case not only the total axial strain $\varepsilon_{ax,tot}$ can be derived, but also its volumetric $\varepsilon_{ax,vol}$ and deviatoric part $\varepsilon_{ax,dev}$ can be analyzed. Both components reflect different rutting mechanisms in the field and thus may be used for an advanced characterization of the permanent deformation behavior of HMA giving valuable information for mix design optimization.

A second research question is how HMA reacts to cyclic compression tests (CCT) in terms of viscoelasticity. To study not only the reaction in axial direction but also perpendicular to the vertical axis in radial direction, strain gauges are attached directly on the HMA specimens around their circumference to ensure high-quality readings. A comprehensive test program is run at mixes with paving grade and polymer-modified bitumen with temperature and frequency sweep. It is analyzed whether the test setup (i.e. whether specimens are glued to the load plates or not) impacts the measured viscoelastic material reaction. While the material viscosity in terms of phase lags is not affected by the setup, the dynamic modulus is at a lower level (-10% to -20%) when unglued specimens are used, as it is the case for CCTs. The impact of time, temperature, binder content and air void content on the viscoelastic material behavior is described. In all CCTs carried out for this study the phase lag between axial loading and axial deformation is smaller than the phase lag between axial loading and radial deformation. It is found after a comprehensive investigation that this difference is material-inherent. As a direct result, the dynamic Poisson's Ratio and its elastic and viscous component are introduced, as well as the dynamic shear modulus and its components. From the results of the test program mentioned above it can be stated that the presently used values for the Poisson's ratio of HMA of 0.30 to 0.35 are only valid for intermediate temperatures and low frequencies or elevated temperatures and high frequencies. The dynamic Poisson's Ratio is affected significantly by test temperature and frequency, as well as by mix design parameters like the binder content and the void content. The dynamic shear modulus $|G^*|$ exhibits a higher temperature and frequency sensitivity than the dynamic modulus $|E^*|$. $|G^*|$ decreases more strongly with increasing temperature and increases more strongly with increasing frequency than $|E^*|$.

On the basis of these findings an analytical model is developed, which predicts the viscoelastic material behavior of HMA from viscoelastic binder characteristics and volumetric characteristics of the mix. The *B(inder)-A(sphalt mix) Model* consists of nine parameters, three of them are dependent on the binder type, the other six parameters are linked to the volumetric characteristics of the mix. The model describes the macroscopic, viscoelastic material parameters of the mix under compression, from dynamic modulus, phase lags and dynamic shear modulus to dynamic Poisson's Ratio. The model parameters may not have a direct physical relation but the *B-A Model* has the vast advantage to predict the HMA behavior over a large range of frequencies/temperatures. Furthermore a conclusive link between the viscoelastic behavior of HMA and the permanent deformation behavior (linear and logarithmic creep rate) is established. In connection with the *B-A Model* this represents an additional powerful tool to predict rutting resistance of mixes.

A further objective of the thesis is to simulate the state of stress within a pavement structure under traffic by the TCCT in a more realistic way. In the standard TCCT the confining pressure is held constant, which does not reflect the situation in a pavement where a cyclic confining pressure is assumed from dynamic wheel loads. Thus, an enhanced TCCT is introduced with cyclic confining pressure, which also takes into account the viscoelastic material reaction to loading by considering the radial phase lag for the confining pressure. Results from standard and enhanced TCCTs are compared and it can be stated that a permanent deformation benefit can be activated, when the material is tested by taking into account its viscoelastic properties. From the results it is assumed that this benefit is mainly caused by a reduced deviatoric strain component. Mixes perform significantly better in terms of long-term behavior. The creep rates decrease by 1/6 up to 1/2 when the standard TCCT results are compared to results from enhanced TCCTs. The results can account for a more specific mix design optimization of HMAs taking into consideration the application of an HMA and the boundary conditions in terms of traffic and climate.

Summarizing the findings it can be concluded that cyclic compression tests have an excellent potential not only for the characterization of the resistance to permanent deformation but also to deal with the viscoelastic material behavior of HMA under compressive loading. Since relevant material reaction does not only occur in axial but also in radial direction, the material behavior can be described in an advanced way.

KURZFASSUNG

Ziel dieser Dissertation ist eine umfassende Beschreibung des Materialverhaltens von Asphalt unter zyklisch-dynamischer Druckschwellbelastung. Zu diesem Zweck wird der triaxiale Druckschwellversuch (TCCT), der zur Zeit hauptsächlich für die Ermittlung des Verformungswiderstandes bei hohen Temperaturen nach (EN 12697-25, 2005) eingesetzt wird, auf verschiedene Anwendungsmöglichkeiten hin untersucht. Daraus ergeben sich vier wesentliche Fragestellungen.

Im ersten Teil der Arbeit wird eine alternative Bewertungsmethode für Standard-TCCT nach (EN 12697-25, 2005) entwickelt. Bisher wird die Kriechkurve durch eine lineare Funktion im quasi-linearen Bereich dieser Kurve angenähert. Die Steigung der Linearen, die lineare Kriechrate f_c, wird als Beurteilungskriterium für die Beständigkeit gegen bleibende Verformungen herangezogen. Allerdings wird der quasi-lineare Teil der Kriechkurve in der Europäischen Prüfnorm nicht näher definiert. Dies führt dazu, dass die ermittelte Kriechrate von der Festlegung des Beginns und des Bereichs des quasi-linearen Kriechens abhängig ist. Bei der Auswertung einer Vielzahl an Standard-TCCTs wurde erkannt, dass die viskoelastische Materialreaktion nach einer bestimmten Anzahl an Lastwechseln konstant bleibt. Ein neues Verfahren wird vorgestellt, mit dem dieser Zeitpunkt ermittelt werden kann. Nach dieser Einschwingphase stellt die Kriechkurve in der log/lin Darstellung eine Gerade dar, sie kann also durch eine logarithmische Funktion beschrieben werden. Zwischen der linearen Kriechrate f_c und der logarithmischen Kriechrate b besteht eine ausgezeichnete Korrelation. Weiters wird gezeigt, dass Prüfergebnisse deutlich genauer in Hinblick auf das Verformungsverhalten bewertet werden können, wenn nicht nur die axiale, sondern auch die radiale Verformung während des Versuchs aufgezeichnet wird. Dadurch kann nicht nur die gesamte Axialdehnung $\varepsilon_{ax,tot}$, sondern auch deren volumetrischer $\varepsilon_{ax,vol}$ und deviatorischer Anteil $\varepsilon_{ax,dev}$ ermittelt werden. Beide Komponenten stehen für die unterschiedlichen Ursachen von Spurrinnenbildung in der Praxis. Die gewonnen Daten müssen also für eine zukünftige Optimierung der Mischgutzusammensetzung abgestimmt auf den Einsatz des Mischguts herangezogen werden.

Eine zweite Fragestellung betrifft das viskoelastische Materialverhalten von Asphalt in zyklischen Druckschwellversuchen (CCT). Da nicht nur die Reaktion in axialer Richtung, sondern auch in der radialen Ebene untersucht werden soll, werden zunächst Dehnungsmessstreifen als adäquates System zur Messung der Umfangsdehnung eingeführt. Diese werden direkt auf die Probekörper rund um die Mantelfläche am Umfang appliziert. Ein umfangreiches Prüfprogramm an Mischgut mit Oxidationsbitumen und polymermodifiziertem Bitumen mit Variation der Prüftemperatur und -frequenz wird präsentiert. In einem ersten Schritt wird untersucht, ob die Materialreaktion davon abhängt, ob der Probekörper fest mit dem Laststempel verbunden ist oder nicht. Dabei konnte festgestellt werden, dass die Materialphasenwinkel unabhängig von der Konfiguration sind. Probekörper, die nicht fest mit den Lastplatten verbunden sind, reagieren jedoch weicher, der dynamische Modul ist um 10% bis 20% niedriger. In einem weiteren

Schritt werden Einflüsse von Temperatur, Frequenz, Bitumen- und Hohlraumgehalt auf das viskoelastische Materialverhalten analysiert. Alle Prüfergebnisse zeigen, dass der axiale Phasenwinkel zwischen axialer Belastung und axialer Reaktion kleiner ist, als der radiale Phasenwinkel zwischen axialer Belastung und radialer Reaktion. Nach einer eingehenden Untersuchung auf mögliche Störquellen für dieses Phänomen wird schließlich die These aufgestellt, dass der Unterschied in den Phasenwinkeln materialinhärent ist. Direkt damit zusammenhängend wird die dynamische Querdehnzahl $|v^*|$ mit ihrem elastischen und viskosen Anteil und der dynamische Schubmodul $|G^*|$ mit seinen Anteilen eingeführt. Es wird gezeigt, dass die derzeit häufig verwendeten Annahmen für die Querdehnzahl von Asphalt von 0.30 bis 0.35 nur dann zutreffen, wenn das Material im mittleren Temperaturbereich bei niedrigen Frequenzen oder bei hohen Temperaturen und hohen Frequenzen belastet wird. Die dynamische Querdehnzahl wird wesentlich von Temperatur, Frequenz und Mischgutzusammensetzung beeinflusst. Anders als der dynamische Modul $|E^*|$ weist der dynamische Schubmodul $|G^*|$ eine stärkere Temperatur- und Frequenzabhängigkeit auf, sinkt also stärker mit steigender Temperatur und steigt schneller mit steigender Frequenz an.

Auf der Basis der vorgenannten Ergebnisse wird im Weiteren ein analytisches Modell entwickelt, dass das viskoelastische Materialverhalten von Asphalt aus dem viskoelastischen Verhalten des verwendeten Bitumens und volumetrischen Kenngrößen des Mischguts vorherzusagen vermag. Das *B(inder)-A(sphalt) Modell* beinhaltet neun Parameter von denen drei von der Bitumenart und die anderen sechs von volumetrischen Kenngrößen des Mischguts abhängen. Das Modell beschreibt alle wichtigen viskoelastischen Kennwerte des Asphalts, vom dynamischen Modul und den Phasenwinkeln über den dynamischen Schermodul bis zur dynamischen Querdehnzahl. Zwar zeigen die Modellparameter keinen direkten physikalischen Zusammenhang, aber das Modell hat den großen Vorteil, das Materialverhalten über eine große Spanne an Temperaturen und Frequenzen vorhersagen zu können. Zusätzlich wird ein eindeutiger Zusammenhang zwischen dem viskoelastischen Materialverhalten ($|G^*|$) und dem Verformungsverhalten (lineare bzw. logarithmische Kriechrate) hergestellt, wodurch das *B-A Modell* auch den Widerstand gegen bleibende Verformungen zu beschreiben vermag.

Eine vierte Fragestellung betrifft die Simulation der Verkehrslastspannungen in einem Oberbau durch TCCTs. Im Standard-TCCT wird der radiale Seitendruck konstant gehalten. Das entspricht jedoch nicht der Annahme, dass in einem Oberbau durch dynamische Radlasten auch viskos verzögerte, dynamische Radialspannungen auftreten. Daher wird ein modifizierter TCCT eingeführt, bei dem der Probekörper nicht nur axial, sondern auch radial zyklisch-dynamisch belastet wird. Dieser modifizierte TCCT berücksichtigt auch die viskoelastische Materialreaktion, in dem der radiale Phasenwinkel für die zyklisch-dynamische Seitendrucksteuerung berücksichtigt wird. Beim Vergleich von Standard- und modifizierten TCCTs wird sichtbar, dass sich der Verformungswiderstand von Asphalt deutlich erhöht, wenn die viskoelastische Materialreaktion berücksichtigt wird. The resultierenden Kriechraten verringern sich um 1/6 bis 1/2, wenn der modifizierte TCCT eingesetzt wird. Die Ergebnisse können dazu beitragen die Mischgutoptimierung noch besser auf einen spezifischen Anwendungsfall unter Berücksichtigung von Randbedingungen wie Verkehr und Klima anzupassen.

Die Ergebnisse der Arbeit bestätigen insgesamt, dass zyklische Druckschwellversuche großes Potenzial haben, einerseits die Verformungsbeständigkeit von Asphaltmischgut detailliert zu beschreiben, andererseits jedoch auch das viskoelastische Materialverhalten im Druckbereich ausgezeichnet abzubilden. Durch Erfassung der Materialreaktion in axialer und radialer Richtung kann das Verhalten umfassend beschrieben werden.

Contents

I	**INTRODUCTION**	**15**
I.1	Background	16
I.2	Motivation and Objectives	22
I.3	Research Approach	24
II	**TEST EQUIPMENT AND DATA EVALUATION**	**29**
II.1	Test Machine	29
II.1.1	Control and Record Unit	30
II.1.2	Software for Test Procedures	32
II.1.3	Triaxial Cell	32
II.1.4	New Pneumatic Device for the Application of Cyclic Dynamic Confining Pressure	32
II.2	Displacement Sensors	34
II.2.1	Approaches to Measure Radial Deformation on Cylindrical Specimens	35
II.2.2	Introduction of Strain Gauges to Obtain Radial Strain	38
II.2.3	Application of Strain Gauges	39
II.2.4	Validation of Measurement by Strain Gauges	41
II.3	Data Evaluation and Analysis	45
II.3.1	Signal Processing	46
II.3.2	Evaluation Software	48
II.3.3	Statistical Analysis	52
II.3.4	Details on the Advanced Approximation Function	53
III	**MATERIALS AND SPECIMEN PREPARATION**	**61**
III.1	Materials	61
III.2	Specimen Preparation	63
IV	**AN IMPROVED ASSESSMENT OF THE RESISTANCE TO PERMANENT DEFORMATION**	**67**
IV.1	Evaluation of Standard TCCTs according to EN 12697-25	67
IV.2	Towards an Alternative Assessment of the Resistance to Permanent Deformation	71
IV.2.1	Test Program	71
IV.2.2	Alternative Assessment of the Resistance to Permanent Deformation in the TCCT	73
IV.2.3	Quasi-3-d Deformation Behavior of HMA in the TCCT	81
IV.3	Conclusions	91
V	**VISCOELASTIC MATERIAL PARAMETERS DERIVED FROM CCTS**	**95**
V.1	Characteristics of Viscoelastic Materials	95
V.1.1	Behavior of Viscoelastic Materials under Cyclic Dynamic Loading	97
V.1.2	Viscoelastic Material Parameters	100
V.1.3	Advanced Viscoelastic Material Parameters	103

V.2	Test Program	106
V.3	Influences of the Connection between Specimen and Load Plates	108
V.3.1	Test Setup	109
V.3.2	Analysis of Signal Data from Force Sensor	111
V.3.3	Analysis of Signal Data from Deformation Sensor	113
V.3.4	Analysis of Material Parameters	117
V.3.5	Conclusions	120
V.4	Linear vs. Nonlinear Viscoelastic Behavior of HMA	121
V.4.2	Conclusions	139
V.5	Viscoelastic Material Parameters Derived from the Compressive Domain	140
V.5.1	Impact of Binder Content	140
V.5.2	Impact of Air Void Content	149
V.5.3	Conclusions	158
V.6	A Study on the Difference between Axial and Radial Phase Lag	159
V.6.1	Influence of Anisotropy of HMA	160
V.6.2	Uniform Radial Deformation of Cylindrical Specimen	165
V.6.3	Impact of the Measuring System	171
V.6.4	Conclusions	178
V.7	Advanced Viscoelastic Material Parameters	178
V.7.1	Dynamic Poisson's Ratio	178
V.7.2	Dynamic Shear Modulus	188
V.7.3	Conclusions	197
VI	**A VISCOELASTIC MODEL OF THE BEHAVIOR OF HMA UNDER COMPRESSIVE LOADING**	**199**
VI.1	Introduction	199
VI.1.1	The Time Temperature Superposition Principle	200
VI.1.2	Master Curve Fitting	202
VI.2	Test Program	207
VI.3	Modeling the Viscoelastic Behavior of Binder and HMA	209
VII	**INTRODUCING AN ENHANCED TCCT WITH CYCLIC DYNAMIC CONFINING PRESSURE**	**247**
VII.1	Approach	247
VII.2	Test Program	248
VII.3	Results from Enhanced TCCTs	248
VII.4	Conclusions and Outlook	267
VIII	**RESUME AND PERSPECTIVE**	**269**
IX	**ABBREVIATIONS AND SYMBOLS**	**277**
X	**BIBLIOGRAPHY**	**283**
XI	**INDEX OF FIGURES AND TABLES**	**289**

A	**ANNEX: THE PRINCIPLE OF STRAIN GAUGES**	**299**
A.1	Basic Information	299
A.2	Principle of Measurement	301
A.3	Characteristics of Strain Gauges	302
A.4	Recording Strain	307
A.5	Calibration of the Wheatstone Bridge Circuit with SGs	309
A.6	Compensation of Disturbance Values	311
B	**ANNEX: MODELING THE VISCOELASTIC BEHAVIOR OF BINDER AND MASTIC**	**313**
B.1	Bitumen → Mastic	313

I INTRODUCTION

The situation in road engineering and construction has changed dramatically since the Marshall method was implemented into the mix design guides and procedures after WW II. Today planners and engineers face an increasing number of heavy goods vehicles (HGVs) traveling the road network. At the same time, limited financial and natural resources increase the need to optimize the efficient use of materials and pavement structures and therefore to construct long lasting pavements. In order to fulfill these demands, all boundary conditions must be taken into consideration, when designing hot mixture asphalt (HMA) on project level. The traffic-situation as well as the climate in the region are important factors when it comes to the optimization of the mix design. From that it seems logical that the traditional Marshall mix design procedure is not capable of meeting the demands of modern pavement engineering since this method does neither take into account the reaction of HMA to low temperatures, nor does it assess the fatigue performance. Improved test methods have been developed to test HMA in the laboratory in a way that simulates traffic and climate conditions occurring in the field as realistically as possible. Today a number of these so called performance-based test procedures for HMA are implemented in harmonized European standards:

− The low-temperature performance and resistance to top-down cracking due to high cooling rates in combination with traffic loading can be tested by the thermal stress restrained specimen test (TSRST) and the uniaxial tension stress test (UTST) at low temperatures. Together with tests to describe relaxation and creep of the HMA the rheological behavior of the material at low temperatures can be obtained. The procedures are set by (EN 12697-46, 2009).

− The stiffness behavior (EN 12697-26, 2004) and fatigue performance (EN 12697-24, 2007) can be described by a 2- and 4-point-bending beam (2PBB, 4PBB) test, the indirect tensile test (ITT), the direct tension (DT) and the direct tension and compression test (DTC). This characterization corresponds to the cracking phenomenon due to fatigue in the field. In addition viscoelastic parameters such as the dynamic modulus $|E^*|$ and the phase lag φ are derived from these test methods.

− The resistance of HMA to permanent deformation at high temperatures (rutting) is characterized by cyclic compression tests (CCT) either uniaxially (UCCT = uniaxial cyclic compression test) or in a triaxial way (TCCT = triaxial cyclic compression test) according to (EN 12697-25, 2005).

All these test methods have been developed in the last three decades and are still subject to optimization in terms of the test setup itself as well as the evaluation and analysis of test data. The Christian Doppler Laboratory (CD-Lab) for Performance-Based Optimization of Flexible Pavements at the Vienna University of Technology, which was established in 2002 for a period of seven years, worked towards the development, optimization and standardization of performance-based test methods on European and national level. Today, Austrian standards permit HMA mix specification according to functional requirements based on the performance concept. Increasing experience in performance based testing reveals that there is still potential for optimizing and enhancing the test procedures to simulate the situation in the field in a more realistic way and to assess key material parameters of HMA. This thesis works towards enhancements in the assessment of the behavior of HMA tested by means of CCT.

I.1 Background

Assessment of the Resistance to Permanent Deformation

A comprehensive pioneering work in the field of TCCT on HMA started in the 1970s at the Belgian Road Research Center (BRRC). Having the main objective to develop a model to describe the permanent deformation behavior of HMA, TCCTs with varying test conditions were carried out to study the impact of different boundary conditions systematically. The derived BRRC-model (Francken, 1977) is still used for a first approximation of the permanent axial strain due to triaxial loading, and the test program was one major source for the development of the European standard for TCCT (EN 12697-25, 2005).

Two German research reports from the 1980s mark another starting point for triaxial cyclic material testing on HMAs. (Jaeger, 1980) tested different HMAs at high temperatures with a sinusoidal axial compressive loading and constant radial confining pressure. Axial as well as radial strain were measured continuously and stored in an analogue way (on paper prints). The permanent axial strain was reconstructed by a 5-parameter model introduced by (Krass, 1971) for creep behavior of HMA. It was shown that the magnitude of axial loading has the most significant impact on the permanent axial strain, followed by temperature and frequency of loading. This is true for those parts of the specimens that are near the loading plates. For the central parts of the specimen temperature has a more dominant influence on the permanent deformation than loading. According to the test results, it was suggested to lower the stress introduced into pavements in the field by trucks by enforcing use of twin tires. This research was followed by (Weiland, 1986). Advanced technologies opened the possibility to digitalize the data, store, process and therefore evaluate and analyze test data in more detail. Amplitudes of stress and strain could be visualized and thus phase lags between those signals could be looked at for the first time. The phase lags incorporate a large scatter due to the (still poor) quality of the digital test data. The results show that there is a difference between the phase angle of axial loading and axial deformation vs. axial loading and radial deformation. They also indicate that temperature has a dominant impact on the magnitude of the phase lag. Influences of the loading frequency could not be analyzed since the tests were all carried out at one frequency only. In addition to a standard asphalt concrete (AC 11) used for surface layers, another mix (AC 22) was tested with a larger maximum aggregate size usually found in binder and base layers. (Weiland, 1986) used an 8-parameter model expanded from the 5-parameter model by Krass. It basically consists of one element for the spontaneous and two elements for the time dependent parts of the deformation.

In the 1990s (von der Decken, 1997) presented an extensive study on TCCTs with cyclic confining pressure on different HMAs and at different temperatures. Interestingly enough, it was found that a constant phase lag between axial loading and radial deformation exists which is independent of mix design and temperature. It was set to 36°. The test data were analyzed by approximating them by means of a sinusoidal function on oscillation level. The creep curve (permanent axial deformation vs. number of load cycles) was described by a function consisting of a constant, a power and an exponential term. Another part of the study deals with correlations between permanent deformation in the field (rutting) and in different test methods. It was found that the best correlation between field and lab performance with regard to deformation resistance can be achieved using the TCCT with cyclic confining pressure.

The TCCT was implemented into the series of harmonized European Standards for testing of HMA in 2004 as EN 12697-25 to assess the resistance to permanent deformation at high temperatures (rutting). The standard test procedure consists of a cyclic dynamic axial loading to simulate a tire passing a pavement structure and a radial confining pressure to consider the confinement of the material within the pavement structure. The axial loading $\sigma_A(t)$ can either be shaped as a sinusoidal function (Figure I-1 a) or a block-impulse (Figure I-1 b). The standard states that the confining pressure σ_C can either be held constant or oscillate dynamically without providing more specific information. The TCCTs carried out for this thesis are loaded by a sinusoidal axial at a constant confinement and cyclic dynamic radial loading, respectively.

However, all types of TCCTs are carried out by loading the specimen in a purely compressive state in axial direction. Therefore the specimen develops accumulated axial strain; part of this deformation is permanent. Thus, the main output of the TCCT is the accumulated, permanent axial strain ε_n vs. the number of load cycles n. Figure I-2 shows a schematic example of the creep curve.

For type testing the test is usually carried out at one temperature (e.g. 50°C), one loading frequency (e.g. 3 Hz) and one state of stress. Most of the laboratories that incorporated the TCCT into their test procedures use constant confining pressure, especially since the test control gets even more complex with cyclic dynamic confining pressure. Furthermore, essential questions about the viscous radial phase lag between axial loading and radial reaction has not been solved satisfactorily yet. Is the radial phase lag different from the axial phase lag? What test and mix parameters influence the radial phase lag and to which extent?

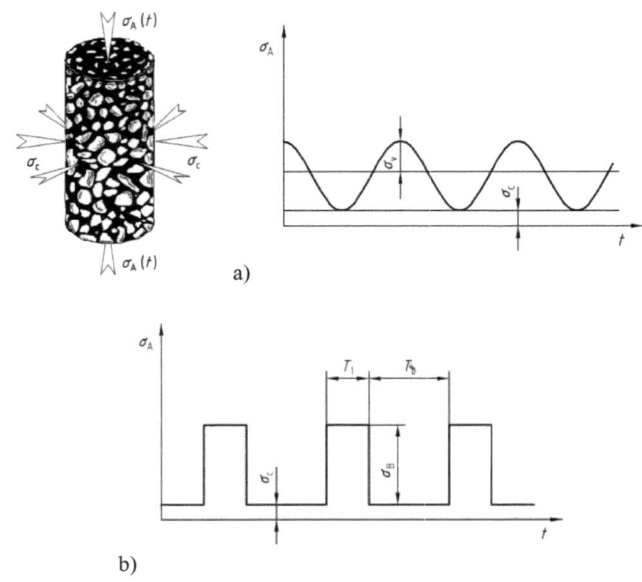

Figure I-1. Loading conditions in the TCCT according to (EN 12697-25, 2005); a) sinusoidal shaped axial loading and b) block-impulses as axial loading, both with constant confining pressure.

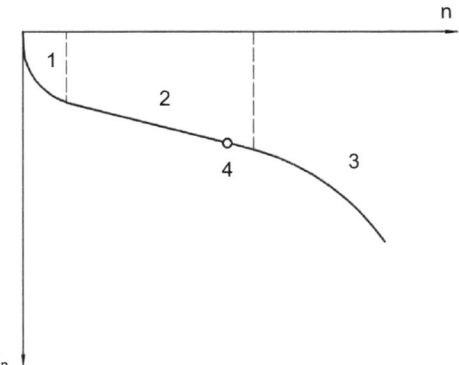

Figure I-2. Standard result of a TCCT according to (EN 12697-25, 2005); accumulated axial strain ε_n vs. number of load cycles n.

A recent, comprehensive study on the assessment of the permanent deformation behavior of HMA with the TCCT can be found in (Kappl, 2007). The dissertation looks at different test procedures and standards to derive deformation resistance parameters of HMA and evaluates the methods from small scale wheel tracking tests to full scale accelerated pavement tests. It gives detailed information about the technical details of necessary hardware to carry out TCCTs according to (EN 12697-25, 2005) and describes the basics of test data evaluation to obtain viscoelastic material parameters as well as parameters for the assessment of the permanent deformation behavior. In addition, the thesis contains an extensive testing program where a number of asphalt mixes with different aggregate and binder types were tested in the TCCT according to (EN 12697-25, 2005). An attempt was made to link the results of TCCTs to various binder parameters or results of simple test methods like the Marshall test. The main findings were that there is a strong correlation between the softening point ring and ball and the cumulated, axial strain after 25,000 load cycles at 50°C. Also, the air void content could be linked to the creep rate resulting from the standard TCCT. The quotient of Marshall stability S and the Marshall flow number F, as well as the flow number F showed significant correlations to the cumulated, axial strain.

The standard procedure for the assessment of the resistance to permanent deformation of HMA in the European standards is given by (EN 12697-25, 2005) for TCCTs. Two methods are provided: Either the quasi-linear part of the creep curve is approximated by a linear function – the slope of the linear function being the benchmark parameter – or the complete creep curve is approximated by a power function and both functional parameters are used to describe the permanent deformation behavior. Problems especially with the first method using a linear function were presented by (Kappl, 2007) since the creep curve has in fact no linear part and the standard does not provide clear regulations on how to determine the quasi-linear part of the creep curve.

In the United States the flow number concept is rather used as a benchmark for the permanent deformation behavior (Witczak, et al., 2002). The flow number is obtained from the point where the secondary creep phase (with quasi-linear creep) ends and the tertiary phase with a progressive creep rate starts.

(Van Dijk, et al., 1975) adapted the energy dissipation criterion used for standard fatigue tests, which are carried out without producing permanent deformation, to testing conditions that aim to provoke permanent strain in the HMA specimen. (Widyatmoko, et al., 1999a) developed a method to incorporate not only the viscous dissipated energy but also dissipated energy due to plastic deformation. By plotting the dissipated energy per cycle vs. the number of load cycles an unambiguous parameter can be derived where the end of the secondary creep phase is located (N_1). In a second paper, (Widyatmoko, et al., 1999b), a method is described to assess the permanent deformation behavior by the dissipated energy approach. Hot Rolled Asphalts (HRA) with different unmodified and modified binders were tested in UCCTs. It is stated that increasing the stress level and/or the test temperature increases the dissipated energy per cycle. A normalized dissipated energy w_{norm} is introduced to make the assessment independent from the stress level. By deriving the load cycle N_1 from test data an unambiguous parameter is presented that indicates where the quasi-linear part of the creep curve ends. The normalized dissipated energy correlates well with the strain rate in the region where the dissipated energy per cycle is constant (secondary creep phase).

Triaxial testing is not only suitable for assessing the permanent deformation behavior. (Weise, et al., 2008) presents a study on testing the fatigue behavior of HMA with triaxial tensile testing. Cylindrical specimens are subjected to cyclic axial compression and cyclic radial tension by introducing negative confining pressure. The phase lag between axial and radial loading was set to 36° and kept constant for all tested mixes and temperatures and frequencies. Two SMA mixes and two AC mixes were compared containing unmodified and modified binder. The results were also compared to standardized fatigue tests, the indirect tensile test (ITT) and the direct tension test (DTT) according to (EN 12697-24, 2007). According to (Weise, et al., 2008), significant correlations between all three tests were found.

Viscoelastic Behavior of Bitumen, Mastic and HMA

As early as in 1954 successful attempts were made to develop "a general system describing the visco-elastic properties of bitumen" by (Van der Poel, 1954). This system already included the time-dependent behavior of bituminous binders but was limited to unmodified binders.

A number of papers deal with the linearity criterion for the viscoelastic behavior of binders and HMAs. Since the viscoelastic parameters, e.g. the complex modulus, phase angles, and also the time-temperature superposition principle (TTSP) rely on the linear viscoelastic behavior of a certain material, it is important to know where the limits for material testing in terms of strain and stress are to ensure the derivation of parameters from the linear viscoelastic domain. (Airey, et al., 2002) and (Airey, et al., 2003) tested different unmodified and modified binders and HMAs to derive linearity criteria for binders and mixes. The limit for viscoelastic behavior was reached when the dynamic modulus decreased to 95% of its initial value. This criterion was also used for a SHRP study (Anderson, et al., 1994). It was shown that the linearity criterion for both binders and asphalt mixes are strain dependent at high stiffness values as well as at low stiffness values for polymer-modified binders (PmB). The strain criterion for mixes was found to be around 10^{-4} m/m whereas the limit for binders is around 10^{-2} m/m and up to 10^{0} m/m for the polymer network.

Usually the dynamic modulus $|E^*|$ and its elastic and viscous part are derived from HMA testing. Furthermore, the phase lag between loading and reaction in one direction can be obtained. Since cyclic compression tests (CCTs) result in relevant deformation not only in direction of loading (axial) but also in the plane perpendicular to the vertical axis (radial), this fact offers the opportunity to study the quasi-3-d behavior of HMA and extend the dynamic material parameters for viscoelastic materials from the dynamic modulus $|E^*|$ to the dynamic Poisson's Ratio $|v^*|$ and further to the dynamic shear modulus $|G^*|$. (Di Benedetto, et al., 2007) presents a study on the 3-d behavior of bituminous binders, mastic and an HMA. DTCs were carried out on all three constituents and the axial and radial deformation was recorded for a range of temperatures and frequencies. (Di Benedetto, et al., 2007) introduces the complex Poisson's Ratio v^* as well as the complex shear modulus G^*, but it does not provide the mathematical expression for the elastic and viscous part of these complex parameters. It has been shown that $|v^*|$ of binder and mastic decrease from around 0.5 at 0°C down to around 0.35 at 25°C. The phase angle between axial and radial deformation is always negative indicating that the radial deformation lags behind the axial deformation. (Di Benedetto, et al., 2007) does not contain data for tested HMAs. In addition the 2S2P1D (2 springs, 2 parabolic elements, 1 dash pot) model is extended from the 1-d case to the 3-d case in (Di Benedetto, et al., 2007). Eleven parameters are necessary to fit the model. A comparison between material parameters from the 2S2P1D model and from material testing is given but the goodness of fit of the model (i.e. the deviation between material parameters derived from modeling vs. testing) is just provided in qualitative means not providing any numbers.

When it comes to modeling the behavior of asphalt mixtures in terms of viscoelastic material parameters the majority of research is focused on mechanistic modeling where the model parameters used to describe the behavior of HMA are interpretable by physical means.

Well-known and widely used models were introduced in the 1960s by two dissertations: (Huet, 1963) introduced a model which can be interpreted as an extended power-law model. (Sayegh, 1965) expanded Huet's model with another spring put into a parallel branch. Thus, also recovery of the material could be described. The 2S2P1D model (Olard, et al., 2003) is an extension of Sayegh's model to enable the model to describe the behavior of both, bitumen and asphalt mixes. The constitutive relations in all mentioned models cannot be solved analytically and have therefore be dealt with by means of numerical algorithms.

A more holistic approach was taken by the Christian Doppler Laboratory for Performance-based Optimization of Flexible Road Pavements at the Vienna University of Technology. By introducing the multiscale approach for HMA (Blab, 2007) links between different levels of observation (bitumen, mastic, mortar, asphalt mixture) were investigated. The objective of the multiscale model for HMA is the determination of macroscopic material parameters on the basis of the mix design and the constituents' properties which, later on, can be used as input for the structural analysis of flexible pavements. These parameters are obtained by means of appropriate upscaling procedures, bridging the scales from finer levels to the macroscale. Hereby, the volumetric parameters of the mix and the viscoelastic properties of bitumen, filler, and aggregate serve as input parameters, allowing to cover the wide range of asphalt mixtures resulting from the different applications, characterized by different mix design and constituents. Different approaches have been presented successfully in the last years, e.g. (Lackner, et al., 2006) on the general procedure, (Aigner, 2010) on the multiscale approach for modeling of the stiffness behavior at higher temperatures and (Fuessl, 2010) on the same approach to describe stiffness and strength behavior at low temperatures.

I.2 Motivation and Objectives

From the state of the art in material science presented in the preceding section, the motivation and objectives of this thesis are to review the triaxial cyclic compression test (TCCT) in the way it is carried out now according to (EN 12697-25, 2005) in various ways and show how the test data can be evaluated, analyzed and interpreted to obtain more advanced information about the material behavior than just creep characteristics, which are the standard results according to (EN 12697-25, 2005). Figure I-4 presents a road map of the thesis to quickly guide the reader through the thesis. In detail the motivations of this thesis are:

- One main motivation is to critically review the method of analyzing the standard results of the TCCT, i.e. the creep curve, today. An enhanced procedure for the characterization of the resistance to permanent deformation will be developed to gather more reliable results from the standard TCCT. Right now, the creep curve is used to derive parameters which characterize the rutting resistance. The radial strain is not taken into consideration in this standard approach. An enhanced method of test evaluation will be introduced taking into account the axial as well as the radial deformation.

- The nature of the TCCT leads to a quasi-3-d state of strain. This is a difference to most of the other performance based test procedures. In the CCT strain occurs not only parallel to the cylindrical axis (in the following referred to as axial strain ε_{ax}) but also strain in the plane orthogonal to the cylindrical axis (radial strain ε_{rad}) occurs. The axial deformation is used for the standard data evaluation to obtain the permanent axial strain whereas the radial strain is not taken into consideration. Furthermore, the test data is only used to present the creep curve vs. number of load cycles. No viscoelastic parameters are derived. From a scientific point of view the question how HMA reacts to cyclic dynamic compressive loading is highly interesting. First of all, axial viscoelastic material parameters ($|E^*|$, $\varphi_{ax,ax}$) can be derived from the compressive loading domain. Furthermore, if the radial strain is recorded with sufficient quality to make sure the data is reliable and repeatable, the dynamic Poisson's Ratio $|v^*|$ can be determined as well as the dynamic shear modulus $|G^*|$. If the tests are run with a temperature and frequency sweep, the quasi-3-d material behavior can be described in temperature and time domain. This is not only valuable information to understand the material behavior better, it is also an important factor for further use in material modeling and simulation (e.g. constitute material models in combination with FEM).

- Material testing on the macroscopic level consumes time and material resources. This is true for specimen preparation and handling as well as for the test procedures. If macroscopic key material parameters could be obtained from testing on a lower scale, in case of HMA on the level of bitumen, in combination with volumetric parameters of the mix, time and cost for material characterization could be substantially reduced. This is of crucial importance if performance based test methods should be implemented for day-to-day standard testing throughout Europe in a large number of laboratories in the future. Thus, another motivation of this dissertation is to establish correlations between material parameters of bitumen, volumetric parameters of the mix and macroscopic material parameters from TCCT of HMA.

- Another critical topic in the standard TCCT today is that the radial confining pressure is kept constant. Two reasons for that are to simplify the test control on the one hand and on the other hand because the interrelation between the axial loading and the radial reaction to this loading has not been studied in a thorough way yet. Thus, the state of stress in a pavement structure under a passing tire is not simulated realistically by the TCCT standard procedure today. In the pavement a dynamic wheel load produces a dynamic radial response due to the confinement of each point within the structure. Figure I-3 compares the stress and strain situation in a pavement structure under a passing tire and in the standard TCCT. Since the state of stress in the field is not incorporated in the test procedure, the test results contain a certain deviation to the real situation. It is therefore aimed to introduce a cyclic dynamic radial confining pressure into the test procedure of the TCCT because it is assumed that this will bring a better correlation of the test results and the rutting occurring in the field. Since a viscous phase lag $\varphi_{ax,rad}$ between axial loading and radial reaction is presumed, this material parameter has to be studied; its magnitude and evolution with the number of load cycles has to be investigated. As a consequence a cyclic confining pressure can be introduced with the analyzed radial phase lag in order to compare results of the standard TCCT with constant confining pressure and the enhanced TCCT with cyclic dynamic confining pressure. On the basis of these results it may be discussed whether this enhanced TCCT is worth to be incorporated in the European standard procedure.

In addition the findings of the research will strongly underline the advantages of the TCCT to other tests available to address the high temperature behavior of HMA like the wheel tracking test (WTT). (Gabet, et al., 2011) contains a detailed study on the WTT and analyzes influences on the scattering of results from WTT, e.g. the type of tires or the preparation of specimens.

The objectives that shall be reached by this thesis are:
- Develop a more reliable evaluation routine for standard TCCTs to describe the creep behavior of HMA and thus overcome the drawback of the present way of evaluating the creep curve according to (EN 12697-25, 2005).
- Get more information from standard TCCT data by taking into account the axial as well as the radial deformation for test evaluation and obtaining the volumetric as well as the deviatoric part of the axial deformation.

- Describe the evolution of viscoelastic material parameters with temperature and test frequency for HMA under cyclic dynamic compressive loading with an emphasis put on the radial phase lag $\varphi_{ax,rad}$, the dynamic Poisson's Ratio as well as the dynamic shear modulus $|G^*|$.
- Develop a model to predict the viscoelastic and permanent deformation behavior of HMA from viscoelastic material parameters of the binder and volumetric mix design characteristics.
- Implement cyclic dynamic confining pressure into the TCCT taking into account the viscoelastic characteristics of the mix to simulate the state of stress within a pavement structure in a more realistic way.

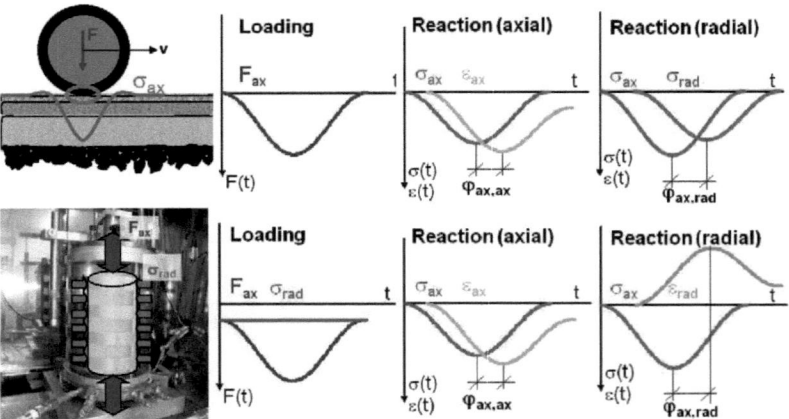

Figure I-3. Stress/strain situation under a passing tire in a pavement structure (top) and in the standard TCCT according to (EN 12697-25, 2005) (bottom).

I.3 Research Approach

To reach the objectives of this thesis stated above the following research approach was taken:

- The first question that arose at the beginning of the investigations was whether the test machine is capable of producing reliable and reproducible results for the test setups that are necessary to gather data for each of the objectives of this thesis. Some shortcomings of the control and record unit of the test machine were found, as well as of the hydraulic circuit for the confining pressure. Thus, new equipment was installed to meet the demands of the research objectives. A second important task was to find an adequate measuring device to record radial strain with sufficient quality. Strain gauges (SGs) were found to be an excellent choice to measure circumferential strain directly on the surface of HMA specimens. The circumferential strain can be converted to radial strain. After some initial tests with this new measuring device and the validation of the measurement with SGs, they were introduced for routine testing. These issues are presented in **chapter II** together with information on the data evaluation and analysis routine. Additional information about the technique of SGs can be found in **Annex A**.

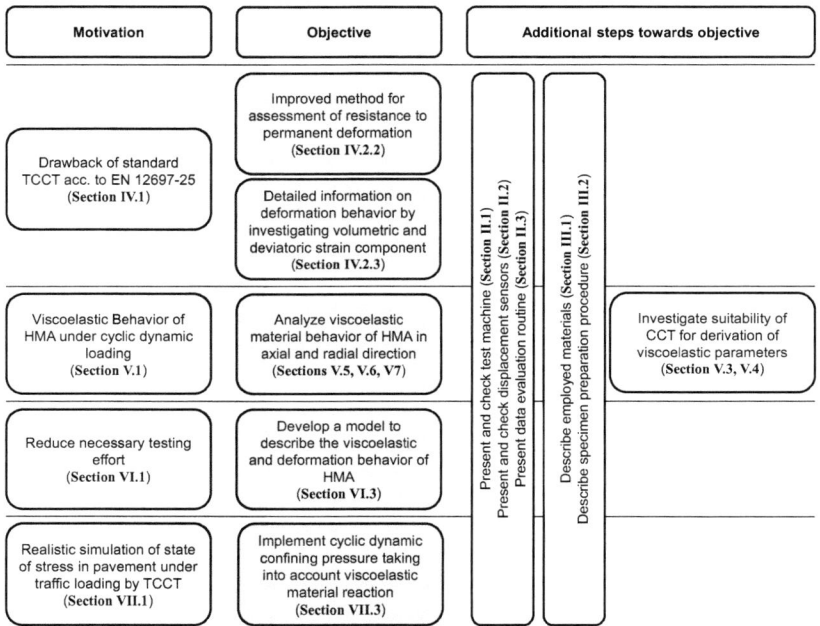

Figure I-4. Road map to guide the reader through the thesis.

- As a next step, materials for the test program were chosen. The majority of the tests was carried out on two different mixes commonly used on the Austrian road network for surface layers. An asphalt concrete with a maximum nominal aggregate size of 11 mm (AC 11 surf) mixed with two different binders, a bitumen 70/100 and a polymer-modified bitumen PmB 25/55-65, as well as with a mineral (porphyrite) commonly used in Austria for surface layers. **Chapter III** contains detailed information on the employed materials and presents the process of specimen preparation according to the respective European standards.
- In a critical review of the European standard for TCCT (EN 12697-25, 2005) a problematic point was isolated when it comes to the evaluation of results. To benchmark different mixes, the creep curve as the main result of a TCCT has to be described by analytical means. The standard asks to put a linear function into the quasi-linear part of the creep curve failing to define the quasi-linear range of this curve. As a consequence, the resulting parameter can be set arbitrarily by the user within a certain range. After looking at comprehensive test data from an older project carried out in the CD-Lab and trying new approaches, an alternative

assessment of the resistance to permanent deformation is provided in **chapter IV** by taking into consideration the viscoelasticity of the material. Also the radial strain data was taken into account to analyze the volumetric and deviatoric part of the specimen's deformation. Impacts of gradation and volumetric properties of the mix and binder type on these two parts of the deformation are presented and discussed.

- Is the CCT a valid test method to derive viscoelastic material parameters from, and does the test setup in terms of whether the load plates are firmly connected to the specimen or not impact the results? Is it possible to test HMA in CCTs in the linear viscoelastic domain at different temperatures and if so, how do viscoelastic material parameters evolute with test temperature and time (frequency)? These were further questions that came up during the course of the research. **Chapter V** tries to give conclusive answers. It also pursues the question where the difference between the axial and radial phase lag has its origins and shows that CCTs with high quality measurement of radial strain can give information on advanced viscoelastic parameters like the dynamic Poisson's Ratio and the dynamic shear modulus.

- Since testing of HMA consumes a significant amount of time and material resources current research is often aimed at developing models that allow the user to derive material parameters of a certain HMA quickly and easily. Since a comprehensive test program had been carried out for chapter V of this thesis with a variation of test temperature und frequency, as well as of volumetric properties of the mix and the binder type, an approach was taken towards a model to describe the behavior of HMA in a state of compressive loading. Therefore some additional tests on bitumen were run with the dynamic shear rheometer (DSR). As a result the *B(inder)-A(sphalt) Model* is developed in **chapter VI**. It enables the user to predict the viscoelastic behavior of an HMA from viscoelastic properties of the bitumen together with volumetric properties of the HMA.

- Taking into account the findings from the preceding chapters another key motivation of this thesis was to simulate the state of stress that occurs within a pavement structure under dynamic traffic loading in a more realistic way than with the standard TCCT today. This may be achieved by substituting the constant confining pressure from the standard TCCT according to (EN 12697-25, 2005) by a cyclic confining pressure taking into consideration the viscoelastic properties of the material. This procedure is presented in **chapter VII**. In a first step, standard TCCTs were carried out to gather information about the phase lag between axial loading and radial reaction. This phase lag was then used in the enhanced TCCT with cyclic confining pressure to set the lag between axial and radial cyclic loading. Figure I-5 shows this approach. Results from standard and enhanced TCCT are compared and discussed. Furthermore the *B-A Model* was extended with a powerful addition. Not only the viscoelastic properties but also the permanent deformation behavior (in terms of the creep rate) can be predicted by the model.

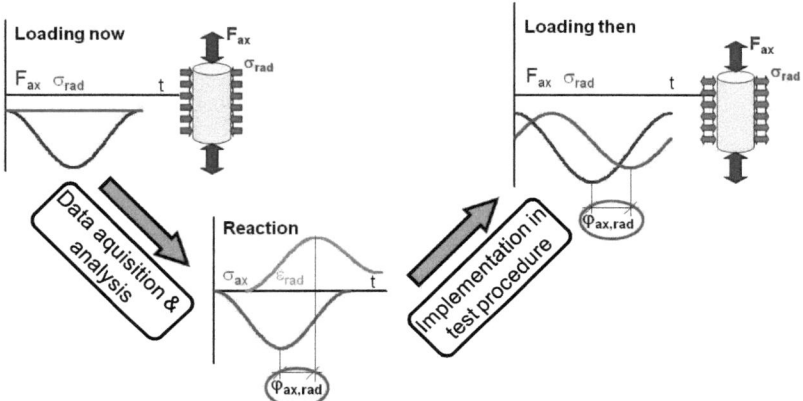

Figure I-5. Approach to achieve an enhanced TCCT with cyclic dynamic confining pressure.

II TEST EQUIPMENT AND DATA EVALUATION

This chapter provides information about
- the test machine and its relevant parts employed for CCTs of HMA specimens within this thesis (section II.1),
- devices used to record deformation in axial as well as in radial direction with an emphasis put upon SGs and the validation of this measuring device (section II.2),
- the evaluation routine for test data to derive meaningful results from signal processing, the evaluation software and the statistical analysis (section II.3).

II.1 Test Machine

In 2002 three new test machines were installed in the laboratory of the Research Center for Road Engineering to cover the complete range of performance based tests on HMA. For the investigation of the material behavior at high temperatures (permanent deformation – rutting) a test stand was acquired consisting of a robust loading frame and load table, a triaxial test cell including measuring sensors, two hydraulic aggregates and a control and record unit. The load table is situated within a temperature chamber. The European Standard for the assessment of the permanent deformation behavior (EN 12697-25, 2005) was the basis for the selection of the components. The triaxial test machine is a servo-hydraulic device with two hydraulic circuits for tests in the tensile and compressive domain. Axial loads can be static or dynamic and cyclic dynamic for a wide range of frequencies. The radial confining pressure controlled by the 2nd circuit can be static as well as cyclic dynamic for low frequencies (up to 0.1 Hz). Figure II-1 shows the components and equipment of the test machine. The most important characteristics and parameters of the test machine are listed in Table II-1. (Kappl, 2007) described each component of the machine in detail. Most of the equipment has not been changed since then. Still, some important parts have been updated. Thus the test machine and its components are presented in a brief overview here. Details on the new equipment are shown as well.

The test machine can be labeled as a universal device; test procedures can be programmed by the user. With additional equipment (Marshall device, load plunger for mastic asphalt,...) various static, dynamic and cyclic dynamic tests can be run. Among them are
- static, uniaxial creep tests,
- static penetration tests,
- uniaxial, cyclic dynamic compression tests (UCCT),
- triaxial, cyclic dynamic compression tests (TCCT) with constant or oscillating confining pressure (at low frequencies) and
- cyclic dynamic tension and compression tests (stiffness tests).

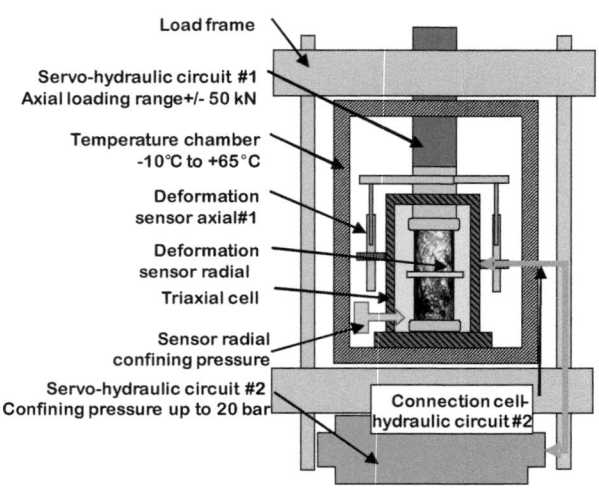

Figure II-1. Components and equipment of the 2-circuit triaxial test machine (Type LFV63/50).

Table II-1. Characteristics of the test machine LFV63/50.

Parameter	Value
Max. force axial (static)	± 63 kN
Max. force axial (dynamic)	± 50 kN
Cylinder Stroke	± 125 mm
Max. test frequency axial	0 to 30 Hz
Max. amplitude axial	+/- 0,5 mm
Max. working pressure of the hydraulic aggregate #1 (axial)	250 bar
Max. flow rate of hydraulic oil in aggregate #1 (axial)	7 l/min
Volume of oil tank of aggregate #1 (axial)	50 l
Max. static confining pressure	20 bar
Max. working pressure of hydraulic aggregate #2 (radial)	25 bar
Max. flow rate of hydraulic oil in aggregate #2 (radial)	4,5 l/min
Volume of oil tank of aggregate #2 (radial)	17 l
Range of temperature chamber	-20°C to +65°C

II.1.1 Control and Record Unit

The principal task of a control and record unit of a testing machine is to control the equipment – in this case the hydraulic circuits that actuate the axial load plunger and the confining pressure – according to the test routine. In addition the record unit should transform analogue measuring signals to digital data which is stored for further evaluation and analysis. A sensitive area is the analogue digital converter (ADC) responsible for the transformation. The first generation of control and record unit which was presented and used by (Kappl, 2007) consisted of a multiplexing unit in combination with one ADC for a total of eight signal ports. The problem with the combination of multiplexer and ADC is that the analogue signals of eight ports are transformed serially. If each transferred data item got its own time stamp directly at the time of the conversion this setback could be compensated. Unfortunately, this was not the case for this control and record unit. A data set of eight signals was transformed with one time stamp when in fact only one of the signals was actually recorded at this point in time. The signal converted after this first one is actually the analogue signal from a short time later. This time span is exactly the time that the converter needs for one conversion and the multiplexer to switch to the next port. The principle is also presented in Figure II-2. So only one signal is converted correctly in time domain. The other signals have a certain phase lag that is induced by the control unit due to the multiplexing ADC. If the data is used for analysis of viscoelastic material properties, these induced phase lags overlap with the actual material phase lags and result in incorrect material parameters.

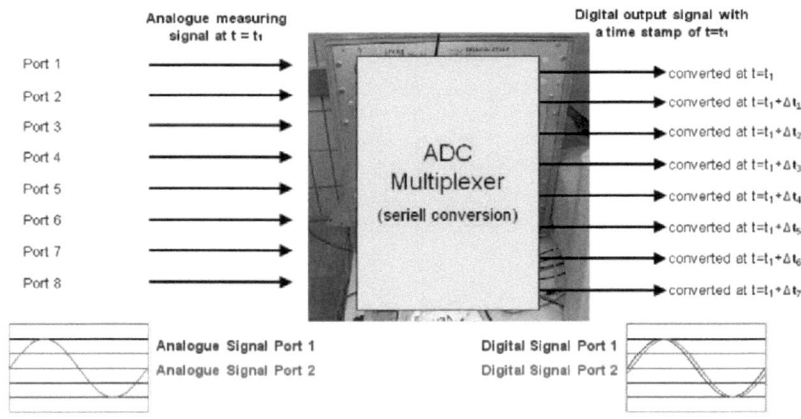

Figure II-2. Phase lag induced by control unit due to multiplexing ADC.

If the machine induced phase lag is constant for each port, it can be compensated by just subtracting it from the derived material phase lag. In the case of the control unit used by (Kappl, 2007) this was not the case. To study viscoelastic properties using this test machine, the machine-induced phase lag biased the results.

The control and record unit was replaced by two new controllers WDC 580 (DOLI Elektronic Gmbh, 2010), one for each hydraulic circuit. Besides other advantages compared to the old unit (e.g. a higher sampling rate of 2.5 kHz), the main reason for choosing this type of controller was that each signal now has its own ADC and therefore no machine-induced phase lags are produced. To prove this assumption, a short and simple test was carried out. A hollow cylindrical aluminum specimen was subjected to a UCCT with a frequency range from 0.1 Hz to 20 Hz at room temperature (around 23°C). The diagram in Figure II-3 presents the 5% and 95% quantiles as well as the median value (50% quantile) of the phase lag $\varphi_{ax,ax}$ between axial loading and axial deformation. Since the material is elastic, no such delay between axial loading and deformation should occur. The phase lag ranges from around $+1°$ to $-1°$. It can therefore be assumed that no significant phase lag is induces by the control and record unit.

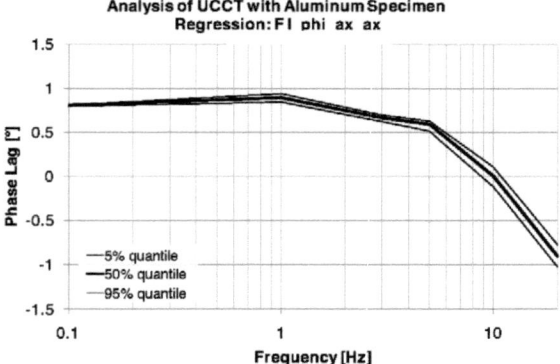

Figure II-3. 5%, 50% and 95% quantiles of $\varphi_{ax,ax}$ for the aluminum specimen at 23°C.

The phase lags that occur in the aluminum test are rather related to the imperfect test data (noise) and the resulting inaccuracy of the evaluation procedure. A scattering of results for phase lags of 2° caused by inaccuracies of the signal recording and evaluation must be taken into account.

II.1.2 Software for Test Procedures

Furthermore, a new control software was installed to simplify programming of new test routines. "GeoSYS" was developed by WILLE Geotechnik and provides a state of the art graphic user interface (GUI) for programming, controlling and recording. The programming language contains all basic logic functions. Thus arbitrary test control commands can be connected to complex test routines. Data from sensors can be recorded with a user-defined sampling rate of up to 2.5 kHz (2.500 samples/s) and is stored in standard ASCII-files.

II.1.3 Triaxial Cell

The triaxial cell is the heart of the test device. As shown in Figure II-4 it consists of a base with eight pressure resistant cable outlets, three supporting rods and the top with a guide jacket for the load plunger and de-aeration vents. The steel jacket is placed around the cell so that the cell can be filled with water and pressurized.

II.1.4 New Pneumatic Device for the Application of Cyclic Dynamic Confining Pressure

To realize cyclic confining pressure also at higher frequencies (> 0.1 Hz) with a well defined time shift to the loading of the axial plunger new equipment has been developed together with Wille Geotechnik. It replaces the old 2nd hydraulic circuit of the testing machine with a completely revised technique based on pneumatics where compressed air is used to activate a stiff membrane. The volume within the membrane is filled with water which controls the water pressure within the triaxial cell.

Figure II-5 shows a sketch of the new device. The system works with compressed air which is connected to a pressure transmitter. This pressure transmitter can be described as a high-end shock absorber also used in trucks. The pressure transmitter is filled with water. The transmitter has a connection to the triaxial cell itself. The actual control mechanism is a valve that allows more or less compressed air to go through. The more compressed air is put onto the pressure transmitter the more water is pressed into the triaxial cell. If the triaxial cell is filled with water and the system is water-tight then the pressure within the cell is changed by the volume of water pressed into the cell from the pressure transmitter. The triaxial cell is equipped with a pressure gauge to measure the pressure in the cell. The gauge is connected to the control unit of the test machine. The control unit activates the valve in the pneumatic device according to the signal of the gauge to reach the pressure given by the user. A linear variable differential transformer (LVDT) below the pressure transmitter records the location of the transmitter and is therefore a safety device to keep the shock absorber membrane (i.e. the pressure transmitter) within safe operation limits.

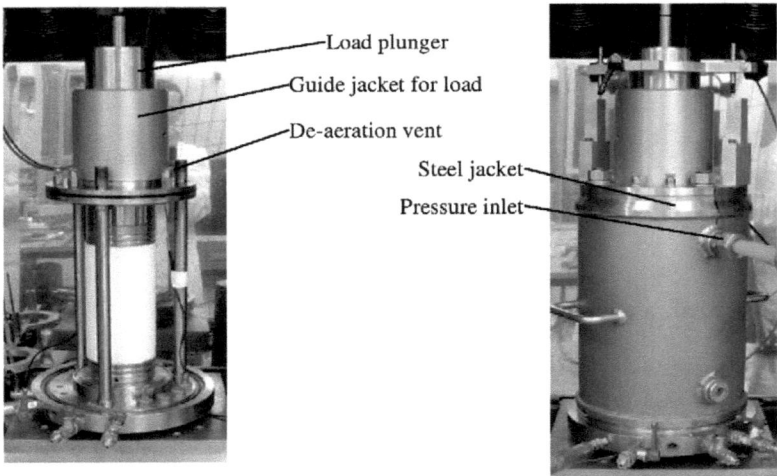

Figure II-4. Main elements of the triaxial cell.

The role of the user is to set a static or dynamic target value for the pressure within the cell and the control unit regulates the pressure with a frequency of 5 kHz. Thus, exact static or dynamic confining pressures are realizable in a wide range of frequencies and amplitudes.

The device needs a compressed air supply with at least 300 l/min and a working pressure of around 8 bar. This ensures that a cyclic confining pressure with amplitudes up to 300 kPa (from 150 kPa to 750 kPa) at frequency up to 3 Hz can be generated by the system. The quality of the results in terms of confining pressure is strongly dependent on the stiffness and compressibility of the system. The stiffer the system including connecting tubes and the less compressible the system is the more reliable are the results. Thus plastic tubes reinforced with a metal grid were used as connectors from the pressure transmitter to the cell and special care was taken to completely de-aerate the system before testing. So far the control for the system has been optimized for the test setup used within this thesis (amplitudes ranging 50 kPa to 100 kPa from a lower radial confining stress of 150 kPa at 3 Hz) but may be adapted to other frequencies and amplitudes with little effort.

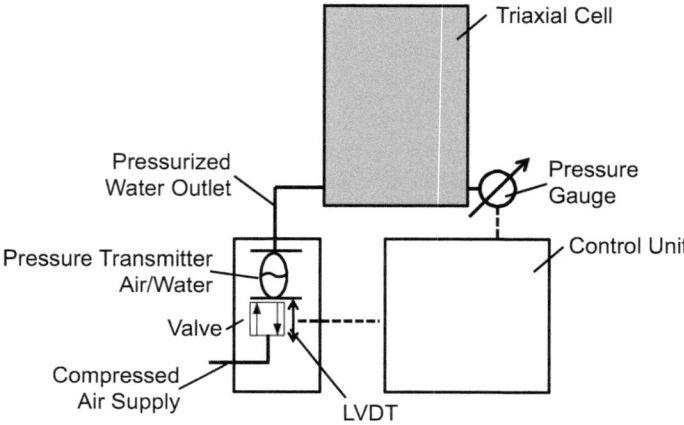

Figure II-5. Principle of the pneumatic device to apply cyclic confining pressure.

II.2 Displacement Sensors

The most important basis for correct and reliable evaluation of test data is to gather reliable signal data for the force applied to the specimen and the resulting deformation of the specimen. In this research, the measurement of axial deformations is realized by two LVDTs recording the movement of the load plunger at both of its sides. Figure II-6 illustrates the setup. The mean value (MV) of both signals is used for further evaluation. More information on the principle of LVDTs can be found e.g. in (Macrosensors, 2011).

When it comes to recording radial deformation or strain of cylindrical specimens under cyclic dynamic loading, various methods have been developed in the last decades. All are perfectly suitable for an isolated spectrum of applications. None of the reviewed methods were capable of meeting the requirements for the measurements for the research presented in this thesis:

– Recording of the circumferential or radial deformation directly on the specimen
– Maximum cumulative strain > 1%
– Exact readings of the radial deformation on oscillation level from 0.1 Hz to 30 Hz

- Constant quality of recorded data under steady oscillations
- Temperature range from 0°C to 60°C
- Measuring device fits in the triaxial cell with its limited space
- Quick and easy application on the specimen

Thus a common and robust measuring method by means of the SG was introduced into the recording of circumferential strain, which can easily be transferred to radial strain.

Figure II-6. Two LVDTs recording the axial deformation on both sides of the load plunger.

II.2.1 Approaches to Measure Radial Deformation on Cylindrical Specimens

In the process of validating possible measuring devices for radial deformation, a number of already existing systems were reviewed. A comprehensive overview of measuring devices for radial deformation is given in (Partl, 1983).

Extensometer-based Device

One example is a device consisting of an extensometer connected to both ends of a chain clamped around the circumference of the specimen. For example, (Trautwein, 2005) used this device in his research. The extensometer records the linear difference in distance between the two ends of the chain. The lateral strain ε_q is calculated from:

$$\varepsilon_q = \frac{\Delta C}{C_0} = \frac{\Delta r}{r_0} \tag{2.1}$$

ΔC .... Change in circumference
C_0 .... Initial circumference
Δr .... Change in radius
r_0 .... Initial radius

Figure II-7 shows the setup of the device on the specimen. The space required by the system is not provided within the triaxial cell and therefore not available for the purpose of the presented research work.

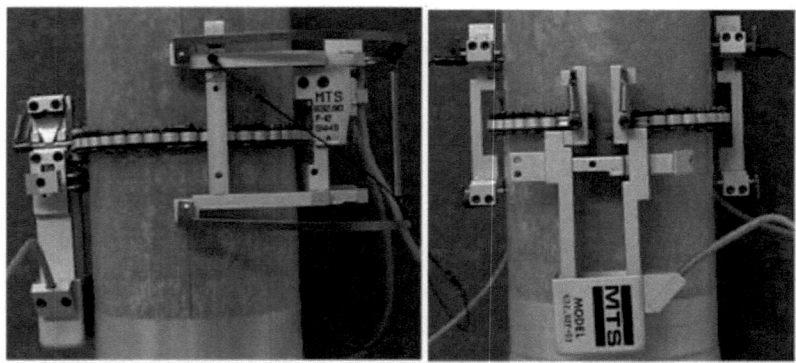

Figure II-7. "Chain Extensometer" attached to a cylindrical specimen. (Trautwein, 2005)

LVDT-based Measuring Frame

The measuring frame with an LVDT to record the deformation was used by (Kappl, 2007) mainly because it fits into the triaxial cell and is water proof. Figure II-8 shows the device attached to a specimen. The frame consists of two semicircular bows; one of them holds an LVDT. A plunger is attached to the other bow. Both parts are held together by two springs. The frame is attached to the specimen outside of the latex membrane which protects the specimen from the pressurized water within the cell. The two springs guarantee a pretension and keep the device in place. When the diameter of the specimen changes due to radial deformation, the two bows are moved apart and the LVDT records a change in length.

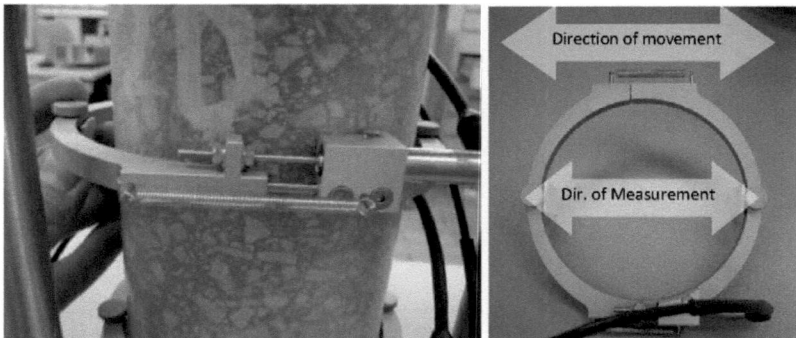

Figure II-8. Measuring frame on LVDT basis.

The main disadvantage of this device is that only changes in the diameter in one direction can be recorded. If the specimen is not deformed uniformly, the system may not detect the maximum change in diameter. Also the device records the deformation outside the membrane, so the influence of this latex membrane (thickness = 0.3 mm) and its contribution to the total measured deformation is unknown. If the user is only interested in the accumulated radial deformation, this device is a perfectly capable tool. Yet, if phase lags between axial loading and radial deformation should be derived from a test, the quality of the signal of this device is not sufficient to carry out any reliable analysis. Figure II-9 shows an example of signal data. A cyclic dynamic test was carried out and the data from force, axial deformation and the LVDT-based radial deformation sensor were recorded. The force and axial deformation sensor show a sinusoidal shape whereas the radial sensor data does not result in any similar shape.

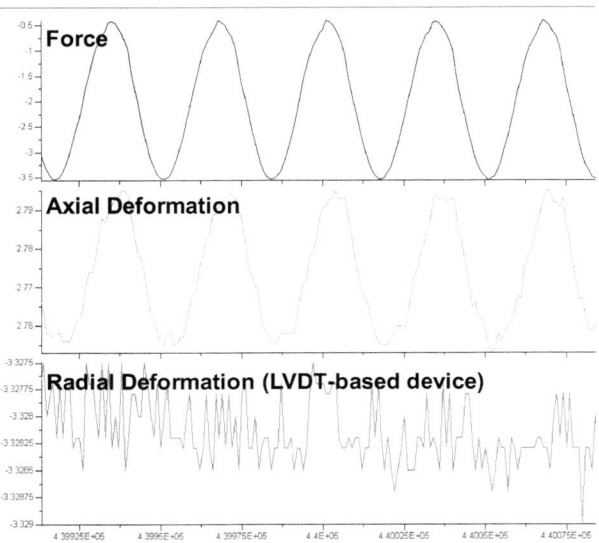

Figure II-9. Signals from force, axial deformation, and radial deformation (LVDT-based) sensor in a cyclic dynamic test.

In the further course of the research it was found that even this LVDT-based measuring frame is capable of recording the radial deformation on oscillation level (and section V.6.3 will prove that), but the necessary effort to keep the device smooth-moving is high and the other setbacks, especially that it measures outside of the latex membrane, cannot be overcome.

Other Devices

Literature provides other devices, many of them based on the principle of SGs. But they are either too spacious to fit in the triaxial cell or do not represent state of the art.

In (Kappl, 2007) for example a prototype of a measuring device consisting of a number of steel springs attached together was constructed, that fit around the specimen. Each steel spring had two SGs attached to it, one on the inside and one on the outside. The main problem was that it could not be achieved to seal all SGs in a way to keep the connectors of the SGs water proof throughout the test.

(Krebs, et al., 1982), (Partl, 1983) and (Huschek, 1983) used devices based on the principle of SGs for their measurements of the radial deformation in the 1980s. A wire was wrapped tightly around the circumference of a specimen. A change in diameter due to loading resulted in an elongation of the wire and a decrease of its cross section. Thus its resistance changed as well. The resistance of the wire was recorded and correlated to the radial strain of the specimen.

II.2.2 Introduction of Strain Gauges to Obtain Radial Strain

SGs are one of the most commonly used measuring systems in the area of experimental analysis of strain. Not only can they be used for numerous different applications, they are also capable of recording strain with high precision. This makes them perfect to determine related physical parameters like forces, momentums or pressures. The basic principle of SGs is simple: The electric resistance of a conductor changes when its dimensions change. The change in length (=strain) of an SG alters its resistance which is recorded. Thus the strain (and not a change in length) is directly connected to the measured value. Or, in other words, a non-electrical, mechanical parameter is transformed to an electrical value which can be used for analogue signal processing.

Annex A provides an overview on the basics of SGs. It is mostly based on a detailed book focusing on the practical use of SGs: (Hoffmann, 1987). For more information the reader may be referred to this literature.

Within the research program carried out for this thesis, SGs are directly glued to HMA specimens to investigate the radial deformation behavior. Figure II-10 shows an SG attached to an HMA specimen. The SGs measure the circumferential strain, which can easily be transferred to radial strain. The radial strain data is used for evaluation and analysis. For this reason it is stated in the further course of the thesis that the radial strain is obtained from SG data.

Figure II-10. End of an SG glued to an HMA specimen.

II.2.3 Application of Strain Gauges

This following section specializes on the application of SGs within this research project. The process is presented as an overview; details can be taken from the internal work instruction developed within the quality management system of the institute's lab (AA542, 2010).

Usually SGs are attached to an object over its full length to ensure that strain is transferred from the object in tension and in compression. Using HMA specimens leads to a special situation because the stiffness of the object (HMA) is distinctively below the stiffness of the adhesive used to glue the SG to the object. If the SG was glued over its full length to an HMA specimen, the stiffness of the adhesive would prevent any deformation in this area which would represent a perfect reductio ad absurdum. It was therefore decided to glue only both ends of the SG to the HMA specimen. Since the radial deformation will only be positive in a purely compressive test, this method of attachment sufficiently ensures the transfer of strain to the SG.

The first step for the application of an SG is to mark half the height of the specimen around its circumference to set the axis where the SG should be attached. Depending on the configuration of the SG two different procedures are presented in the following:

Application of a Single Strain Gauge Directly on the Specimen

If a single SG should be used to derive radial strain from the recorded data, the SG is placed around the cylindrical surface of the specimen and temporarily fixed by adhesive tape to make sure it stays in place as perfectly orthogonal to the cylinder axis as possible. Both ends of the SG should be in a homogeneous area not directly above one large aggregate or a large area with mastics to make sure stress transmission into the glued area is in a homogeneous region. The adhesive area is cleaned by a mixture of ethanol and acetone to ensure the bond between specimen and adhesive (Figure II-11). Then the two component adhesive is applied to an area that exceeds the edges of the SG (Figure II-11) in a thin layer. The area should cover around 15 x 15 mm. After this the SG is put upon the still soft glue and held with the pressure of the palm for around two minutes. It is important to note that from the experience gathered in this thesis, neither the cleaning mixture nor the glue affect the bitumen or mastic of the specimen.

The same procedure is carried out with the other end of the SG. The SG must be laid tightly around the specimen. After 24 h the adhesive is dried enough to apply a silicone based seal to protect the SG and the connectors from moisture.

Applications of Two Strain Gauges attached Directly to Each Other

The other possible SG setup is to use two SGs with a length of the measuring grid of 150 mm and attach them around the total circumference of the cylindrical specimen without gluing them directly on the surface of the specimen.

The first step is to glue the ends of both SGs together using basically a high-end superglue. Both ends should overlap by about 5 mm (Figure II-12). It must be taken special care on the fact that both SGs are in line. After a couple of seconds the glue has dried enough to attach the system of two SGs temporarily around the specimen's circumference by adhesive tape as shown in the left picture in Figure II-12.

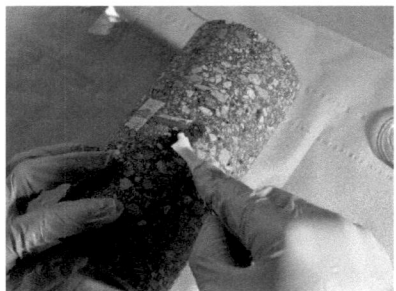 

Figure II-11. Cleaning of the adhesive area (left) and attaching the SG to the glued area (right).

 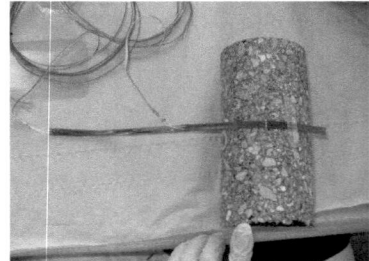

Figure II-12. Overlapping ends of two SGs (left) and attaching two SGs temporarily to the specimen (right).

The two SGs are passed around the specimen, and again both ends are glued together using the advanced superglue (Figure II-13). The SGs must be wrapped around the specimen tightly. After 24 h the SGs are coated by a silicone based gel to protect them against moisture.

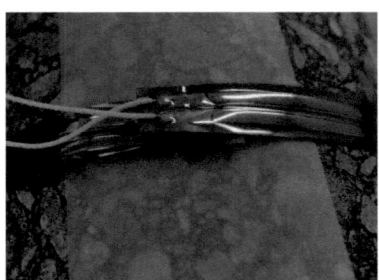 

Figure II-13. Overlapping ends of two SGs before (left) after being glued together (right).

II.2.4 Validation of Measurement by Strain Gauges

The measurement of strain with SGs was introduces for a series of measurements on HMA specimens for the first time in the laboratory of the Research Center for Road Engineering at the Vienna University of Technology. To proof that the measuring device consisting of SG and glue produces correct and reliable readings, a number of tests were carried out comparing the data recorded from SGs and a parallel used, reliable measuring device (LVDTs). The tests were carried out statically as well as dynamically.

Tests with Axial Strain Measurements

In a first step the readings of the SGs were compared to those of LVDTs when SGs are attached to the specimen in axial direction along the cylinder axis. For the tests two SGs 100 mm were glued to both sides of the cylindrical specimen (diameter = 100 mm, height = 200 mm) according to the procedure in section II.2.3. The specimen was then glued to two load plates because the SGs are only attached to the specimen at their ends, and thus only tensile strain can be measured by the SGs. Figure II-14 shows the specimen with the two load plates and the SG in axial direction. The specimen was produced from an asphalt concrete with a maximum nominal aggregate size of 11 mm (AC 11) with a polymer-modified binder PmB 25/55-65. The binder content is 5.3% (m/m) and the specimen contains 4.1% (v/v) air voids.

Figure II-14. HMA specimen glued to the load plates with an SG attached on both ends in axial direction.

The test was carried out in two steps at 30°C. First a constant axial tensile stress of 0.10 N/mm² was applied to the specimen for 200 sec followed by a recovering period of 175 sec without any loading. Figure II-15 shows the applied force and the reaction of the specimen in terms of axial strain as MV of the signal from both LVDTs and both SGs respectively. Comparing both signals it can be observed that the difference between them is negligible. The SGs recorded an accumulated strain in the loading phase of 0.2994% and the LVDTSs a strain of 0.3089%. The SGs measured 3.08% less strain than the LVDT. It has to be kept in mind that the LVDTs do not measure directly on the specimen but on top of the load plunger. So the LVDTs record a mean value of deformation over the complete height of the specimen whereas the SGs measure within the center part of the specimen over a length of 100 mm. Thus, only a small part of the deviation between both systems is directly related to the systems itself, the other part is linked to the different range of measurement.

The total recovered strain recorded by the SGs $\Delta\varepsilon_{ax}$ is 0.1772%, and for the LVDTs it is 0.1627%. The relative difference between SGs and LVDTs comes to 8.91%. Again, it must be pointed out that only part of the difference is directly related to the measuring systems. In the deformation as well as in the recovery deformation phase a non-uniform strain distribution over the length of the specimen is probable.

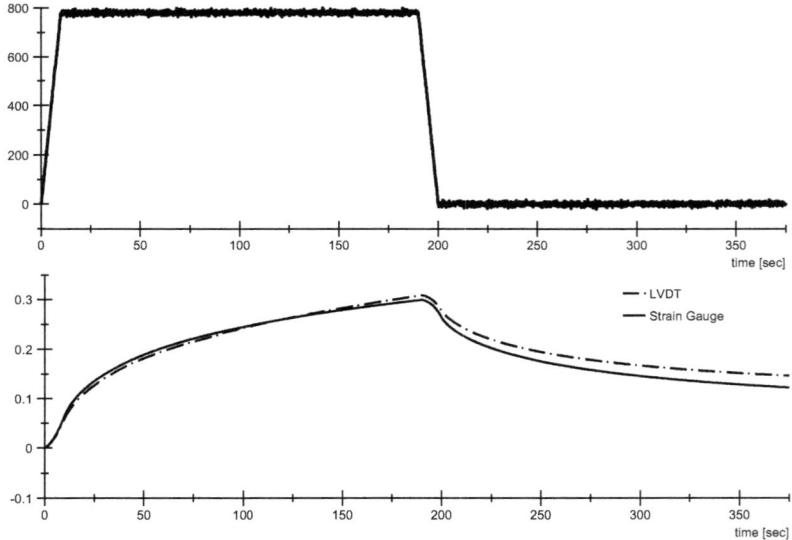

Figure II-15. Static test to compare LVDT vs. SG signal data at 30°C; applied tensile force (top) and reaction of the specimen in terms of axial strain (bottom).

Following this 1st step with constant force, a second phase was started in which the applied axial loading was cyclic dynamic in a sinusoidal way without confining pressure (UCCT). Force-controlled oscillations with a mean stress of -0.04 N/mm² and an amplitude of 0.14 N/mm² were applied. Ten load cycles were carried out at 0.1 Hz and another ten load cycles at 0.5 Hz. Figure II-16 presents the applied load and the reaction at the 0.1 Hz frequency packet. The figure shows the first load cycles of the test and the control unit still adapts the force to the given values by the user. Thus, the extrema of the force are not constant for the first load cycles.

Table II-2 shows the results in terms of strain amplitudes of the LVDT-signals and the SG-signals. Since the amplitude rather than the absolute values are important for the further research, these values were taken as a benchmark. The SGs result in 7.9% to 8.9% higher amplitudes than the LVDTs. Part of the difference may also be related to the fact that the oscillations approximated the 0% strain level. That is the level where the SGs might be too relaxed to record correctly since they are not form-fit in the compressive axial domain due to the fact that they are only glued to the specimen at their ends.

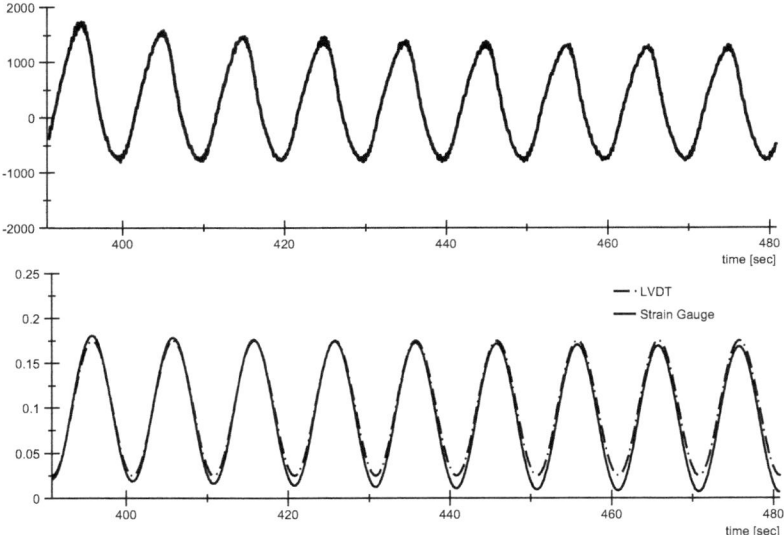

Figure II-16. Cyclic dynamic test to compare LVDT vs. SG signal data at 30°C; applied tensile force (top) and reaction of the specimen in terms of axial strain (bottom) at 0.1 Hz.

Tests with Radial Strain Measurements

A second force-controlled test regarding the evaluation of the SG measuring system was carried out with the following setup. An HMA specimen (AC 11 PmB 25/55-65) with a binder content of 5.3% (m/m) and 4.3% (v/v) air voids was used for the investigation. One SG 150 mm was glued to the specimen in the standard way described in section II.2.3. In addition the LVDT-based device to record radial deformation (see section II.2.1) was also attached to the specimen in the same height as the SG to be able to compare both recorded signals. As it will be shown in a later section (V.6.3), the LVDT-based device is capable of recording not only accumulated radial deformation but also radial deformation on the level of oscillation with sufficient quality. This is true if it is maintained extensively to keep the LVDT as smooth moving as possible and the springs free from rust.

Table II-2. Axial strain amplitudes of LVDT and SG for 0.1 Hz and 0.5 Hz at 30°C.

Load Cycle	Frequency [Hz]	Strain Amplitude [cm/m=%]		SG/LVDT [%]
		LVDT	SG	
1	0.1	0.1502	0.1623	108.1
2	0.1	0.1500	0.1621	108.1
3	0.1	0.1502	0.1623	108.1
4	0.1	0.1500	0.1620	108.0
5	0.1	0.1501	0.1620	107.9
6	0.1	0.1503	0.1622	107.9
7	0.1	0.1502	0.1622	108.0
8	0.1	0.1501	0.1619	107.9
9	0.1	0.1500	0.1619	107.9
11	0.5	0.1503	0.1632	108.6
12	0.5	0.1504	0.1637	108.8
13	0.5	0.1505	0.1633	108.5
14	0.5	0.1502	0.1636	108.9
15	0.5	0.1500	0.1634	108.9
16	0.5	0.1501	0.1631	108.7
17	0.5	0.1502	0.1632	108.7
18	0.5	0.1504	0.1632	108.5
19	0.5	0.1502	0.1633	108.7

A UCCT at 30°C with a frequency range from 0.1 Hz to 30 Hz was carried out with a mean axial compressive stress of 0.25 N/mm^2 and an amplitude of 0.15 N/mm^2. The recorded data were evaluated with the standard procedure which will be described in section II.3.2. The diagram in the following present the 5% and 95% quantiles as well as the median value (50% quantile) of results from the LVDT-based device and the SG.

The diagram in Figure II-17 presents the comparison of both measurement devices in terms of amplitude. It gives the ratio between the strain amplitude measured with the SG and LVDT. Obviously the SG records smaller amplitudes than the LVDT-based device. The relative difference between both measuring systems is around or below 10%. It must be taken into consideration that one of the downsides of the LVDT-based devices is the fact that it only measures on two points of the circumference and in only one direction. The SG on the other hand measures continuously around the circumference. Together with the lower data quality compared to the SG data, this difference can be explained.

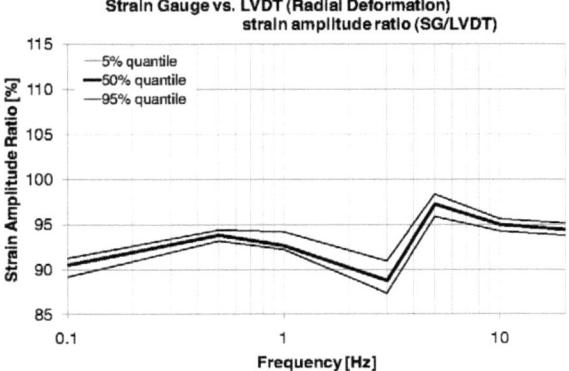

Figure II-17. 5%, 50% and 95% quantiles of the strain amplitude ratio SG/LVDT at 30°C.

When comparing results from validation of SGs in axial direction (Figure II-16 and Table II-2) to results from validation of SGs in radial direction (Figure II-17), it becomes obvious that the axial strain obtained from SGs is larger than the strain from LVDTs, whereas the radial strain obtained from SGs is smaller than the strain from LVDTs. It is assumed that this difference is due to the different LVDT-measuring devices. The axial deformation is measured by two LVDTs on top of the load plunger (Figure II-6), the radial deformation is recorded by the measuring frame with LVDT shown in Figure II-8.

II.3 Data Evaluation and Analysis

In order to derive reproducible and reliable material parameters from unprocessed test machine data, the test data has to be evaluated and analyzed carefully especially at higher test frequencies (f > 1 Hz) since the derived material parameters from test data react highly sensitive to the method of evaluation. (Pellinen, et al., 2003) presents a study that analyzes the effect of different evaluation methods for the determination of dynamic modulus and phase lag of HMAs in cyclic dynamic tests. To show the complexity of the way from raw machine data to a single material parameter the following sections give an insight in signal processing. It starts from conversion from analogue to digital data, to the evaluation of this digital output data by means of an evaluation software, and finally the statistical analysis of the results.

The procedure of data evaluation presented in this chapter is kept general. Details on the derivation of certain parameters (e.g. viscoelastic material parameters) are given in the respective chapters.

II.3.1 Signal Processing

A physical signal in its simplest definition is a way to transport information. Two dimensional physical signals are characterized by the distribution of an amplitude value (y) vs. time (x) or any other physical quantity.

In general, signal processing contains all steps that aim to extract or systematically reduce information on a process from a recorded signal. If analogue signals should be used for electronic data processing, the most important part of signal processing is the transformation from analogue to digital data. An analogue signal is identified by the fact that an amplitude value is defined for any arbitrary point in time. This value is continuously variable within a defined range. Diagram a) in Figure II-18 shows an example of an analogue signal. Analogue signals are commonly electric signals (i.e. voltage or amperage). Problems in the processing may occur due to non-linearity of transducers or more often due to noise. Noise in physics is disturbance value with a wide, non-specific range of frequencies.

As stated above, analogue signals cannot be used for electronic data processing and must therefore be discretized in time and amplitude domain to a digital signal in time and amplitude domain. The analogue signal is sampled at fixed time intervals. Each sampled point in time is given an amplitude value from a discretized range by an analogue-digital converter (ADC). Diagram d) in Figure II-18 shows an example of a digital signal. Obviously information is lost in the process of converting analogue to digital signals. The double discretization results in two independent errors. The quantification in time domain leads to the sampling error. Details of the analogue signal in between sampled time intervals are not stored in the digital file. The quantification of the amplitude causes the quantization/rounding error. In theory the intervals of discretization in time and amplitude domain can be reduced to a minimum. However, in reality the quantization in time domain is limited by the operation speed of the microprocessor. The same is true for the amplitude domain. The higher the resolution of the analogue-digital converter, the more time it needs for converting. (Best, 1991)

The noise of the analogue signal mentioned above also effects the converted digital signal. Figure II-19 presents the effect graphically. It shows the digitalized, recorded signal of an axial load cell. The measured signal contains a certain amount of noise. If this unprocessed data is used for further evaluation of material parameters, like the dynamic modulus and phase lag (e.g. by using a peak-finding algorithm) the noise of the signal will have a significant effect on the material parameters and results will at best be imprecise or incorrect at worst.

A quick and simple method has to be used to eliminate or reduce the noise without affecting the measured signal in time domain. One way is regression analysis in combination with a suitable approximation function. In this process an analytical function is determined to fit the measured data so that the residua between the approximation function and the signal data become a minimum. Different regression methods are suitable for different problems. The smooth (red) curve in Figure II-19 shows how the measured signal was approximated by a simple sinusoidal approximation function. Hereby the noise is canceled out not effecting further evaluation. One important function of the evaluation software, which is presented in the next section, is the mentioned mathematical smoothing of the measured signals.

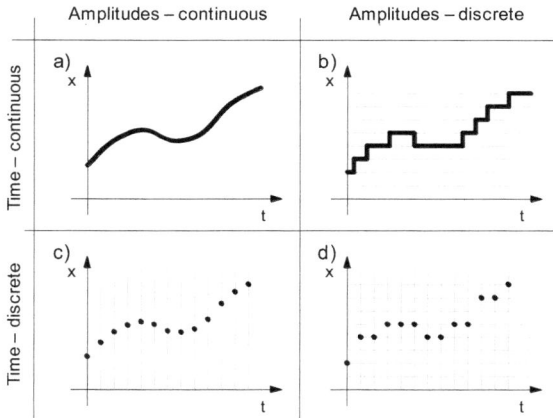

Figure II-18. Classification of signals: a) Continuous signal in time and amplitude domain = analogue signal; b) signal continuous in time domain and discrete in amplitude domain; c) signal discrete in time domain and continuous in amplitude domain; d) discrete signal in time and amplitude domain = digital signal. According to (Best, 1991)

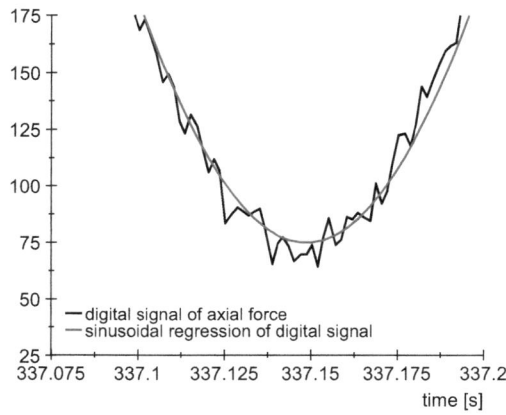

Figure II-19. Example of a digitalized signal with clear effects of noise (black) and analytical regression function by sinusoidal approximation (red).

It can be seen in Figure II-19 that the recorded signal is slightly asymmetrical and this asymmetry is not captures by the approximation. The approximation function used in this case is a simple sine. It will be shown in section II.3.4 that a more advanced approximation function can handle this kind of asymmetry.

II.3.2 Evaluation Software

For the evaluation of the test data from the unprocessed raw data and signal processing by different approximation functions to the identification of material parameters (i.e. dynamic modulus, phase lags,...) all tests carried out for this thesis were processed with a robust evaluation software that was developed at the Research Center for Road Construction, Vienna University of Technology in 2005 and has been improved ever since. The evaluation software can be characterized as robust since it runs stable and finds an optimum in terms of approximation quality if the test data resembles a sine-like shape. This was proven by (Kappl, 2007): A sinus signal was created for which all functional parameters are known. This sinus data was evaluated by the software. The parameter of the sine regression resembled the parameters of the created signal.

The routine can handle all kinds of cyclic dynamic tests from stiffness and fatigue tests with different setups (4PBB, DTC, ITT) to CCTs. It consists of five modules. The following sections will provide information to lay a comprehensible basis for the interested reader. In addition, an advanced approximation function used for data regression is taken care of in more detail. Figure II-20 gives an outline of the data evaluation by the software routine. An additional module for the evaluation of static low-temperature tests (TSRST, UTST) is available.

Determination of Frequency Packets

To make sure that the sum of squared errors in the identification of parameters for the approximation function converges to a global minimum, the initial estimates of the parameters have to be as close as possible to the final values. Therefore the raw test data file (in ASCII format) is divided into packets with identical frequencies of loading. This is crucial since the regression is sensitive to the initial value of the frequency. The control/record unit of the test machine records a virtual counter for load cycles that is increased by 1 after each load cycle. By the first and last time step within a load cycle the periodic time T_p is determined and thus the frequency of the load cycle by the reciprocal value $1/T_p$.

Determination of Initial Values and Parameters of Approximation Function

The evaluation routine provides different approximation function:

$$f(t) = a_1 + a_2 \cdot \sin(2\pi \cdot f \cdot t + a_3) \qquad (2.2)$$

$$f(t) = a_1 + a_2 \cdot \sin(2\pi \cdot f \cdot t + a_3) + a_4 \cdot t \qquad (2.3)$$

$$f(t) = a_1 + a_2 \cdot \sin(2\pi \cdot f \cdot t + a_3) + a_4 \cdot t + a_5 \cdot \sin(4\pi \cdot f \cdot t + a_6) \qquad (2.4)$$

a_1 .... Offset of the fundamental oscillation
a_2 .... Amplitude of the fundamental oscillation
a_3 .... Phase lag of the fundamental oscillation
a_4 .... Gradient of the linear tem
a_5 .... Amplitude of the 1st harmonic oscillation
a_6 .... Phase lag of the 1st harmonic oscillation
f .... Frequency of the oscillation
t .... Time

These functions will be described in greater detail in section II.3.4. For a short identification they are referred to as

– the approximation F for equation (2.2) since this represents a simple sine (the fundamental oscillation),
– approximation $F+L$ for equation (2.3) for the added linear term,
– and approximation $F+L+1H$ for equation (2.4) for the additional 1st harmonic oscillation.

The user can decide how many load cycles are merged to one data block for approximation. The minimum number of load cycles for one approximation block is two for reasons of stability of the evaluation routine. For evaluation of tests carried out for this thesis the number of load cycles per approximation block was set to 3. One approximation block will be dealt with as one data set.

Regardless of which approximation function is used, the first iteration always starts with the simple function F. The initial values $a_{i,0}$ are set to:

$$a_{1,0} = 0$$
$$a_{2,0} = \frac{\max_{RD} - \min_{RD}}{2} \qquad (2.5)$$
$$a_{3,0} = 2\pi \cdot f$$

max_{RD} ... Maximal amplitude value of the raw data within the approximation block
min_{RD} .... Minimal amplitude value of the raw data within the approximation block
f .... Frequency from the determination of frequency packets

The residue between function (2.2) and the test data is aimed to become a minimum. By using the method of least squares together with Newton's method this optimization task is solved after a number of iterations. Since Newton's method works based on the first derivative, the sinusoidal terms require the system of equations to be expanded as a Taylor series to obtain a linear system of equations. More details on the algorithm can

be found in (Blab, et al., 2009) and for details on the mathematical procedures used within the evaluation routine a number of mathematical textbooks exist, e.g. (Zeidler, 2004). The iteration process is carried out until one of the three stop criterions is true:
- The number of iterations is equal to 50 (this value was found sufficient in a number of test runs by (Kappl, 2007)),
- the error of the last iteration is smaller than a user-defined error R_l or
- the difference between the last error and the error before that is smaller than a user-defined value (default: $\Delta R_l = 10^{-10}$).

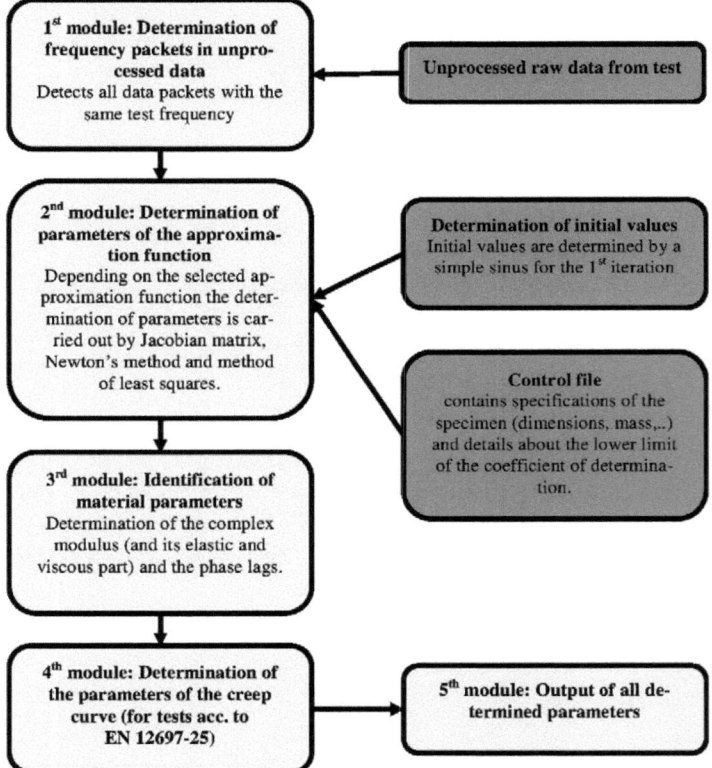

Figure II-20. Principle of the evaluation software for the determination of material parameters and test results of cyclic dynamic material tests.

For the more advanced approximation function $F+L$ and $F+L+1H$ the initial values are set after a first iteration as shown in (2.5) with the simple sine to:

$$a_{1,0} = a_{1,1} \text{ (from 1}^{st}\text{ iteration with simple sine)}$$
$$a_{2,0} = a_{2,1} \text{ (from 1}^{st}\text{ iteration with simple sine)}$$
$$a_{3,0} = a_{3,1} \text{ (from 1}^{st}\text{ iteration with simple sine)}$$
$$a_{4,0} = 0 \qquad (2.6)$$
$$a_{5,0} = \frac{a_{2,0}}{10}$$
$$a_{6,0} = a_{3,0} + \frac{2\pi}{2}$$

To quantify the quality of the regression, the coefficient of determination R^2 is introduced. For a nonlinear regression, R^2 is defined as (Sachs, et al., 2009):

$$R^2 = 1 - \frac{A}{Q_y} \text{ with } A = \sum_{i=1}^{n}(y_i - \hat{y}_i)^2 \qquad (2.7)$$

A is the sum of squared errors of the measured value y_i and the regression values $\hat{y}_i$, and Q_y is the covariance

$$Q_y = \sum_{i=1}^{n}(y_i)^2 - \frac{1}{n}\left(\sum_{i=1}^{n}y_i\right)^2 \qquad (2.8)$$

y_i............Measured value at x_i
$\hat{y}_i$............Regression value at x_i
n............Number of values

R^2 ranges between 0 (no correlation between measured and regression values) to 1 (perfect correlation between measured and regression values). It can therefore be used as a quality parameter how well the approximation function fits the test data.

The result of the 1st and 2nd module according to Figure II-20 is a set of regression parameters $a_{i,k}$ for i from 1 to 6 for data from sensor k. The evaluation software can do a regression on data from five individual sensors at once. A force sensor and at least one deformation sensor is the obligatory minimum. In addition two other deformation sensors and the mean value of two deformation sensors can be evaluated. Also the time stamp, the number of the first load cycle of an approximation block and the recorded temperature for this approximation block is given in an output file. Table II-3 provides an example of the first columns of an output file. The 1st column sets the time stamp, the 2nd column shows the first load cycle of the approximation block, followed by the frequency of the block and the temperature for this approximation block. The next six columns provide the regression parameters $a_{1,F}$ to $a_{6,F}$ for the force sensor. The last column depicted in Table II-3 presents the coefficient of determination for the force sensor. By looking at the data the information can be extracted that the test temperature was 30°C, the frequency of loading was 0.1 Hz for the first 25 load cycles, followed by a frequency of 0.5 Hz.

Table II-3. Example of an output file from the evaluation routine.

Time	LC	Freq	Temp	a1,F	a2,F	a3,F	a4,F	a5,F	a6,F	Rsqr,F
83.236	1	0.1	30	-1.993649	0.9558319	-0.02523143	-0.00006462	0	0	0.9998828
113.236	4	0.1	30	-1.995526	0.9829656	-0.02660261	0.00009752	0	0	0.9995138
143.236	7	0.1	30	-1.992864	1.0027760	-0.02824109	-0.00006323	0	0	0.9998084
173.236	10	0.1	30	-1.994632	1.0164530	-0.03138920	0.00000012	0	0	0.9999025
203.236	13	0.1	30	-1.992120	1.0275290	-0.03420649	-0.00000047	0	0	0.9998981
233.236	16	0.1	30	-1.992029	1.0329320	-0.03698802	-0.00003379	0	0	0.9998834
263.236	19	0.1	30	-1.992980	1.0371310	-0.03941389	0.00001020	0	0	0.9998975
293.236	22	0.1	30	-1.992798	1.0395680	-0.04203499	-0.00000418	0	0	0.9999062
338.662	27	0.5	30	-1.995721	1.0425800	-0.02353552	0.00014248	0	0	0.9998961
344.662	30	0.5	30	-1.992668	0.9588136	-0.02431338	-0.00004240	0	0	0.9998932
350.662	33	0.5	30	-1.992334	0.9852713	-0.02506380	-0.00040967	0	0	0.9995307

The regression parameters are used in the following to derive different material parameters (e.g. the dynamic modulus $|E^*|$). The procedure will be shown in section V.1.2.

II.3.3 Statistical Analysis

Evaluation of test data leads to a number of parameters, e.g. viscoelastic material parameters. Usually more than one data item is derived from one specific part of a test with constant test conditions, e.g. in terms of test temperature and frequency to make sure whether the results follow a certain trend or whether they are stable, and how large is the scattering of results for constant test conditions. One approach is to look at each individual result and compare it to the rest of result data for the respective test conditions. This is time consuming for the researcher and confusing for the reader since diagrams get overcrowded and the essential statements drawn from the results cannot be stated clearly. In addition, the quality of data must be described by objective parameters. These statistical parameters allow results from one test to be compared to results from other tests in terms of quality of the raw data and scatter of results.

Basic statistical parameters, like the mean value (MV) or the variation and standard deviation (SD) are taken as sufficiently explained by literature, e.g. (Sachs, et al., 2009). Another parameter especially to describe the quality of an approximation, the coefficient of correlation R^2, has already been presented in the preceding section II.3.2.

For a number of analyses, quantiles will be given to characterize the scattering of results. Basically, they describe a percentual probability by which a given value is undercut. For example, the 95% quantile of a value x means, that 95% of the results will be below this value x or that 5% of the results will be higher than x. In a mathematical way it can be stated that

$$F^{-1}(p) := \inf\{x \in \mathbb{R} : F(x) \geq p\} \tag{2.9}$$

$F^{-1}(p)$....p-quantile, p between 0 and 1
x............Random variable
F...........Cumulative distribution function

Figure II-21 shows a graphic example for a normal distribution. The p-quantile represents a value of a sample with p% of the sample below and (1-p)% above the value.

For the statistical analysis of the CCTs carried out in this thesis, each frequency packet was look at separately. To account for a certain time that is needed in cyclic dynamic tests to reach a steady state, the complete data set was not used for statistical analysis. Depending on the number of data blocks within one frequency packet, the following share of data blocks was used:
- Less than 40 data blocks: last 50% of data blocks
- 40 to 100 data blocks: last 20 data blocks (20% to 50%)
- More than 100 data blocks: last 20% of data blocks

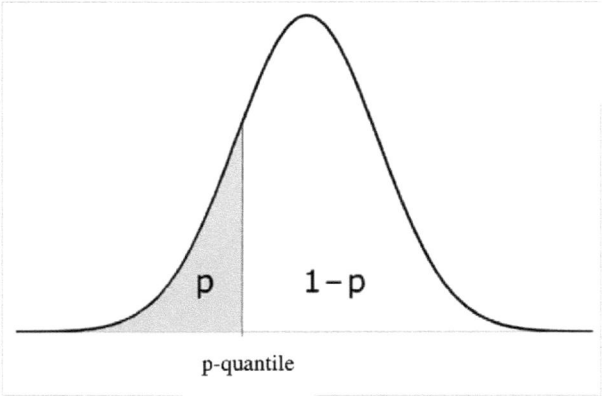

Figure II-21. Graphic example of the p-quantile for a normal distribution.

II.3.4   Details on the Advanced Approximation Function

As shown in (Blab, et al., 2009) standard TCCTs with an axial sinusoidal load and a constant confining pressure result in axial deformations of the specimen that cannot be described with a standard sine according to (2.2). In fact this statement can be generalized for all force-controlled cyclic dynamic tests in the compressive domain with or without confining pressure. It is necessary to take into account the accumulating compressive axial strain by adding a linear term as presented in (2.3). Figure II-22 shows a graphic example of the approximation of test data from a CCT with the $F+L$ function. In the top diagram the test data of axial stress (σ_{ax}) and strain (ε_{ax}) vs. time is shown for two oscillations. In addition the approximation function for ε_{ax} is depicted. At closer examination it is obvious that for specimens from CCTs the reaction in terms of deformation will not follow a simple sinusoidal function even when a linear term is added. At the point of maximum loading the test data of deformation is broader and flatter whereas on the point of minimum loading the deformation appears narrower with a larger peak. The diagram on the bottom shows the stress-strain relationship for the same oscillations. It reveals more clearly that the test data differs from the $F+L$ approximation or rather that the function is not able to fit the test data with satisfactory quality. Especially in the loading and unloading phase, the shape of the test data shows a distortion compared to the sine approximation.

For these reasons (Kappl, 2007) introduced an advanced approximation function for the regression of TCCTs. Basically a new sinusoidal term was added representing the first harmonic of the oscillation characterized by the double frequency of the fundamental oscillation. The $F+L+1H$ function is presented in (2.4).

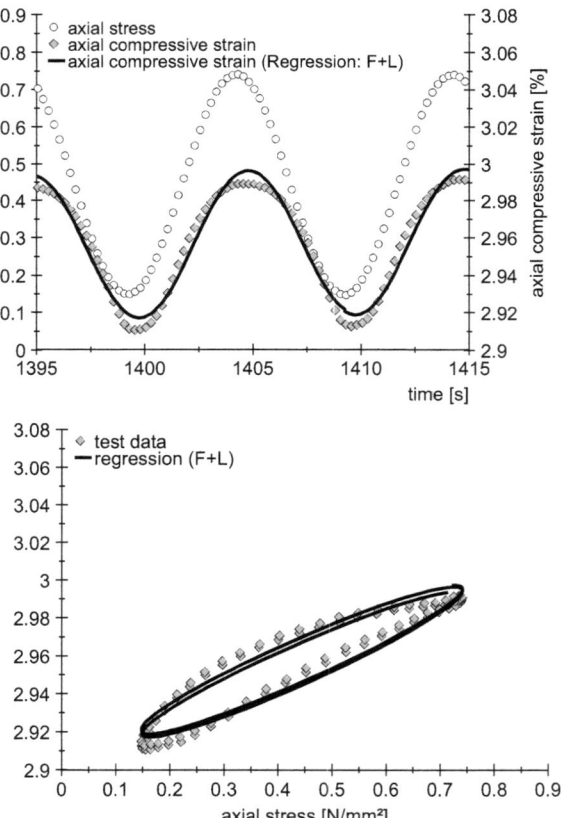

Figure II-22. Example of CCT test data with axial stress (σ_{ax}) and strain (ε_{ax}) and the analytical approximation $F+L$ – in time domain (top) and as a stress-strain relationship (bottom).

Figure II-23 presents the same data as Figure II-22 but with the $F+L+1H$ approximation. By comparing the figures it becomes clear, that the advanced approximation fits the deformation data in a better way than the standard sine approach. This is true for the extrema as well as the loading and unloading phase. Thus, it is stated in (Blab, et al., 2009) that the material parameters (e.g. the dynamic modulus and the phase lags) will represent the true values of the test data better if the $F+L+1H$ function is used for regression. However, the impacts of the 1st harmonic term on the approximation and the shape of the function were not considered. The following section is aimed towards a better understanding of the advanced approximation function, at least in a phenomenological way. Section V.3 will critically review the advanced approximation function, i.e. whether it is correct to use a more complex approximation function to fit the test data better in a mathematical way without mechanical reason for the use of the more complex approximation function.

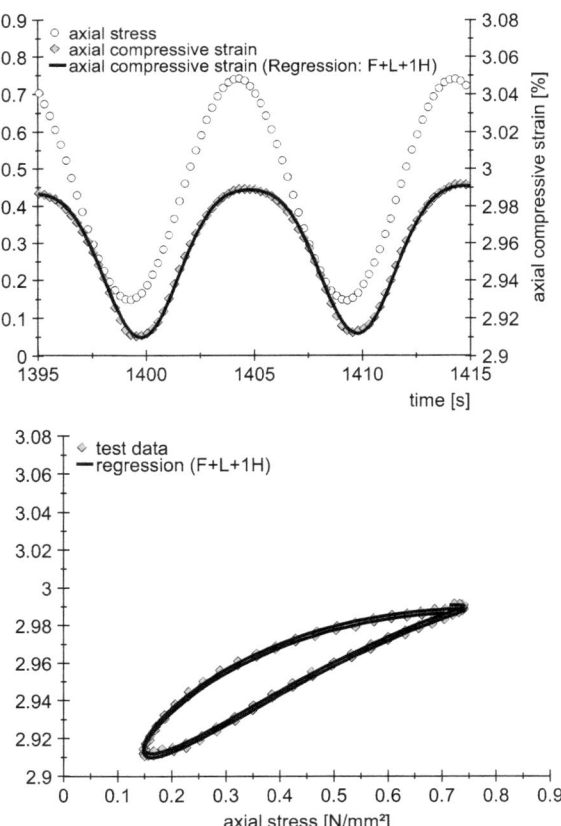

Figure II-23. Example of CCT test data with axial stress (σ_{ax}) and strain (ε_{ax}) and the analytical approximation $F+L+1H$ – in time domain (top) and as a stress-strain relationship (bottom).

To understand the impact of the 1st harmonic term on the shape of the advanced approximation function, this chapter provides a detailed analysis of characteristics of this function.

For this, it may be assumed that data from two sensors are available for evaluation. Both sensors show a sinusoidal behavior (e.g. force and displacement from a cyclic dynamic test) and one sensor lags behind the other sensor (e.g. due to viscoelastic material behavior). Both sensors can be evaluated with the *F+L+1H* function. It is important to know that due to the nature of this function not only one uniform phase lag can be obtained for each oscillation but that the phase lag itself is a function of time. Thus phase lags can be obtained from four different well-defined amplitude function values:

- at the point of the minimum of the oscillation (=*min*),
- at the mean value between minimum and maximum coming from the min (=*MV+*),
- at the point of the maximum of the oscillation (=*max*) and
- at the mean value between maximum and minimum coming from the max (=*MV*).

Figure V-3 shows an example of these functional values.

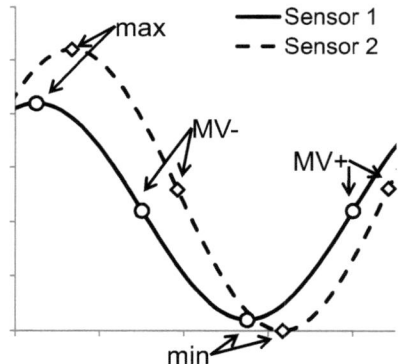

Figure II-24. Definition of function values for the calculation of the four different phase lags.

It was found that the ratio of the amplitude of the 1st harmonic to the fundamental oscillation as well as the shift factor between the 1st harmonic and the fundamental influence the shape of the regressed function in terms of incline between the extremal values and the shape of the extrema respectively, as well as values of the extrema. The amplitude ratio is referred to as

$$AR = \frac{a_5}{a_2} \qquad (2.10)$$

AR.........Ratio of amplitude of 1st harmonic to fundamental oscillation [-] and the shift factor between both sinusoidal terms

$$\gamma = a_6 - a_3 \tag{2.11}$$

γ............ *Shift factor between 1st harmonic and fundamental oscillation [°].*

To systematically study the impact of the 1st harmonic a theoretic example with the following input parameters is given. The offset (a_1), the phase lag of the fundamental oscillation (a_3) as well as the linear term (a_4) are set to 0. The amplitude of the fundamental oscillation (a_2) is 1, the amplitude of the 1st harmonic (a_5) is varied between 0.1 and 0.3 to demonstrate the impact of this parameter. Since a_1 is 1, a_5 is equal to the amplitude ratio AR according to (2.10). The phase lag of the 1st harmonic (a_6) varies within a range of −180° to 180° to show the impact of this parameter. This particular range was chosen because the test results are usually within this area. Since a_3 is 0, a_6 is equal to the shift factor between both sinusoidal terms γ according to (2.11). The frequency f is set to 0.1 Hz. The parameter set is also shown in Table II−4.

Table II−4. Input data used for analysis of the advanced approximation function F+L+1H.

Parameter	Values	Details
a_1	0	Offset
a_2	1	Amplitude of fundamental oscillation
a_3	0	Phase lag of fundamental oscillation
a_4	0	Gradient of linear term
$a_5 = AR$	0.1 to 0.3	Amplitude of 1st harmonic
$a_6 = \gamma$	−180° to 180°	Phase lag of 1st harmonic
f	0.1 Hz	Frequency

Figure II-25 shows nine diagrams with a systematical variation of the shift factor γ between fundamental and 1st harmonic. Starting from the top left at −180° γ is raised by +45° in each diagram to +180° at the bottom. The amplitude ration AR is kept constant at 0.2. The following statements regarding the change of the function are made in comparison to the standard sinusoidal oscillation which is also shown in the diagrams. At −180° the maximum of the function is shifted towards the right and the minimum towards the left. Thus the mean gradient of the unloading phase is increased, and at the same time the mean incline of the loading phase decreased. The function is stretched symmetrically around the base line in amplitude direction. Therefore there is no phase lag between the standard sine and the advanced function at the mean amplitude value. In other words, the advanced function does not influence the behavior of the oscillation in terms of phase lags at the mean amplitude values (Φ_{MV-} and Φ_{MV+}) compared with the standard sine approach.

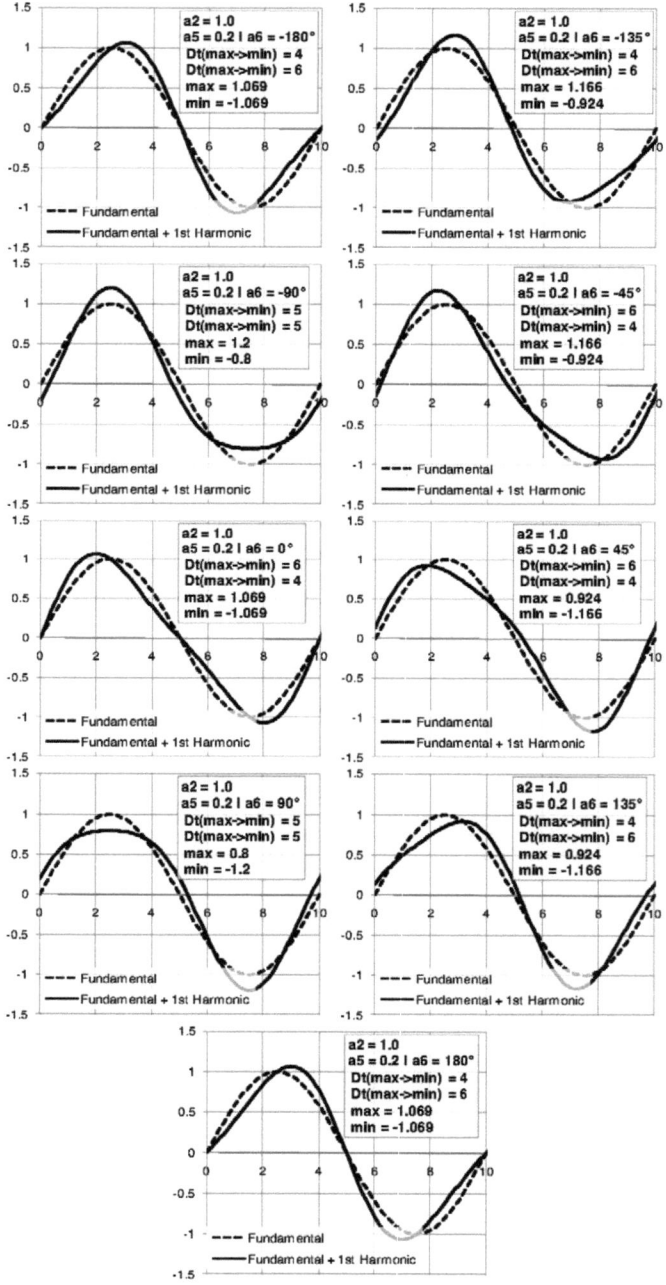

Figure II-25. Variation of shift factor between 1st harmonic and fundamental from $a_6 = -180°$ (upper left) to $+180°$ (bottom).

At −135° the mean gradient of loading and unloading phase are still the same as for the first case. But the function is now stretched asymmetrically so that the maximum of the advanced function is larger than the maximum of the standard sine whereas the minimum is smaller. So there is a phase lag in all four points, the two extremal and the mean amplitude values.

At −90° the situation is inversed to the situation at −180°. The mean gradient of the loading and unloading phase is the equal. The extrema in time domain coincide with the extrema of the standard sine. The difference in the extremal amplitude values is at its maximum. The maximum is raised by the full amplitude ratio AR, the minimum is decreased by AR. There is no phase lag between the extrema of the standard and advanced function.

Again, put in other words, the advanced function starts to shift from the standard sine at the mean amplitude values from −180° on with an increasing time lag to −135° and −90°. For the extremal amplitude values the time lag is constant between −180° and −135°. It starts to get smaller from −135° on and the values overlap at −90° in time domain.

The situation at −45° is analogue to −135° in terms of the values of the extrema and reversed in terms of the mean gradient of loading and unloading phase. The same is true for 0° and −180°. 45° shows analogue behavior as −45° in terms of mean gradients and the behavior is reversed in terms of values of the extrema. 90° is the reversed −90° case and at 135° analogy can be found between to 45° in terms of extremal values, whereas the mean gradients are reversed. Finally at 180° after a full 360° the cycle starts again.

At a shift factor of −180° the 1st harmonic's influence is dominant in the loading and unloading phase leaving the shape of the extrema unaffected. From −180° to −90° the impact of the 1st harmonic shifts to the other extreme. At −90° the shape around the extrema are clearly deformed. In terms of shape of the function the changing influence of the 1st harmonic on the deformation of the sine continues with a turning point every 90°.

Table II-5 shows information about phase lags between the advanced and the standard approximation function at the two extremal amplitude values and the MV in the unloading (MV−) and the loading (MV+) phase. The signs in the table indicate the phase lag qualitatively. A "+" means that the advanced function is ahead of the standard function. 0 indicates that no phase lag occurs.

What becomes clear is that the advanced function can be used to check the shape of a sine quickly. When test data from two sensors (e.g. load and deformation) are analyzed with the standard function only one constant phase lag can occur in one analyzed load cycle. With $F+L+1H$ function the phase lag changes within one load cycle. Thus this approximation is able to fit the material response in a more detailed way which will be shown in the further course of the thesis.

Table II-5. Phase lags between advanced and standard approximation function at different functional values.

Shift Factor γ	Phase lag of advanced vs. standard function @				Change of phase lag of advanced vs. standard function @			
	max	MV-	min	MV+	max	MV-	min	MV+
-180°	+	0	-	0				
					0	-	0	+
-135°	+	-	-	+				
					-	0	+	0
-90°	0	-	0	+				
					-	0	+	0
-45°	-	-	+	+				
					0	+	0	-
0°	-	0	+	0				
					0	+	0	-
+45°	-	+	+	-				
					+	0	-	0
+90°	0	+	0	-				
					+	0	-	0
+135°	+	+	-	-				
					0	-	0	+
+180°	+	0	-	0				

III Materials and Specimen Preparation

To produce the composite material HMA, aggregates (particle size > 0.063 mm) including a certain amount of filler (particle size < 0.063 mm) and bituminous binder are mixed based on mix designs. The following sections present relevant characteristics of each component and the mix designs used. In addition, the process of specimen production and preparation is presented.

III.1 Materials

Aggregates

The coarse aggregate used for mixes is a porphyrite from the Lower Austrian quarry "Loja". This stone is commonly used in the Eastern part of Austria for surface layers. The relevant characteristics according to the European Standard (EN 13043, 2002) are shown in Table III-1 taken from the statement of compliance that is issued by the manufacturer.

Table III-1. Characteristics of "Loja" according to (EN 13043, 2002)

Parameter	Additional Information	Test Procedure	0/2	2/4	4/8	8/11
Aggregate size	---	---				
Particle size distribution	---	EN 933-1	$G_F85.$ $G_{TC}20$	G_C 90/15	G_C 90/15	G_C 90/15
Content of fines	< 0,063	EN 933-1	f_{10}	f_1	f_1	f_1
Fines quality	---	EN 933-9	MB_F NPD	MB_F NPD	MB_F NPD	MB_F NPD
Particle shape	> 2 mm	EN 933-4	---	SI_{15}	SI_{15}	SI_{15}
Percentage of crushed and broken surfaces	---	EN 933-5	$C_{100/0}$	$C_{100/0}$	$C_{100/0}$	$C_{100/0}$
Angularity D < 2 mm		EN 933-6	$E_{CS}35$	---	---	---
Resistance to fragmentation	LA 8/11	EN 1097-2	LA_{20}	LA_{20}	LA_{20}	LA_{20}
Resistance to polishing	PSV 8/11	EN 1097-8	PSV_{50}	PSV_{50}	PSV_{50}	PSV_{50}
Resistance to wear	---	EN 1097-1	M_{DE} NPD	M_{DE} NPD	M_{DE} NPD	M_{DE} NPD
Particle density	ρ_{rd} [Mg/m^3]	EN 1097-6	2.81-2.87	2.81-2.87	2.81-2.87	2.81-2.87
Water absorption	---	EN 1097-6	$WA_{24}1$	$WA_{24}1$	$WA_{24}1$	$WA_{24}1$
Resistance to freezing and thawing	if applicable	EN 1367-1	F_1	F_1	F_1	F_1
Compatibility between aggregate and bitumen	Number of Stones Coverage in %	EN 12697-11 Part B	≤ 1 ≥ 80	≤ 1 ≥ 80	≤ 1 ≥ 80	≤ 1 ≥ 80

Filler/Fines

The filler used for the research is powdered limestone ($CaCO_3$.). The residual moisture content is 0.1% (m/m) or lower. The particle size distribution ensures that at least 80% (m/m) are finer than 0.09 mm.

Binders

Two different bituminous binders were used. A standard, unmodified bitumen 70/100 (penetration between 70 and 100 $^1/_{10}$ mm) and an SBS-polymer modified bitumen PmB 25/55-65 (penetration between 25 and 55 $^1/_{10}$ mm, softening point > 65°C). The characteristics of both binders are presented in Table III-2.

Table III-2. Characteristics of bitumen.

Parameter	Test Procedure	70/100	PmB 25/55-65
Penetration [0.1 mm]	EN 1426	84.0	46.0
Softening Point Ring & Ball [°C]	EN 1427	46.8	73.0
Fraaß Breaking Point [°C]	EN 12593	-17	max. -10 [a]
Ductility at 13°C [cm]	ÖNORM C 9218	-	min. 30 [a]
Elastic Recovery at 25°C [% relative] [a]	EN 13398	-	min. 50 [a]
max. increase of Softening Point [K] [a]	EN 1427	9	-
min. Softening Point after hardening [°C] [a]	EN 1427	45	-
min. residual Penetration [0.1 mm] [a]	EN 1426	46	-
min. Solubility [% (m/m)] [a]	EN 12592	99.0	-
min. Dynamic Viscosity at 60°C [Pa·s] [a]	EN 12596	90	-
min. Kinematic Viscosity at 135°C [mm²/s] [a]	EN 12596	230	-
max. Content of Paraffins [% (m/m)] [a]	EN 12606-1	2.2	-
BBR m-value (-18°C) ageing A [-]	EN 14771	0.338	0.320
BBR S-value (-18°C) ageing A [MPa]	EN 14771	260	65
BBR m-value (-18°C) ageing C [-]	EN 14771	0.274	0.243
BBR S-value (-18°C) ageing C [MPa]	EN 14771	356	264
DSR |G*| (+64 C) ageing A [Pa]	EN 14770	1243	8290
DSR δ (+64°C) ageing C [°]	EN 14770	87.9	63.2
G*/sin(δ) (+64°C) ageing A [Pa]	EN 14770	1243	9290
Rotational Viscosity (+135°C) ageing A [MPas]	EN 13302	375	1629
PG exact [°C]	SHRP	63-24	> 82-19
Δ PG [°C]	SHRP	87	> 101

[a] According to specification sheet of the bitumen manufacturer

Mix Designs

The mix used for this research was an AC 11 surf (in the course of the thesis abbreviated as AC 11) which is an asphalt concrete with a largest nominal aggregate size of 11 mm. This represents a typical mix used for surface layers on Austrian roads. The target grading curve is the Fuller Parabola. During the research project, different batches of the particle size groups of aggregates were used. With new batches of the same particle size groups the grading curve of the mix can change slightly. It was tried to approximate the initial grading curve in a best possible way. Still minor deviation could not be avoided. Therefore 10 slightly different grading curves were produced which are presented in Figure III-1. However, it was taken care of the fact that the produced grading curves are within the limits given by the national standard (ONR 23580, 2009) for the mix. The upper and lower limits are depicted in the diagram in Figure III-1 as well.

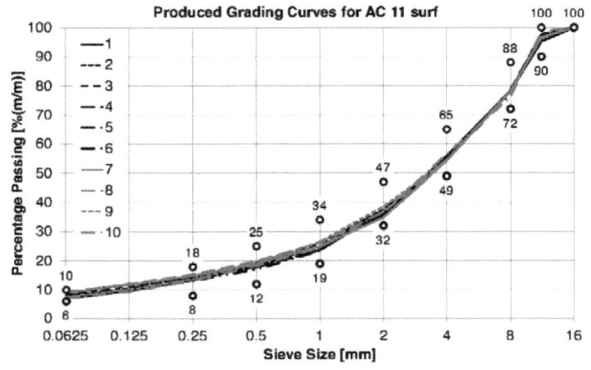

	Percentage Passing [%(m/m)]									
	Grading Curve									
Sieve Size [mm]	1	2	3	4	5	6	7	8	9	10
0.063	6.9	7.6	8.5	9.2	9.2	8.1	9.0	9.2	7.6	6.9
0.125	9.8	10.6	11.4	11.9	11.9	10.6	11.9	12.2	10.2	9.4
0.25	13.5	14.5	14.5	14.4	14.4	13.2	14.9	15.4	13.8	13.2
0.5	18.2	19.3	19.1	17.9	17.9	17.1	19.1	19.9	19.1	18.3
1	25.0	26.2	25.9	23.7	23.7	23.5	25.0	26.2	26.5	24.5
2	35.6	37.4	36.9	36.6	36.6	37.5	34.8	36.7	38.6	36.8
4	55.7	54.5	54.5	54.6	54.6	54.5	54.3	54.5	54.5	56.5
8	78.1	78.1	78.1	78.4	77.3	78.1	78.1	78.1	78.1	76.1
11.2	95.3	97.6	97.6	97.6	96.5	96.6	96.6	96.6	96.5	98.6
16	100.0	100.0	100.0	100.0	100.0	100.0	100.0	100.0	100.0	100.0

Figure III-1. Produced grading curves for the AC 11 surf

III.2 Specimen Preparation

The complete process of specimen preparation from mixing and compaction to coring and cutting was carried out in accordance to the respective European Standards (EN).

The mix was produced in a reverse-rotation compulsory mixer according to (EN 12697-35, 2007). The mix drum as well as the mixing device are heated to ensure correct mix and compaction temperatures. The pre-heated aggregates and filler are mixed for 1 minute before the pre-heated bitumen is added to the mix. Aggregates, filler and bitumen are mixed for an additional 3 minutes. After the mixing process the material is compacted in a segment roller compactor according to (EN 12697-33, 2007). Figure III-2 shows the compactor used in the lab. Slabs compacted by the device have a base area of 50x26 cm and a variable height of up to 22 cm. The radius of the segment of 55 cm corresponds to the size of standard roller compactors used in the field. It was shown in

(Airey, et al., 2005) that this compaction method reproduces the compaction at construction sites in the best way of all compaction methods given by European Standards at this time.

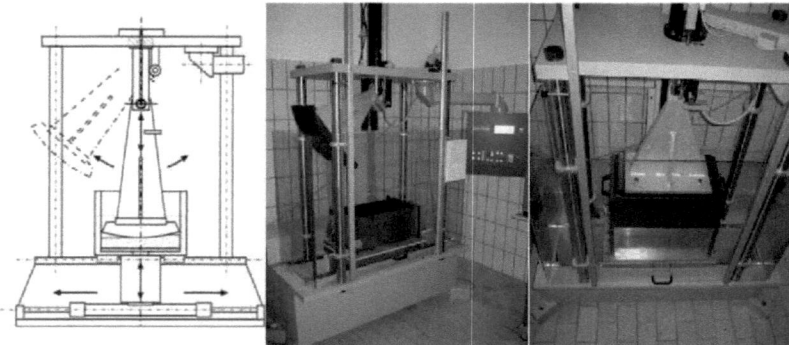

Figure III-2. Segment roller compactor used for slab compaction.

The slabs were usually compacted to a target height of 13.0 cm in a displacement-controlled way. The complete slab was compacted in two layers. Single-layered compaction leads to large scatter of the density between upper and lower parts of the slab (Höflinger, 2006). Since the bulk density is known as well as the target content of air voids, the target density can be derived. The target density and the target volume of the slab define the necessary mass for compaction.

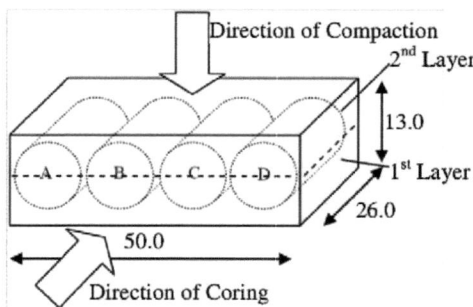

Figure III-3. Principle of specimen direction within HMA slab.

From the slabs, four specimens are cored out with a diameter of 100 mm. Compared to the direction of the compaction force, the direction of coring is orthogonal to the compaction. The obtained specimens are then cut to a height of 200 mm. The reason for this specimen orientation is mainly practical. If the specimens were cored out so that the height of the specimen is in direction of the compaction force, the slab would have to be at least 22 cm in height. This would almost double the mass of the slab to around 80 kg and thus the slabs would be hard to handle in the lab. Of course, the stress/strain situation in a pavement is simulated better when the specimens are cored so that its height is in direction of the compaction force.

For this research, SGs were attached to most of the specimens to derive the radial strain from the recorded data. The procedure for attaching SGs is described in section II.2.3 in short, as well as in (AA542, 2010) in more detail.

Before the specimens are finally tested they are stored at the test temperature for at least 4 h but no longer than 7 h. This is in accordance to (EN 12697-25, 2005).

IV AN IMPROVED ASSESSMENT OF THE RESISTANCE TO PERMANENT DEFORMATION

The present way of data evaluation and interpretation of results from standard TCCTs according to (EN 12697-25, 2005) is not thoroughly satisfying. Section IV.1 describes the evaluation routine and shows that the EN standard requires approximating the quasi-linear part of the creep curve with a linear function. As a matter of fact there is no linear part of the creep curve. The standard fails to give information how to define the quasi-linear part. The results of the test are dependent on the choice of the load cycle range for the quasi-linear part. Different users may employ different methods to define the quasi-linear range of the creep curve Thus, the results of TCCTs according to (EN 12697-25, 2005) (= resistance to permanent deformation) are repeatable but may not be comparable if different users or laboratories evaluate the same test data. It is therefore seen as crucial to develop an assessment of the resistance to permanent deformation that is well-defined through procedures that are not dependent on the user. This objective is worked on in section IV.2.2.

Furthermore CCTs offer the rare and valuable opportunity among the performance based test methods for HMA to study not only the axial deformation behavior but also deformation perpendicular to this axis in the radial plane. This holds true if strain in both direction is recorded with sufficient quality, which is the case in this thesis. Thus a detailed study into this quasi-3-d state of strain from standard TCCTs according to (EN 12697-25, 2005) is carried out in section IV.2.3.

IV.1 Evaluation of Standard TCCTs according to EN 12697-25

When standard TCCTs according to (EN 12697-25, 2005) are carried out, the main interest in terms of results is the assessment of the resistance to permanent deformation. Therefore the time-strain-curve or creep curve is taken into consideration. To describe the resistance to permanent deformation the axial creep curve or parts of it have to be represented analytically by a suitable regression function.

Creep curves obtained from TCCTs according to (EN 12697-25, 2005) can be divided into three different phases (Figure IV-1):

- The primary phase (1): Within the first phase of a TCCT a certain amount of re-compaction leads to decreasing slope of the curve with increasing number of load cycles.
- The secondary phase (2): The main phase of the TCCT is characterized by a quasi-constant slope of the curve.
- The third phase (3): Usually the standard TCCT does not reach this state where the deterioration of the specimen leads to an increase of the slope of the curve with increasing number of load cycles.

(EN 12697-25, 2005) provides two methods for the regression of the creep curve which allow the assessment of the resistance to permanent deformation quickly and easily. The two methods described in section 5.6 of (EN 12697-25, 2005) are
- the determination of a creep rate f_c by means of a linear function and
- the determination of a regression parameter B by means of a power function.

For type testing of HMA according to (EN 13108-1, 2006), the linear creep rate f_c is the benchmark parameter for the resistance to permanent deformation.

For both approaches the measured axial deformation must be converted to axial strain ε_{ax}:

$$\varepsilon_{ax}(n) = \frac{h_0 - h(n)}{h_0} \cdot 100 = \frac{\Delta h(n)}{h_0} \cdot 100 \qquad (4.1)$$

$\varepsilon_{ax}(n)$......Axial strain at the load cycle n [%]
h_0..........Initial height of the specimen before the TCCT
$h(n)$.......Height of the specimen at the load cycle n

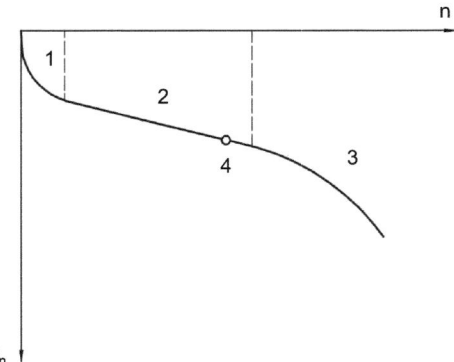

Figure IV-1. Example of a creep curve (cumulative axial deformation ε_n (= $\varepsilon_{ax}(n)$) vs. number of load cycles n) showing the different phases.

The axial strain is determined for the complete test and presented in a load-cycle-strain-diagram with linear scale for both axes. If a secondary creep phase with a quasi-constant incline of the creep curve can be determined from the data in the diagram, an approximation function of the following kind is to be fitted to the quasi-linear part of the curve by using the method of least squares:

$$\varepsilon_{ax}(n) = A_1 + B_1 \cdot n \qquad (4.2)$$

$\varepsilon_{ax}(n)$......Approximated function for permanent axial strain at the load cycle n
A_1..........Regression parameter describing the intersection of the approximation function with the y-axis (offset)
B_1..........Regression parameter describing the incline of the approximation function
n ....Load cycle

An example of the linear regression to the creep curve is shown in Figure IV-3.

The creep rate f_c is determined as the incline B_1 of the linear function in micrometers per meter per load cycles:

$$f_c = B_1 \cdot 10^4 \qquad (4.3)$$

f_c is used to determine the resistance of a specimen to permanent deformation. The smaller its absolute value, the smaller is the increase of permanent deformation vs. load cycles. Thus a smaller absolute value of f_c means that the resistance to permanent deformation is higher. The European Standard states that this method is easy to handle with the disadvantage that it represents the actual creep curve only "to a certain extent". Besides, f_c is clearly dependent on the choice of the interval of load cycles used for the approximation, since the creep curve usually does not show a constant incline but only a quasi-linear domain.

A quasi-linear domain is characterized by the fact that the first derivative of the axial strain ε_{ax} with respect to the number of load cycles approaches a constant value. This means that before starting a regression the range of load cycles has to be found in which the axial strain behaves quasi-linear. Therefore the differential quotient of the axial strain has to be analyzed:

$$\frac{\Delta \varepsilon_{ax}}{\Delta n} = \frac{\varepsilon_{ax,i} - \varepsilon_{ax,i-1}}{n_i - n_{i-1}} \qquad (4.4)$$

Figure IV-2 shows an example of the differential quotient of the axial strain for the complete range of load cycles in the left diagram and from the 5,000th to the 25,000th load cycles in the diagram on the bottom. If only the top diagram is taken into consideration it seems that the differential quotient stabilizes shortly after the beginning of the test. When the scale is changed (lower diagram), it is obvious that the differential quotient never reaches a constant value. The question where the quasi-linear part of the creep curve starts is not answered by the EN standard.

It was therefore decided to set a fix range of load cycles for which the linear approximation is carried out for standard TCCT evaluation in Austria. This range from the 5,000th to the 20,000th load cycle was used for all TCCTs within this thesis.

The second method to approximate the creep curve given by (EN 12697-25, 2005) is a regression with a power function:

$$\varepsilon_{ax}(n) = A \cdot n^B \qquad (4.5)$$

$\varepsilon_{ax}(n)$......Approximated function for axial strain at the load cycle n
A............Regression parameter characterizing the intersection of the function with the y-axis at x=1
B............Regression parameter characterizing the incline of the function in the log-log-scale
n............Load Cycle

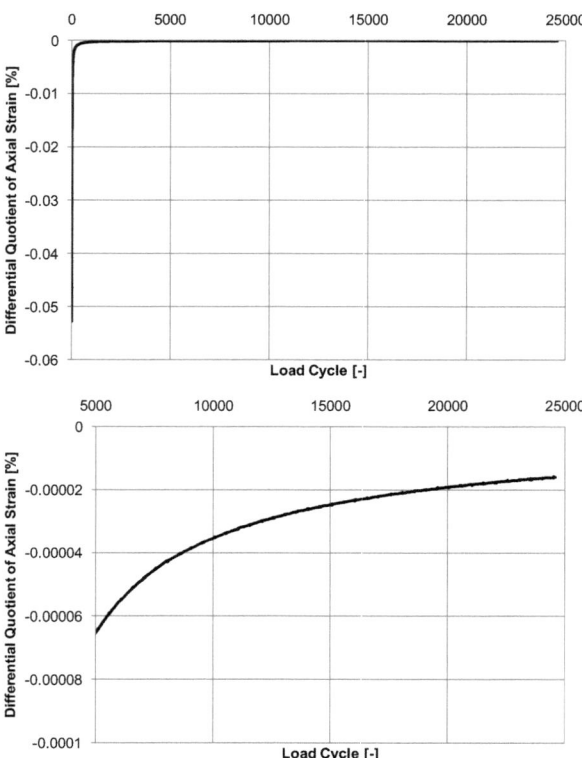

Figure IV-2. Differential quotient of axial strain vs. load cycles for the complete range of load cycles (top) and from 5,000th to 25,000th load cycle (bottom).

An example of the exponential approximation to the creep curve can be seen in Figure IV-3.

The function represents a linear behavior in the log-log scale. The standard gives two parameters to characterize the resistance to permanent deformation. First of all the calculated permanent axial strain at the 1,000th load cycle $\varepsilon_{1000,calc}$

$$\varepsilon_{1000,calc} = A \cdot 1000^B \qquad (4.6)$$

and secondly the exponent B. In most cases the approach with the power function will approximate the complete creep curve with sufficient quality. Therefore the approximation with the power function is carried out from the 1st to the 25,000th load cycles for TCCTs within this thesis.

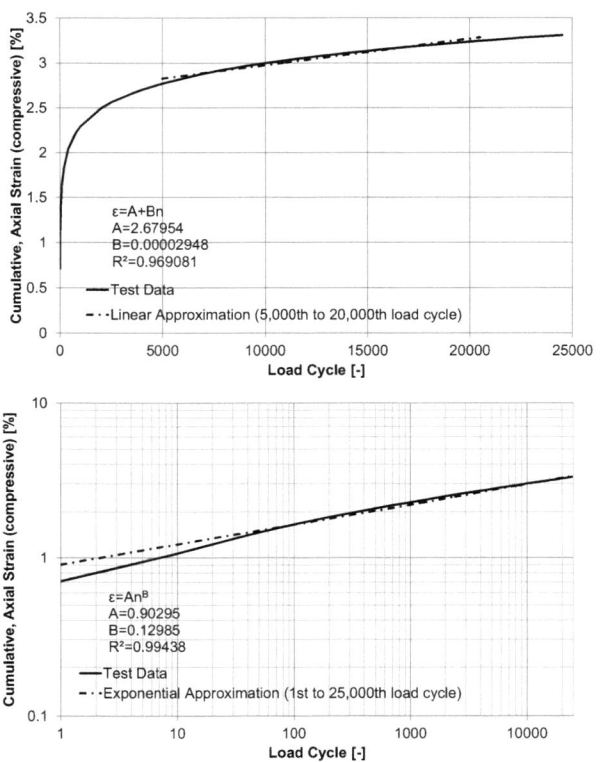

Figure IV-3. Linear (top) and exponential (bottom) approximation to the TCCT test data.

IV.2 Towards an Alternative Assessment of the Resistance to Permanent Deformation

IV.2.1 Test Program

To put the assumptions and findings of the following sections on a sound basis of different mixes, data from a former project (Wistuba, et al., 2007) will be used in the following. The project worked on the performance of HMA and contains numerous results of TCCTs for various mixes. The axial deformation was recorded in the same way as it was done for this thesis, but the radial deformation was measured by the LVDT-based device described earlier in section II.2.1. This has no effect on the quality of the results since only the accumulated radial deformation is relevant for the investigations of this section. For this purpose the LVDT-based device is a perfect tool. Details of the mixes can be found in Table IV-1.

Table IV-1. Mixes used for investigation of alternative assessment of deformation resistance.

Gradation type [-]	Max. aggregate size [mm]	Binder [-]	Aggregate [-]	Binder content [% (m/m)]	Air void content [% (v/v)]	Test temperature [°C]	# of specimens [-]
AC	11	PmB 45/80-65	Diabase	5.6	3.0	50	4
			Steel slack			50	3
						40	3
		70/100	Porphyrite	4.8		50	3
				5.3			
				5.8			
		PmB 25/55-65		5.3			
SMA	11	PmB 45/80-65	Diabase	6.5	3.0	50	3
			Steel slack	5.6			2
		70/100	Diabase	6.5			4
			Steel slack	5.6			2
SMA	11	PmB 45/80-65	Diabase	5.8	10.0	50	3
		70/100	Diabase				3
			Steel slack				1
AC	22	50/70	Limestone	4.5	3.0	40	2

The gradation types contained a standard AC 11 and SMA 11 suited for surface layers, both with a maximum aggregate size of 11 mm. One SMA was designed with a target air void content of 3.0% (v/v), the other one with a higher void content of 10.0% (v/v) used for noise reducing pavements. Two binders were compared, an SBS-modified PmB 45/80-65 (penetration between 45 and 80 and a softening point above 65°C) and a 70/100 bitumen. Also two mineral types were employed in the survey: a diabase and industrially processed steel slack (referred to as LD in the following diagrams for the steel making method "Linz-Donawitz" or basic oxygen steelmaking). Most specimens were tested at standard TCCT conditions of 50°C, the AC 11 with the PmB and steel slack was also tested at 40°C. In addition a typical base layer AC 22 with a 50/70 bitumen and limestone as the mineral aggregate (referred to as Hollitzer in the following diagrams for the quarry the material was taken from) was tested. According to the standard, it was tested at 40°C.

In addition, data from standard TCCT carried out for this thesis are also used for the investigation. AC 11 mixes with two different binders, one with an unmodified 70/100 bitumen and one with an SBS-modified PmB 25/55-65 were tested. For the 70/100 bitumen specimens with three different binder contents (4.8% (m/m), 5.3% (m/m), 5.8% (m/m)) were prepared. For the PmB 25/55-65 specimens with one binder content (5.3% (m/m)) were prepared. The target air void content for all specimens is 3.0% (v/v).

Compaction method and dimensions of the specimens are in accordance to section III.2.

IV.2.2 Alternative Assessment of the Resistance to Permanent Deformation in the TCCT

As mentioned in the introductory remarks of this chapter, the present way of assessing the resistance to permanent deformation from results of standard TCCTs according to (EN 12697-25, 2005) is not entirely satisfying. Especially the linear regression to the quasi-linear part of the creep curve is problematic, since there is no linear part of this curve and the standard does not state how to define this quasi-linear part. Thus the present practice in Austria is to set a fixed number of load cycles between which the linear regression is derived from, in detail between load cycle 5,000 and 20,000. This is not satisfying since it does not account for different behavior of different mixes.

In the course of investigating viscoelastic material parameters derived from TCCTs, which will be presented in chapter V, an interesting relationship between the evolution of the viscoelastic parameters with the number of load cycles and the development of the creep curve was found. When the creep curve is plotted in the log/lin scale, i.e. the number of load cycles in log and the strain in linear scale, it becomes linear from the point where the viscoelastic parameters get stable.

Figure IV-4 shows an example: the evolution of the viscoelastic material parameters with the number of load cycles under standard TCCT conditions, i.e. 50°C, 3 Hz, 25,000 load cycles with an axial sinusoidal compressive stress ranging from 150 kPa to 750 kPa and a constant radial confining pressure of 150 kPa for an AC 11 70/100 with a binder content of 5.6% (m/m) and 1.5% (v/v) air void content. The three diagrams show the dynamic modulus $/E^*/$, the axial phase lag $\varphi_{ax,ax}$ between axial loading and axial deformation and the radial phase lag $\varphi_{ax,rad}$ between axial loading and radial deformation vs. the number of load cycles indicated by the black line. The x-axis presents the number of load cycles in log-scale. From the data it is obvious that there is a phase with changing viscoelastic material parameters at the beginning of each test. The dynamic modulus is at a low level at first, increasing to a stable level, whereas the phase lags are higher for the first load cycles descending to a lower steady level. This shows that the material reacts less stiff and more viscous at the beginning of a CCT.

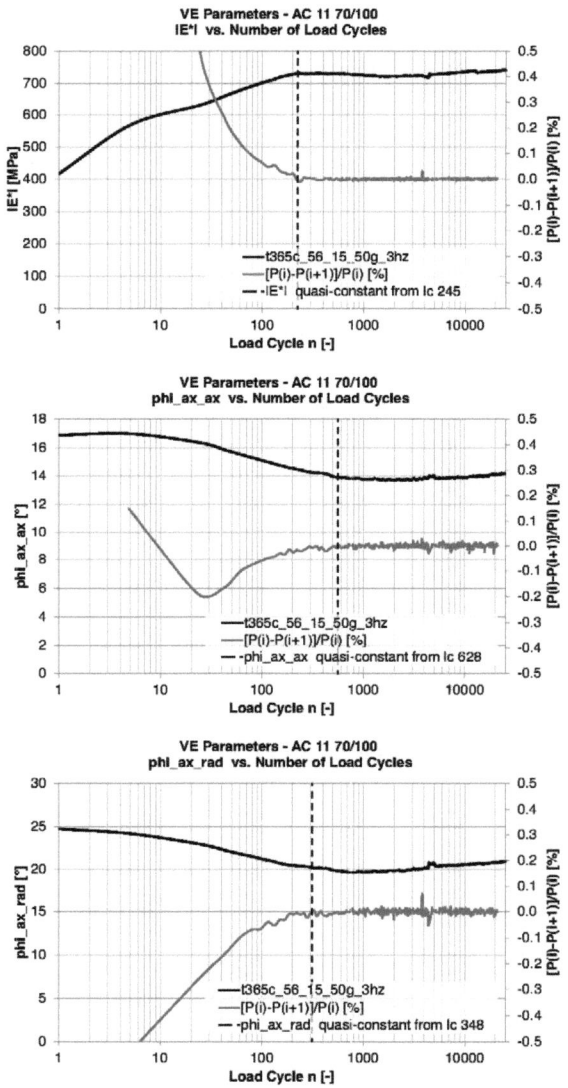

Figure IV-4. Evolution of $|E^*|$ (top), $\varphi_{ax,ax}$ (middle) and $\varphi_{ax,rad}$ (bottom) of AC 11 70/100 with the number of load cycles at standard TCCT conditions (50°C, 3 Hz).

After this phase the parameters are practically constant. To give a rational procedure how to find this point from where on the parameters are stable, the following approach was taken. To smooth the data, the calculations are carried out with the mean value of data points from 20 consecutive load cycles. So, the value for a viscoelastic parameter

$P(i)$ and $i = 1$ would be the mean value of this parameter from load cycle 1 to 20, for $i+1$ from load cycle 21 to 40 etc. The differential quotient of these mean values

$$\frac{P(i+1) - P(i)}{P(i)} \tag{4.7}$$

is derived for the total number of load cycles. When this value approaches 0, it can be stated that the respective parameter P is stable at this range of load cycles. The data is shown in the diagrams in Figure IV-4 as grey lines and the y-scale to the right of the diagrams shows the data range. This differential starts from a positive or negative value and then approaches 0. The point from where the viscoelastic parameters are stable is set at the load cycle when the differential reaches 0 for the first time. This analysis is carried out for all viscoelastic parameters available. For a standard, state of the art test setup for TCCT force and axial deformation are recorded with sufficient quality to derive the dynamic modulus $|E^*|$ and the axial phase lag $\varphi_{ax,ax}$. The load cycle when all evaluated viscoelastic parameters are constant is crucial for the further evaluation of the creep curve. For the mix presented here as an example, the steady state is reached between load cycle 248 for $|E^*|$, load cycle 348 for $\varphi_{ax,rad}$ and load cycle 628 for $\varphi_{ax,ax}$. Thus for this test all viscoelastic parameters are constant after load cycle 628. If this load cycle is now taken and marked in the creep curve of this test, then the behavior of the curve is a linear in the log/lin scale. This fact is shown in Figure IV-5 for the same material and test. Thus a function of the following kind can be used to approximate the creep curve:

$$\varepsilon_{ax}(n) = a + b \cdot \ln(n) \tag{4.8}$$

a .... regression parameter indicating the intersection of the function at load cycle 1
b .... regression parameter indicating the slope of the function in the log/lin scale
= logarithmic creep rate
n .... number of load cycles

To be able to rank the quality of this fit to the fit quality of the regressions that already exist in the standard, a comparison was made, which included the linear function in the specified load cycle range (5,000 to 20,000) and the power function over the complete number of load cycles. In addition the power function in the linear viscoelastic range was also compared to the results of the newly introduced logarithmic function. For this reason Figure IV-6 shows the creep curve again in the log/lin scale (top) and the lin/lin scale (bottom) with the approximation functions mentioned above. From the visual investigation no clear conclusions can be drawn, besides that the power functions used for the complete range of load cycles seems to overestimate the strain in the first part of the creep curve, where the material reacts non-linear viscoelastic. But since the long term behavior is of greater interest it is more important how the regression functions fit the creep curve in the higher range of load cycles. The lower diagram in lin/lin scale also shows that the linear viscoelastic range starts very early after a little more than 2% of the total number of load cycles.

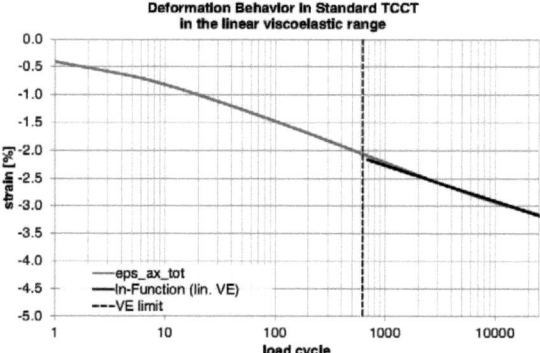

Figure IV-5. Creep curve for AC 11 70/100; the point from where the behavior of the material is linear viscoelastic is marked (load cycle 628).

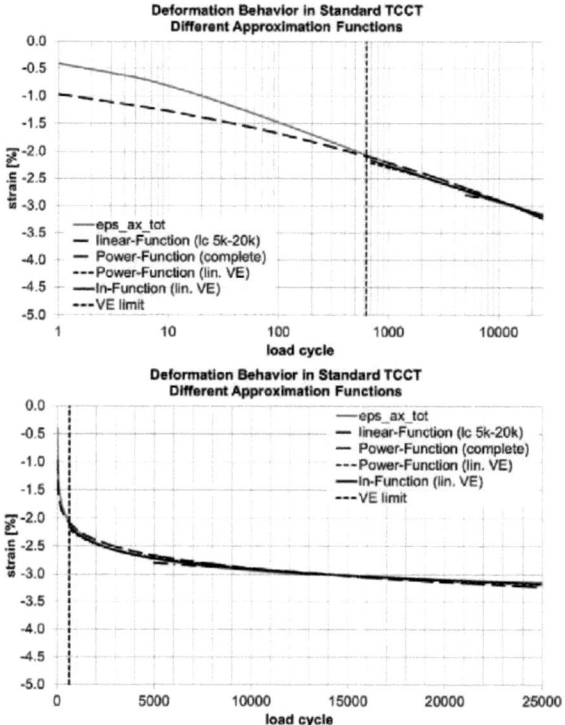

Figure IV-6. Creep curve for AC 11 70/100 and different approximation functions in the log/lin scale (top) and the lin/lin scale (bottom).

To quantify and rank the fit quality of the different approximation functions, the deviations between each regression and the creep curve from test data were calculated for each load cycle and the 2.5%, 50% and 97.5% quantiles of these deviations were obtained. To make sure that the data is comparable the deviations were only compared in the linear viscoelastic part of the creep curve and for the linear regression between the specific load cycles (5,000 and 20,000). Figure IV-7 presents the results in the top diagram. The approximation quality of the linear function implemented in the European standard (equation (4.2)) and the newly introduced logarithmic function (equation (4.8)) are comparable. In fact the *ln* function fits the creep curve even a little better. In 97.5 out of 100 cases the deviations will stay below 1.63% compared to 2.06% for the linear regression. The power functions both show larger deviations.

In the lower table in Figure IV-7 the regression parameters for all four approximation functions are depicted. For the linear function the creep rate f_c $(= b \cdot 10^4)$ is shown instead of the slope b because the creep rate is the benchmark parameter according to the standard. Again, as the long term behavior is the main point of interest the slope of each curve (parameter b) is the major parameter. Interestingly enough this parameter is similar for the linear (-0.219 µm/(m*n)) and the logarithmic function (-0.281 µm/(m*n)).

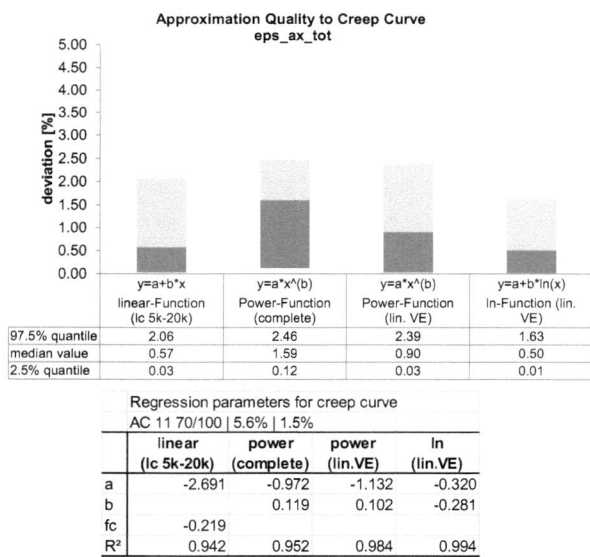

Figure IV-7. Fit quality of different approximation functions (top) and regression parameters (bottom).

From this first analysis with the new logarithmic function it can be stated, that this approach is perfectly able to describe the creep curve. In fact when the material behavior is considered, a logarithmic function seems to be reasonable at least in the first and second phase of the creep curve before the material is subject to micro- and macro-cracks (in the third phase) because with this linear behavior in the log/lin scale, the creep curve will neither reach an asymptotic value, nor does it produce the same permanent strain with every load cycle. The deformation rate will rather slow down. It was also shown that at least for the presented case the creep rate f_c from the linear function and the slope of the *ln* function b exhibit similar values and both approximations result in similar fit quality.

To extend the data basis for this investigation mixes from an earlier project presented in section IV.2.1 were used to compare the different evaluation methods in terms of approximation functions. Unfortunately only the creep curves for the radial and axial deformation were stored over the number of load cycles for the tests. Thus no viscoelastic parameters could be derived from the data and for this reason the viscoelastic linearity limit could not be obtained. Still, the benefit from this large variety of mix designs was considered larger than the setback of limited test data. To be on the safe side, the limit for linear viscoelasticity was set to 500 load cycles for all tests to be able to use the logarithmic approach as well. To safe some space here, the diagrams analogue to Figure IV-7 are not depicted here for each single mix. From the analysis it can be shown that for all materials and tests, the fit quality of the logarithmic regression is similar and in many cases even slightly better than the quality of the linear function. Also the power function always resulted in worse fit quality. More important is the question how the results from linear and logarithmic function are correlated. For this reason Figure IV-8 compares the results derived from all materials with the linear function between load cycle 5,000 and 20,000 in terms of the creep rate f_c to the parameter b from the logarithmic function within the linear viscoelastic range (in this case from load cycle 500 on). The diagram contains the mean values of the results derived from a certain number of single tests for each material and the MV ± SD. For all three values a linear regression analysis was carried out (highlighted in grey for the mean values). Since the slope of the linear is nearly 1 (0.9978) and the offset is very small (-0.022) with a high coefficient of correlation (0.9785) it can be stated that both parameters correlate excellently.

When taking into consideration that the results were derived from different gradation types with different binder types, binder contents and contents of air voids as well as composed of different aggregates and tested at different temperatures, the strong correlation between both evaluation methods is significant. It can be stated positively that there is a well-defined interrelation between the standard linear regression and the newly introduced logarithmic expression. The major advantage of the logarithmic function is that the point from where it is laid into the creep curve can be defined unambiguously and this point can be derived easily from the evolution of viscoelastic parameters with the number load cycles. It is strongly recommended to substitute the linear regression with the presented evaluation routine with the logarithmic approach in the standard (EN 12697-25, 2005), since the linear function produces results that depend on the individual user. The correlation found in this study can be used to convert parameters of material already tested and evaluated with the standard routine.

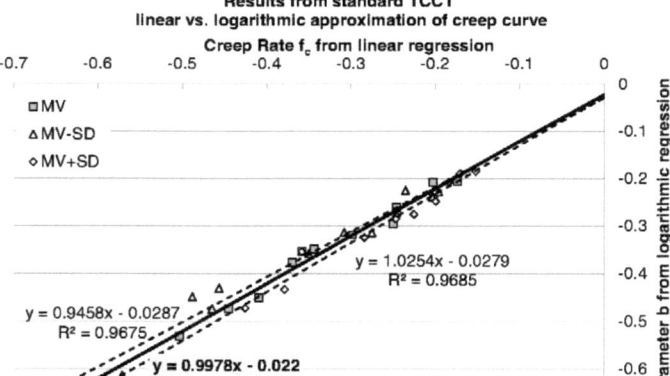

Figure IV-8. Comparison of results from standard TCCT according to (EN 12697-25, 2005) with the linear and logarithmic regression.

For four different mixes, AC 11 70/100 with three different binder contents and one AC 11 PmB 25/55-65, standard TCCTs were carried out for this thesis. They were also evaluated according to the already existing standard approaches as well as with the newly introduced logarithmic approach. The tables in Figure IV-9 sum up the different parameters of the regression to the creep curve for all four mixes. From the coefficient of correlation in the last row of each table it becomes obvious that the new logarithmic regression within the linear viscoelastic part of the test delivers best results.

Figure IV-8 already presented the correlation between the linear regression to the creep curve given in (EN 12697-25, 2005) and the logarithmic function introduced in this thesis. It was shown that both parameters (linear and logarithmic creep rate) correlate strongly. To demonstrate that the mixes tested for this thesis also fit into this picture, Figure IV-10 presents a diagram where the mean values of the creep parameters of the four tested mixes are incorporated into the data. All four points lie within the analyzed data from the other mixes and underline the fact that logarithmic creep parameters can be derived from linear ones.

Regression parameters for creep curve eps_ax_tot
AC11 70/100 Loja 50°C | 4.8% | 3.0%

	linear (lc 5k-20k)	power (complete)	power (lin.VE)	ln (lin.VE)
a	-2.4 ± 0.06	-1.061 ± 0.075	-1.147 ± 0.075	-0.623 ± 0.126
b		0.098 ± 0.006	0.089 ± 0.005	-0.215 ± 0.009
fc	-0.187 ± 0.01			
R²	0.968 ± 0.003	0.98 ± 0.003	0.997 ± 0.001	1 ± 0

Regression parameters for creep curve eps_ax_tot
AC11 70/100 Loja 50°C | 5.3% | 3.0%

	linear (lc 5k-20k)	power (complete)	power (lin.VE)	ln (lin.VE)
a	-3.207 ± 0.191	-1.172 ± 0.048	-1.267 ± 0.074	-0.251 ± 0.012
b		0.12 ± 0.002	0.111 ± 0.001	-0.355 ± 0.023
fc	-0.301 ± 0.014			
R²	0.962 ± 0.001	0.986 ± 0.003	0.993 ± 0.003	0.999 ± 0.001

Regression parameters for creep curve eps_ax_tot
AC11 70/100 Loja 50°C | 5.8% | 3.0%

	linear (lc 5k-20k)	power (complete)	power (lin.VE)	ln (lin.VE)
a	-3.544 ± 0.112	-1.513 ± 0.145	-1.643 ± 0.129	-0.826 ± 0.21
b		0.102 ± 0.007	0.092 ± 0.005	-0.327 ± 0.011
fc	-0.276 ± 0.005			
R²	0.966 ± 0.004	0.98 ± 0	0.994 ± 0.001	0.999 ± 0.001

Regression parameters for creep curve eps_ax_tot
AC11 PmB 25/55-65 Loja 50°C | 5.3% | 3.0%

	linear (lc 5k-20k)	power (complete)	power (lin.VE)	ln (lin.VE)
a	-2.217 ± 0.159	-0.969 ± 0.094	-1.102 ± 0.101	-0.653 ± 0.104
b		0.098 ± 0.003	0.084 ± 0.003	-0.188 ± 0.008
fc	-0.158 ± 0.008			
R²	0.964 ± 0.003	0.958 ± 0.006	0.994 ± 0.002	0.999 ± 0.001

Figure IV-9. Regression parameters from standard TCCT for AC 11 mixes.

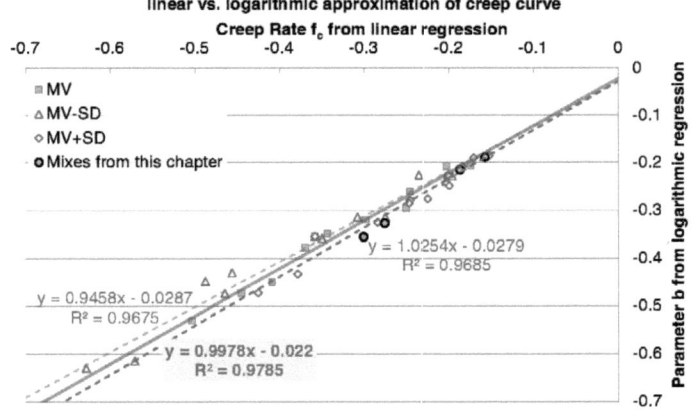

Figure IV-10. Linear vs. logarithmic regression to the creep curve including the mixes tested in this chapter.

IV.2.3   Quasi-3-d Deformation Behavior of HMA in the TCCT

HMA is a composite material made of three constituents, the binder, the mineral aggregate structure and a certain amount of air voids. In terms of the permanent deformation behavior the following hypothesis is raised: When an HMA specimen is loaded so that permanent deformations are provoked by the test setup, these deformations can be caused by two different mechanisms:
- The structure can be compacted in a way that the content of air voids is reduced without any change in its shape. This can be referred to as a purely volumetric strain and the mechanical phenomenon is connected to rutting by pure compression without any doming.
- On the other hand the specimen could just change its shape without any change in its volume; a phenomenon that can be referred to as purely deviatoric. With respect to rutting in the field, this deformation type would be a shear deformation where doming is dominant on both sides of the tire lane.

Usually a CCT will result in a mix of both deformation mechanisms and it is presumed at this point that the ratio between both kinds of deformations is influenced by the mix design and also mainly by its stiffness parameters. Thus this ratio will also change with temperature and frequency. At a low stiffness level where the binder lost most of its bearing ability and reacts more and more like a fluid, aggregates can reposition more easily, and are therefore more mobile. The aggregate skeleton cannot be held in its initial position and thus the deviatoric strain component will get more dominant the lower the stiffness of the mix gets. On the other hand, when the material shows higher stiffness the weakest part are the air voids that will be compressed by the mix constituents and the volumetric part of deformation will be more dominant.

It is also relevant in the field to know how a material reacts in terms of permanent deformation and how the ratio between both deformation types is. Depending on the application of a mix, surface layers are more prone to shear deformation especially in areas where high shear stresses are introduced into the pavement, e.g. at intersections or on airfields where the noseleg of aircrafts often turns without lateral movement. Binder layers are influenced by vertical stresses and are also confined within the pavement structure. Therefore volumetric deformations are of higher interest for these layers.

When both, the axial and radial deformation is recorded with sufficient quality, the two strain components, volumetric and deviatoric can be determined as follows: A cylindrical specimen shall have an initial height h_0, an initial diameter of d_0 and thus an initial volume of

$$V_0 = \frac{d_0^2 \pi}{4} \cdot h_0 \tag{4.9}$$

The change in height Δh_n and in diameter Δd_n with each load cycle n results in the volume

$$V(n) = \frac{(d_0 + \Delta d_n)^2 \pi}{4} \cdot (h_0 - \Delta h_n) \qquad (4.10)$$

If the pure deviatoric part of the deformation (without any change in volume) shall be derived, $V(n)$ in the formula above has to be substituted by V_0 and solved for Δh_n which then becomes $\Delta h_{n,dev}$:

$$\Delta h_{n,dev}(n) = h_0 - \frac{V_0 \cdot 4}{(d_0 + \Delta d_n)^2 \cdot \pi} \qquad (4.11)$$

This change in height can be converted to an axial strain component by dividing it by the initial height h_0. If the total axial strain is referred to as $\varepsilon_{ax,tot}$ then the deviatoric and volumetric strain component can be defined as follows:

$$\varepsilon_{ax,dev}(n) = \frac{\Delta h_{n,dev}(n)}{h_0}$$
$$\varepsilon_{ax,vol}(n) = \varepsilon_{ax,tot}(n) - \varepsilon_{ax,dev}(n) \qquad (4.12)$$

By following this procedure, four different permanent strain components can be derived:

- $\varepsilon_{ax,tot}$ is the total axial strain directly from the measured axial deformation data.
- $\varepsilon_{ax,vol}$ is the volumetric part of the axial strain responsible for the change in volume of the specimen without a change in shape.
- $\varepsilon_{ax,dev}$ is the deviatoric part of the axial strain responsible for the change in shape of the specimen. It has no impact on the volume of the specimen.
- ε_{rad} is the radial strain directly from the measured radial strain data.

Figure IV-11 shows an example for the four strain components for an AC 11 70/100 tested at standard TCCT conditions. The top diagram presents the evolution of the components with the number of load cycles. The dash-dot line with two dots shows the total axial strain, the continuous line the deviatoric, the dash-dot line with one dot the volumetric axial strain and the dashed line the radial strain. The lower diagram shows the ratio of deviatoric and volumetric strain to the total axial strain for certain numbers of load cycles. The heading of the diagram describes the mix shown in the capture. In the case of Figure IV-11 the mix has a target binder content of 5.6% (m/m) and an actual volume of air voids of 1.5% (v/v). From the data shown in the diagram it becomes obvious that the deviatoric part of the strain slightly dominates for the first 100 load cycles. From this point on both axial strain components exhibit a balanced behavior. About half the axial strain is transformed to a change in volume and the other half to a change in the shape of the specimen.

The following diagrams for the mixes mentioned in section IV.2.1 always show the ratio of deviatoric/volumetric axial strain to total axial strain. The bars represent the mean values of a certain number of single tests; the scattering of results is also shown by the standard deviations. The diagrams shown here also contain the total axial strain as curves to compare different gradation types, binders, aggregates and air void contents.

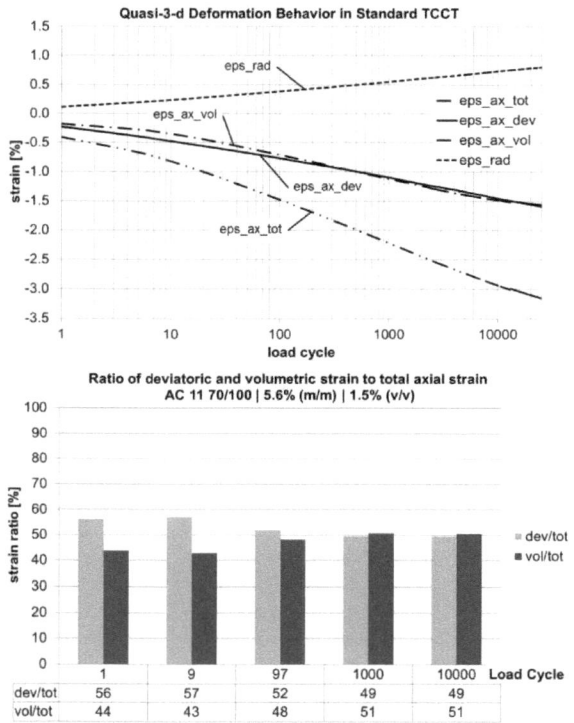

Figure IV-11. Different strain components from standard TCCT for an AC 11 70/100 mix vs. load cycles (top) and ratio of deviatoric and volumetric to total axial strain (bottom) at 50°C.

Figure IV-12 contains results for the AC 11 mix with the PmB 45/80-65. The type of aggregate was varied as well as the temperature of testing. Diagram a) shows the mix with diabase. While the volumetric strain is dominant in the very beginning indicating a certain amount of recompaction, both strain components exhibit equal shares after that. The situation for the extremely dure steel slack stands in sharp contrast at 50°C (diagram b)). There is only a small share of volumetric strain (20% to 30%). Thus in the mix with industrial aggregate hardly any change in volume is activated. Although the total axial strain is larger for the mix with steel slack (-3.7% at 10,000 load cycles) than for the diabase mix (-2.5%), the volumetric strain is comparable for both mixes (around -1.2% at 10,000 load cycles). The type of aggregate used for the mix seems to have an impact rather on the deviatoric strain component than on the volumetric component.

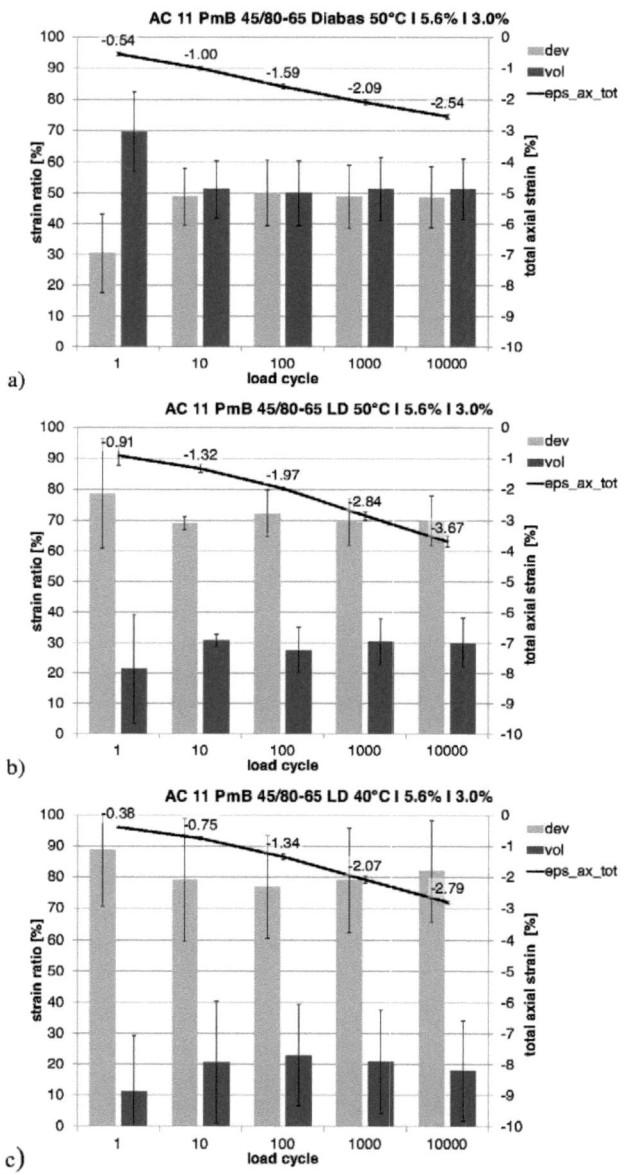

Figure IV-12. Strain ratio of deviatoric and volumetric to total axial strain from standard TCCT for an AC 11 PmB 45/80-65 diabase (a) and steel slack (b) at 50°C and steel slack (c) at 40°C.

When the steel slack mix is tested at 40°C (diagram c)), an even smaller share of the strain is due to volumetric change (around 20%) compared to the test at 50°C. At 10,000 load cycles, only -0.5% volumetric strain and -2.3% deviatoric strain occurs. So, the volumetric strain is cut to less than half of the value at 50°C whereas the deviatoric strain is similar at both temperatures. This shows that the decreasing permanent axial strain with decreasing temperature is caused by a strong decline of the volumetric component.

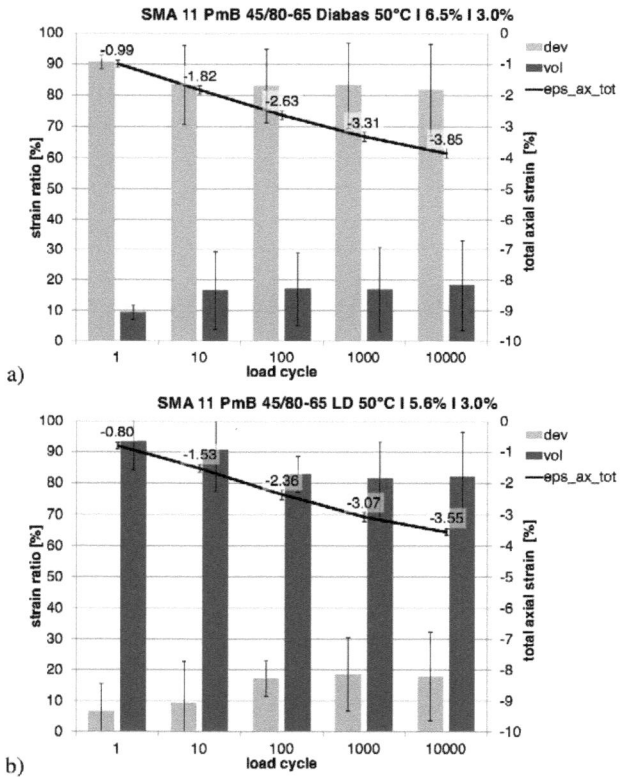

Figure IV-13. Strain ratio of deviatoric and volumetric to total axial strain from standard TCCT for an SMA 11 PmB 45/80-65 diabase (a) and steel slack (b).

Compared to the case in Figure IV-12, the mix in Figure IV-13 is different in terms of gradation type. An SMA 11 where the mastic and split contents of the mix are dominant was tested. SMA mixes are supposed to be highly resistant pavements for heavy trafficked section of the road network. Diagram a) shows the mix with diabase; compared to the AC mix the total axial strain after 10,000 load cycles is higher (-3.9% vs. -2.5%) but since the deviatoric part is dominant here (80%), only -0.75% volumetric strain occurs. This is far less than for the AC mix (-1.3%). The SMA mix will mostly produce shear deformation. Interestingly enough the SMA mix with steel slack (diagram b)) shows the other extreme. More than 80% of the total strain is due to volumetric deformation. Nearly -2.9% volumetric strain can be measured after 10,000 load cycles. Obviously the mix with the industrially processed aggregate steel slack is far less resistant to a change in volume but exhibits only minor deviatoric strain.

Figure IV-14 contains the results for the SMA 11 mixes with the unmodified 70/100 bitumen. The overall axial strain is clearly larger than for the mix with the PmB. This seems to be due to the binder type. The diabase mix exhibits 25% more, the steel slack mix 50% more axial strain. In terms of volumetric and deviatoric components both mixes show the same tendency as the PmB SMAs but the difference between both components is less pronounced. At 10,000 load cycles around 23% volumetric strain occurs for the 70/100 mix (Figure IV-14) compared to under 20% for the PmB mix (Figure IV-13) when diabase is used as aggregate. For the steel slack the same can be said for the deviatoric part.

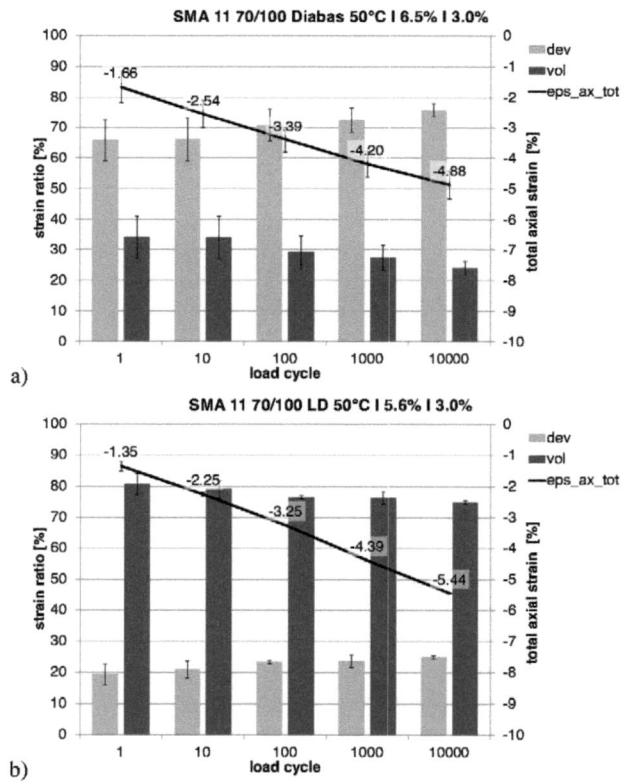

Figure IV-14. Strain ratio of deviatoric and volumetric to total axial strain from standard TCCT for an SMA 11 70/100 diabase (a) and steel slack (b).

For the SMA, the binder does not seem to influence the material behavior in terms of volumetric and deviatoric deformation, but only impacts the total axial strain. The type of aggregate has a strong impact, as it is shown.

A noise reducing SMA 11 with a target air void content of 10% is presented in Figure IV-15. It was composed of PmB and diabase. Compared to the standard SMA 11 (diagram a) in Figure IV-13) the total strain is about 18% larger. The behavior in terms of volumetric to deviatoric strain is clearly different. The volumetric strain component is more distinct here than for the low air void content SMA. At 10,000 load cycles around 42% of the total axial strain are due to a change in volume (compared to under 20% for the low void SMA). -1.9% volumetric axial strain can be measured after this number of load cycles. This is more than 2.5 times the volumetric change of the low void SMA. The content of air voids makes the significant difference in the deformation in this case.

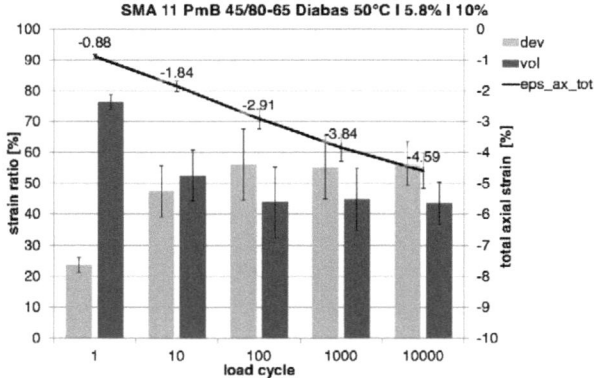

Figure IV-15. Strain ratio of deviatoric and volumetric to total axial strain from standard TCCT for an SMA 11 PmB 45/80-65 diabase with high air void content.

In Figure IV-16, the two diagrams describe the results of the high air void SMAs with unmodified binder and diabase (diagram a)) as well as steel slack (diagram b)). The diabase mix suffers from significantly higher total strain compared to the low void SMA (diagram a) in Figure IV-14) (-6.5% vs. -4.9% at 10,000 load cycles = +33%). Regarding the volumetric/deviatoric deformation behavior, both mixes the low and high void SMAs exhibit similar ratios of around 23% volumetric strain component.

The high void content SMA with steel slack (diagram b)) produces even less total axial strain than the low void SMA (-3.4% to -5.4% at 10,000 load cycles = -37%). In terms of the volumetric strain component the difference is less (-3.1% to -4.1% at 10,000 load cycles = -24%). Still, this behavior is surprising since the mix with higher content of air voids is more stable to permanent deformation. This SMA type enables more aggregate interaction than the low void content SMA and the steel slack seems to be an a optimal material for the high void content SMA since the total axial strain is even smaller than for the low void content SMA.

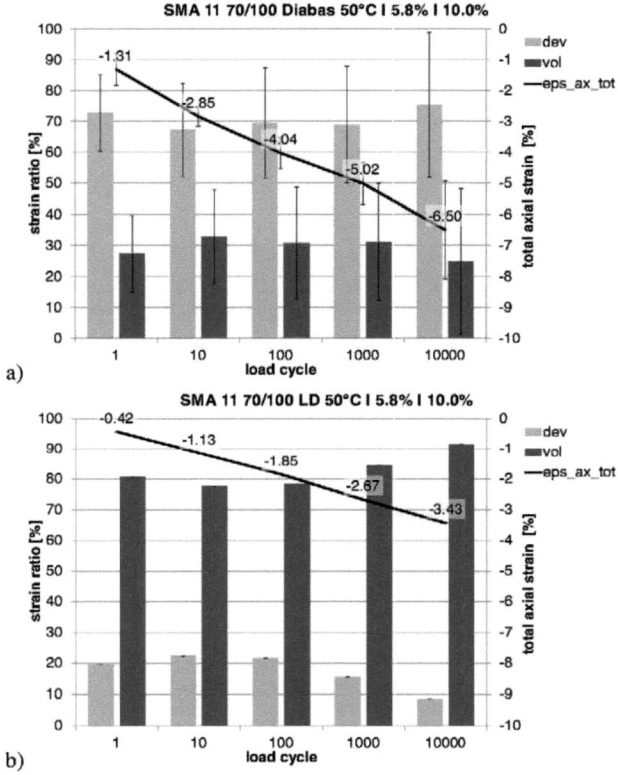

Figure IV-16. Strain ratio of deviatoric and volumetric to total axial strain from standard TCCT for an SMA 11 70/100 diabase (a) and steel slack (b) with high air void content.

Figure IV-17 presents the results of the two binder/base layer mixes AC 22 with SBS-modified binder (diagram a)), the other one with unmodified binder (diagram b)). Since the test temperature and the binder content, the type of aggregate as well as the binder type of the unmodified mix is different from the other presented HMAs, comparisons between the mixes above would not make sense. Especially interesting seems to be the evolution of the strain ratio with the number of load cycles. The volumetric share is very high in the beginning, 82% (-0.26% volumetric strain) for the PmB mix and decreases to around 20% at 10,000 load cycles (-0.48% volumetric strain). Thus the volumetric strain component not even doubles from the 1st load cycle to load cycle 10,000 whereas the deviatoric component starts at -0.06% and shows a value of -1.9% at 10,000 load cycles.

For the mix with the unmodified 50/70, the volumetric strain stays practically constant at around -0.10%. From this it can be stated that the AC mix with the larger maximum aggregate size exhibits a small recompaction at the first loading whereas any further deformation is due to deviatoric strain.

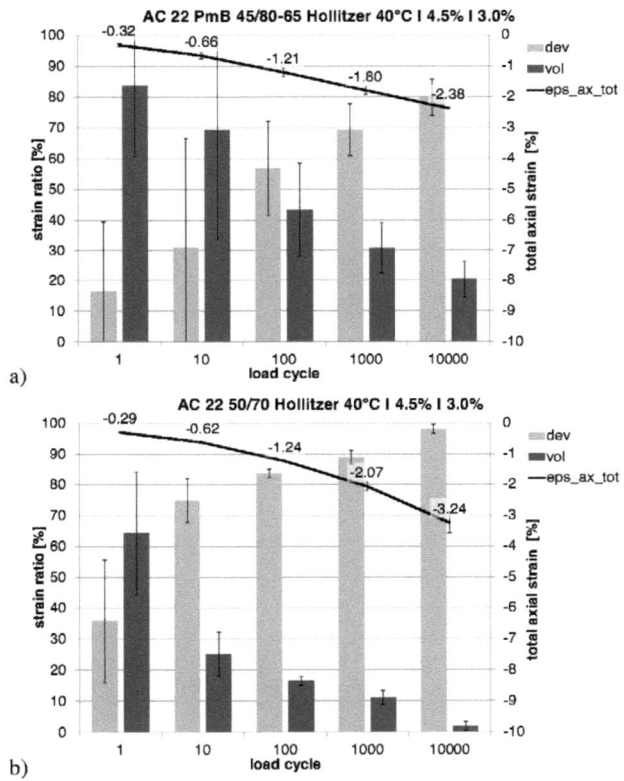

Figure IV-17. Strain ratio of deviatoric and volumetric to total axial strain from standard TCCT for an AC 22 PmB 45/80-65 limestone (a) and AC 22 50/70 limestone (b) at 40°C.

Figure IV-18 shows results for the three AC 11 70/100 mixes tested within this thesis. Interestingly enough the mix with the lowest binder (diagram a)) content exhibits less axial strain than the other two mixes. Also, the ratio between deviatoric and volumetric axial strain at load cycles 1 to 10 changes for the three binder contents. The mix with the lowest binder content shows higher volumetric strain in the beginning (60%), followed by the two other mixes (50% to 55%). After 1,000 load cycles all three mixes reach the same level of about 60% $\varepsilon_{ax,vol}$ to 40% $\varepsilon_{ax,dev}$. This shows that for the low binder content a stronger compaction of the voids in the mix happens at the beginning of the test. For mixes with a higher binder content, the deviatoric component is more dominant in the first part of the test. A higher binder content may work like a lubricant that enable aggregates to slide past each other more easily and thus create stronger reorientation of the aggregate skeleton within the first few load cycles.

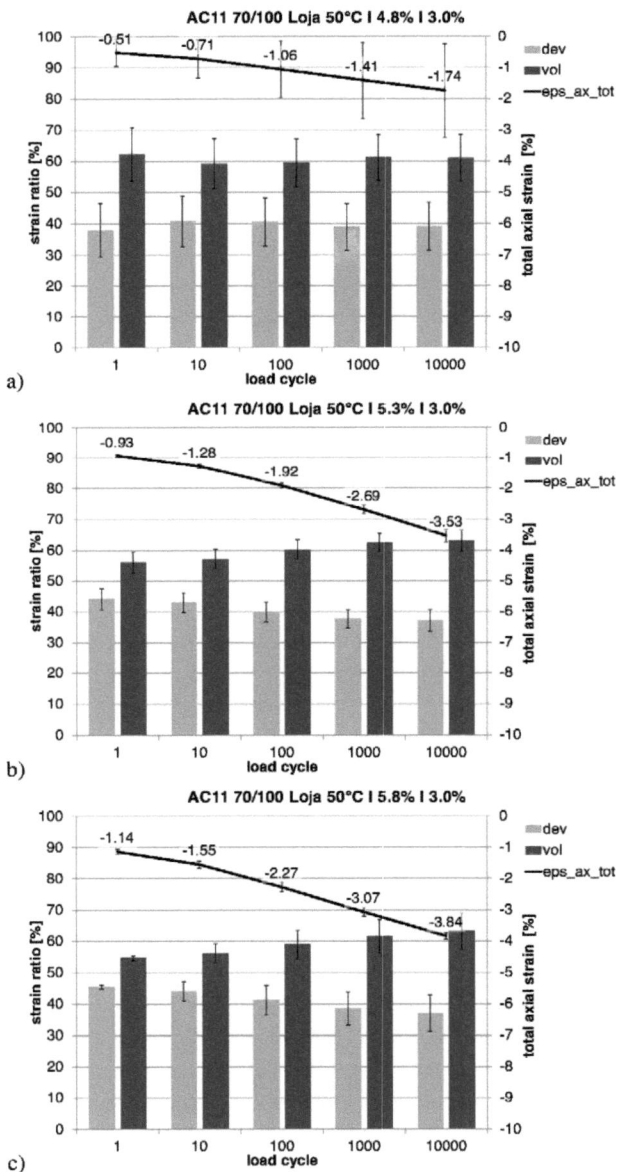

Figure IV-18. Strain ratio of deviatoric and volumetric to total axial strain from standard TCCT for AC 11 70/100 mixes with 4.8% (a), 5.3% (b) and 5.8% (c) binder content.

The PmB mix was tested at one binder content of 5.3% (m/m). The respective results are shown in Figure IV-19. The left diagram shows the strain ratio of deviatoric and volumetric to total axial strain. Compared to the mixes with the paving grade bitumen 70/100, the mix exhibits less total axial strain (around 68% of the 70/100 mixes with the same binder content) and also a more balanced ratio of deviatoric and volumetric strain. Both mixes, the modified and unmodified exhibit similar deviatoric strain. The PmB mix shows around -1.1%, the UmB mix around -1.3% deviatoric strain. This shows that the higher stability of the PmB mix is mainly caused by less compaction of the voids in the mix whereas the shear stability is only slightly affected. The PmB mix suffers only -1.3% volumetric strain whereas the unmodified mix exhibits nearly double this value (-2.2% volumetric strain).

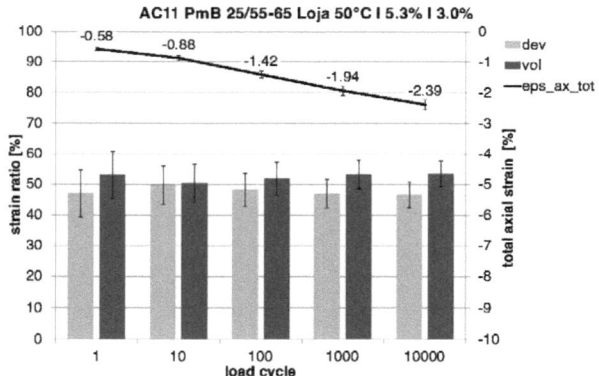

Figure IV-19. Strain ratio of deviatoric and volumetric to total axial strain from standard TCCT for AC 11 PmB 25/55-65 mix with 5.3% binder content.

IV.3 Conclusions

This chapter was aimed towards an enhanced characterization of the rutting resistance with the TCCT and dealt with two major issues:

A first step towards an enhanced assessment of the permanent deformation behavior was to investigate **alternative methods to describe the creep curve** as the main result of TCCTs. In the present version of the standard (EN 12697-25, 2005) the creep curve can either be described by a power function over the complete number of load cycles or by a linear regression within the quasi-linear part of the creep curve. For the characterization of HMAs, the linear creep rate f_c is used as a benchmark in the European Standard for type testing (EN 13108-1, 2006). The drawback with the linear approximation is that the standard does not provide information how to find the quasi-linear part of the curve. Thus, the derived creep rate depends on the arbitrarily chosen data range for linear regression analysis. The investigations within this thesis showed that the creep curve gets linear in the log/lin-scale when the viscoelastic parameters get constant. An alternative method to describe creep curves has been introduced. First, the load cycle, from which on the material is in a constant, linear viscoelastic steady state is determined by a well-defined procedure. Then a logarithmic function is used to approximate the creep curve from that load cycle on. The slope of the logarithmic function in the log/lin-scale (parameter b) (= logarithmic creep rate) is used to describe the permanent deformation behavior. It was shown that the fit quality of the logarithmic function is similar and often even better than the quality of the linear function given by the standard. In addition, the major advantage is that the creep rate is unambiguously defined and does not depend on the user in any way. A correlation analysis between the linear and logarithmic creep rate showed an excellent correlation between both parameters. It is therefore recommended to implement the logarithmic approach into a next version of the standard (EN 12697-25, 2005) to ensure reliable and reproducible results from TCCTs.

Secondly, the **quasi-3-d deformation behavior of HMA** in the TCCT was investigated on a number of different mixes. Since strain gauges were used as a measuring device to obtain radial deformation, both, the axial and radial deformation, could be recorded with high quality. Thus it became possible to study not only the total axial strain $\varepsilon_{ax,tot}$ and the total radial strain $\varepsilon_{rad,tot}$ but also the volumetric ($\varepsilon_{ax,vol}$) and deviatoric ($\varepsilon_{ax,dev}$) part of the axial strain. Two different reasons for permanent deformation, post-compaction of air voids without a change in shape (volumetric deformation) and shear deformation without a change in volume (deviatoric deformation) could therefore be investigated separately. Data from a former project, where a number of different mixes was tested with a variation of the gradation type, the aggregate and binder type as well as the void content and the test temperature were employed for an analysis of the quasi-3-d deformation behavior. To summarize the findings in short, the following can be stated:

- The type of aggregate has a clear impact on the deviatoric part of the deformation. In the underlying case study, a diabase was compared to an industrially produced steel slag. The mix with steel slag resulted in similar volumetric but higher deviatoric strain.
- When the temperature is set to a lower level, the decrease in total axial strain is due to reduced volumetric strain whereas the deviatoric part does not change.
- In terms of gradation type, a standard AC mix was compared to an SMA mix both with 11 mm maximum aggregate size. The results were not conclusive since the SMA mix with natural aggregate resulted in larger deviatoric strain whereas the SMA mix with steel slag suffered from larger volumetric strain compared to the AC mix. It could not be determined whether this is due to the gradation type or the type of mineral aggregate.
- For the SMA mix, the binder type was varied as well. Unmodified 70/100 bitumen was used as well as PmB 45/80-65. From the results, it can be concluded that the binder type affects the total axial strain leaving the ratio of volumetric to deviatoric strain untouched.

- An increase in the void content of a mix leads to an increase of the volumetric part of the deformation and thus a higher total axial strain.
- When the maximum aggregate size was raised from 11 mm to 22 mm the volumetric part of the deformation decreased.

The investigation on the volumetric and deviatoric part of the total axial strain from a number of different mixes reveals a significant potential of the approach to characterize the permanent deformation behavior of HMA at elevated temperatures. It is valuable information for road construction in the field to know how a certain mix reacts to loading. Especially the question of how resistant a mix to both types (volumetric/deviatoric) of permanent deformation is can help optimize mixes for different applications. While surface layers are more susceptible to shear deformation especially in those areas where high shear stresses are introduced into the pavement (intersections, airfields), binder layers are more confined within the structure than upper layers. Thus, volumetric deformations are more relevant for these parts of a pavement.

With only little additional effort to record the accumulated radial strain in the TCCT (an LVDT-based device does the job) a large benefit can be created in terms of results. It is therefore recommended to introduce this approach into a next revision of the European standard (EN 12697-25, 2005) to gather more data and experience on this important matter.

V VISCOELASTIC MATERIAL PARAMETERS DERIVED FROM CCTS

This chapter presents a detailed study on viscoelastic parameters derived from tests in a purely compressive state (CCTs). Because the radial strain is obtained with high quality from data recorded by circumferential SGs, the quasi-3-d state of strain can be taken into consideration in this study and parameters like the Poisson's Ratio or the dynamic shear modulus are analyzed as well. This investigation is carried out for tests with temperature and frequency sweep. In consequence, the evolution of the viscoelastic parameters with temperature and frequency is investigated. The results reveal a detailed insight into the nature of HMA behavior. They are also relevant for further use in material modeling and simulation. Since both, the axial and radial strain were analyzed from a comprehensive test program, it can be shown that there is an obvious material inherent difference between both phase lags, the axial and radial phase lag, that brings the need to incorporate the Poisson's Ratio into the system of dynamic material parameters.

Before the findings of the analysis are presented, a short theoretic summary on the behavior of viscoelastic materials is presented in section V.1. The test program is summarized in section V.2 and section V.3 gives an answer to the question whether the test setup (i.e. whether the specimen is firmly connected to the load plates or not) has an impact on the material reaction and thus, whether standard CCT is a valid test method for the determining viscoelastic material properties. In the following section V.4 test conditions chosen for the tests are looked at in terms of whether the material is in the linear or non-linear viscoelastic domain within these conditions. This investigation is followed by section V.5 which presents an analysis of the viscoelastic parameters, as there are phase lags in axial and radial direction and the dynamic modulus. Results will be shown for tests with temperature and frequency sweep. Section V.6 aims to explain the significant difference between axial and radial phase lags in all CCTs of this study. As one result the dynamic Poisson's Ratio as well as the dynamic shear modulus are introduced, analyzed and discussed in section V.7.

V.1 Characteristics of Viscoelastic Materials

To evaluate and analyze test data from cyclic (and static) material tests in a reliable, repeatable and efficient way, certain mathematical and statistical methods are necessary. A vast number of textbooks and other publications on these topics are available, e.g. (Sachs, et al., 2009) and (Zeidler, 2004). Certain basics on rheology, especially on the theory of viscoelasticity, are required to interpret results from test evaluation. The most important concepts that are used further on in the thesis are presented in this section. For more details on the theory of viscoelasticity (Findley, et al., 1989) gives a detailed overview or summarized in a compact way in (Hofko, 2006). Therefore the following section contains only those mathematical and mechanical concepts and methods that are used and needed in the analysis and interpretation of results of cyclic dynamic material tests.

To describe the mechanical behavior of materials used in civil engineering, a purely elastic approach is often used to save time and mathematical effort. For many common materials (e.g. steel, concrete) this approximation is an efficient way to solve standard design problems, as the materials are stressed within a range in which the material behavior is dominantly elastic. For critical load cases close to the bearing capacity even so called elastic materials reach a state where more complex constitutive relationships occur and have to be taken into account to stay on the save side and design efficiently. These include plastic and viscous components. Basically, constitutive relationships can be divided into four main categories (Krass, 1977):

- **Elastic** deformations are spontaneous deformations of a material under loading that reset spontaneously after the load is removed.
- **Plastic** deformations are also spontaneous deformations under loading that are permanent or irreversible. Usually these deformations occur only when a certain, material-specific yield stress is reached.
- **Viscoelastic** deformations are time-dependent deformations. Viscoelastic displacements approach an asymptotical value and are completely, yet time-dependently reversible after removal of the load.
- **Viscoplastic** deformations are also time-dependent. They also approach an asymptotical value under constant loading, but they are permanent.

Table V-1 gives an overview on these deformation components.

Table V-1. Different deformation components.

Parameter	time-dependent	reversible
elastic	no	Yes
plastic	no	No
viscoelastic	yes	Yes
viscoplastic	yes	No

Bituminous bound materials used in road construction (a prominent representative being HMA) are usually a composite from
- mineral aggregate as the load transferring component,
- bitumen as the binder and
- a certain content of air voids (except for mastic asphalt).

Since bitumen shows time- and temperature-dependent deformation behavior with elastic, viscous and plastic components, the composite HMA develops a complex material behavior as well. Regarding HMA the deformation components can be characterized as follows (Blab, et al., 1999):
- Reversible deformation components: Elasticity and viscoelasticity are responsible for reversible deformation of HMA. At low temperatures and high frequencies of loading the mechanical behavior shifts towards a virtually (linear) elastic, time-independent behavior. Reversible deformations are seen as the dominant

factor for fatigue in terms of bottom-up (fatigue) and top-down (low-temperature) cracking.
- Irreversible deformation components: Plasticity and viscoplasticity are responsible for irreversible deformation of HMA. The higher the temperature and the lower the frequency of loading the more dominant get these components. They are seen as the main factor for rutting.

The different strain components are presented in a graphical example in Figure V-1. The figure shows the strain vs. time for a viscoelastic behavior that is stressed with a constant load in the first part of the curve and then unloaded. The deformation components can be described as follows:

$$\varepsilon_{tot}(t) = \varepsilon_e + \varepsilon_p + \varepsilon_{ve}(t) + \varepsilon_{vp}(t) \quad (5.1)$$

ε_{tot} .... Total strain at time t under loading
ε_e .... Spontaneous, time-independent elastic strain component (reversible)
ε_p .... Spontaneous, time-independent plastic strain component (irreversible)
ε_{ve} .... Time-dependent viscoelastic strain component (reversible)
ε_{vp} .... Time-dependent viscoplastic strain component (irreversible)

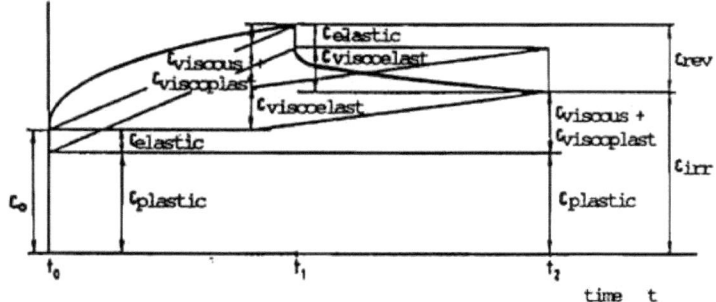

Figure V-1. Illustration of the different deformation components. (Thamfeld, 1990)

V.1.1 Behavior of Viscoelastic Materials under Cyclic Dynamic Loading

In the definitions above the mechanical behavior of materials was described only under constant stress in time domain. Data for rheological models are often obtained from static creep and relaxation tests. The test data is then fitted to the models by use of differential or integral operations. Since HMA is commonly used as a construction material for traffic infrastructure it is mostly stressed by dynamic loading from passing vehicles. For purely elastic materials it may be irrelevant whether a structure is stressed by static or dynamic loading since the material parameters of elastic materials are independent on the loading time or temperature. HMA as a viscoelastoplastic material develops different material behavior with regard to frequency of loading and temperature. Therefore, if HMA structures should be describe mathematically in a correct way and thus designed efficiently, this time/temperature dependency of mechanical parameters has to be taken into account. Identification tests for material parameters must be carried out dynamically to identifiy the time dependent behavior. In addition the temperature has to be varied as well to assess the temperature dependency. According to the viscoelastic theory it does not make a difference whether material parameters are derived from static or dynamic tests. Still, at the present time no satisfying correlations between results from static and dynamic tests have been established.

As a result of dynamic material tests, complex stiffness (modulus) and compliance are derived. The tests within this thesis were carried out with a sinusoidal loading, and therefore the following derivation of the dynamic material parameters are based on cyclic dynamic shapes.

To demonstrate the derivation of the dynamic material parameters, a specimen shall be loaded with a sinusoidal, external force F with a constant amplitude F_0 and an angular frequency of ω. The cyclic dynamic loading can be expressed by $F=F_0\cos(\omega t)$. The external force F induces a stress within the specimen:

$$\sigma = \sigma_0 \cdot \cos(\omega \cdot t) \qquad (5.2)$$

σ ....Sinusoidal stress [N/mm²] with a cycle period $T=2\pi/\omega$ [s]
σ_0 ....Amplitude of the sinusoidal stress [N/mm²]
ω ....Angular frequency $\omega=2\pi f$ [rad/s]
t ....Time [s]

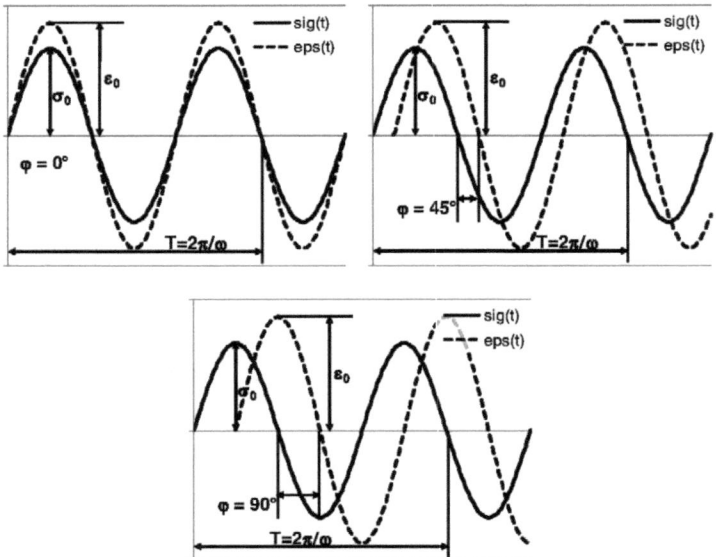

Figure V-2. Stress σ(t) and strain ε(t) under sinusoidal loading for an elastic (upper left), viscoelastic (upper right) and viscous (below) material.

If the specimen shows viscoelastic properties and a sinusoidal loading is introduced, the reaction of the material in terms of strain ε will occur in a sinusoidal shape. As displayed in Figure V-2, the frequency of the material reaction in terms of strain ε(t) is equal to the frequency of loading. Depending on the ratio of elastic to viscous components of the material behavior a phase lag/phase angle φ between stress σ(t) and strain ε(t) occurs. For a purely elastic material (upper left diagram in Figure V-2) both parameters oscillate in phase (φ = 0°). If the material reacts completely viscous (lower diagram in Figure V-2) the phase angle reaches a maximum of φ = 90°. A viscoelastic behavior produces phase lags in between those two boundaries. As an example a situation with a phase angle of φ = 45° is shown in the upper right diagram in Figure V-2.

The material reaction on a stress history according to equation (5.2) in terms of strain results in

$$\varepsilon = \varepsilon_0 \cdot \cos(\omega \cdot t - \varphi) \qquad (5.3)$$

ε_0........... Amplitude of the sinusoidal strain reaction [-]
ω........... Angular frequency [rad/s]
φ........... Material phase lag

For one HMA the phase lag is a function of frequency and temperature of loading. It is independent of the magnitude of loading, if the material is stressed within the linear viscoelastic domain.

By taking advantage of the Euler's formula and differential operators (for details see e.g. (Findley, et al., 1989)) the combination of ε_0, σ_0 and φ can be arranged, so that

$$E^* = \frac{\sigma_0}{\varepsilon_0} \cdot (\cos\varphi + i \cdot \sin\varphi) = E_1 + i \cdot E_2 \qquad (5.4)$$

E^* is the complex modulus and can be divided into a real (E_1) and an imaginary (E_2) part as commonly used in the complex number system. E_1 and E_2 can be presented differently as well:

$$E_1 = \frac{\sigma_0}{\varepsilon_0} \cdot \cos\varphi \qquad (5.5)$$

$$E_2 = \frac{\sigma_0}{\varepsilon_0} \cdot \sin\varphi \qquad (5.6)$$

E_1 is also referred to as the storage modulus and E_2 as the loss modulus. The storage modulus represents the elastic part and the loss modulus the viscous part of the material stiffness. For a perfectly elastic material (with a phase lag of 0°) the viscous or loss modulus is 0, whereas for a purely viscous material the elastic or storage modulus is 0.

E^* is a complex number. Usually the magnitude of E^*, the dynamic modulus $|E^*|$ is given, which is defined as follows:

$$|E^*| = \sqrt{E_1^2 + E_2^2} = \frac{\sigma_0}{\varepsilon_0} \qquad (5.7)$$

Analogous to the approach above the counterpart of the complex modulus, the complex compliance J^* and its loss and storage or real and imaginary part J_1 and J_2 can be derived. The magnitude of the complex compliance is defined as the dynamic compliance

$$|J^*| = \sqrt{J_1^2 + J_2^2} = \frac{\varepsilon_0}{\sigma_0} \qquad (5.8)$$

J_1 and J_2 are

$$J_1 = \frac{\varepsilon_0}{\sigma_0} \cdot \cos\varphi \qquad (5.9)$$

$$J_2 = \frac{\varepsilon_0}{\sigma_0} \cdot \sin\varphi \qquad (5.10)$$

By applying further mathematical methods (for details the reader may again be referred to e.g. (Findley, et al., 1989)) other relationships between the derived parameters can be given:

$$J^* \cdot E^* = 1 \qquad (5.11)$$

$$|J^*| = \frac{1}{|E^*|} \qquad (5.12)$$

$$\tan\varphi = \frac{E_2}{E_1} = \frac{J_2}{J_1} \qquad (5.13)$$

$$J_1 = \frac{1}{|E^*|} \cdot \cos\varphi = \frac{E_1}{E_1^2 + E_2^2} \quad \text{and} \quad J_2 = \frac{1}{|E^*|} \cdot \sin\varphi = \frac{E_2}{E_1^2 + E_2^2} \qquad (5.14)$$

$$E_1 = \frac{1}{|J^*|} \cdot \cos\varphi = \frac{J_1}{J_1^2 + J_2^2} \quad \text{and} \quad E_2 = \frac{1}{|J^*|} \cdot \sin\varphi = \frac{J_2}{J_1^2 + J_2^2} \qquad (5.15)$$

V.1.2 Viscoelastic Material Parameters

With the evaluation software presented in section II.3.2, viscoelastic material parameters are derived from the obtained regression parameters after the 1st and 2nd module of the evaluation. The parameters computed by the software are:

- 4 different phase lags at different functional values (Φ_{max}, Φ_{min}, Φ_{MV-}, Φ_{MV+}) between data from force and the deformation sensors,
- 4 different elastic (E_1) and viscous (E_2) parts of the complex modulus with the 4 respective phase lags for data from the force sensor, as well as from deformation sensors and
- 4 dynamic moduli $|E^*|$ for data from the force sensor, as well as from deformation sensors.

If the maximum number of sensors is used for the evaluation 4x4 = 16 phase lags, 2x4x3 = 24 different values for E_1 and E_2 and 1x4x3 = 12 different $|E^*|$ are determined by the software.

Φ represents the measured phase lag between two signals. It does not represent the material phase lag φ, since inertia effects of moveable parts of the test machine impact the measured phase lag. It is shown below how the measured phase lag Φ is converted to the material phase lag φ. Phase lags are obtained from four different well-defined function values. Figure V-3 shows a schematic representation of these function values. In a first step, the extreme values and respective points in time are derived from the analytical approximation functions (see equations (2.2) to (2.4)). Therefore the 1st derivation of the function is set to zero and the time $t_{k,max}$ and $t_{k,min}$ for the sensor k is calculated. The phase lag $\Phi_{k,l,max}$ between the sensors k and l at the maximum is then obtained by

$$\phi_{k,l,max} = \frac{(t_{l,max} - t_{k,max})}{T_p} \cdot 360° \qquad (5.16)$$

$\Phi_{k,l,max}$... Phase lag between sensor k and sensor l at the maximum with impacts of test machine (inertia effects) [°]
$t_{l,max}$ ...... Time of the maximum function value of sensor l
$t_{k,max}$ ...... Time of the maximum function value of sensor k
T_p .... Periodic time of the respective oscillation

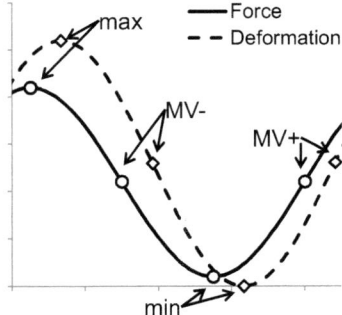

Figure V-3. Definition of function values for the calculation of the four different phase lags.

In an analogous way the phase lag at the minimum is obtained. The other two values are derived from the point where the functions reach the mean value between maximum and minimum. There is a mean value in the unloading phase (MV-) and one in the loading phase (MV+). The times t_{MV-} and t_{MV+} are not obtained analytically. The mean value of the function is calculated from the minimum and maximum of the function. Then the point of time for the mean functional value is approximated by the bisection method. The initial point of time for this approximation is the maximum and minimum of the function respectively. The reason for obtaining phase lags and material parameters at four different functional values are connected to the advanced approximation function $F+L+1H$ presented in section II.3.4. When only the simple sine is taken into account all four mentioned phase lags above will result in equal values. Yet, when a linear term and, even more, the 1st harmonic is added to the approximation function all four values become different.

As a result of the RILEM TC 182 on performance testing and evaluation of bituminous materials, (Di Benedetto, et al., 2001) suggests a method to obtain the dynamic moduli. The European Standard for stiffness tests on HMA specimens (EN 12697-26, 2004) follows this suggestion as the following formulas show:

$$E_1 = \gamma \cdot \left(\frac{F}{z} \cos(\phi) + \frac{\mu}{1000} \cdot \omega^2 \right) \quad (5.17)$$

$$E_2 = \gamma \cdot \left(\frac{F}{z} \sin(\phi) \right) \quad (5.18)$$

$$|E^*| = \sqrt{E_1^2 + E_2^2} \quad (5.19)$$

$$\varphi = \arctan\left(\frac{E_2}{E_1} \right) \quad (5.20)$$

γ............ Shape factor as a function of shape and dimension of the specimen [mm/mm²]
F .... Measured force [N]
z .... Measured deformation of the specimen [mm]
Φ .... Measured phase lag between force and deformation signal
μ .... Mass factor taking into account inertia effects of the moveable parts of the test machine [kg]
ω............ Angular frequency $\omega = 2\pi f$ [rad/s]

The two factors implemented in (5.17) and (5.18), γ and μ are defined for different test types and shapes of the specimen. For the CCTs with cylindrical specimens EN 12697-26 provides the following factors:

$$\gamma = \frac{4h}{D^2 \pi} \quad (5.21)$$

$$\mu = \frac{M}{2} + m \quad (5.22)$$

h .... Height of the specimen [mm]
D........... Diameter of the specimen [mm]
M .... Mass of the specimen [kg]
m........... Mass of the dynamic parts of the test machine [kg]

Obviously γ is just an auxiliary value to transform the force and displacement in (5.17) and (5.18) to stress and strain respectively. μ takes care of the inertia effects. For the test setup of this thesis M as the mass of the specimen is around 4 kg and the dynamic masses of the test machine are around 18 kg. These values have been determined by (Kappl, 2007).

The notation of (5.17) and (5.18) to obtain E_1 and E_2 are misleading. An unambiguous formulation would be

$$E_1 = \gamma \cdot \left(\frac{\Delta F}{\Delta z} \cdot \cos(\phi) + \frac{\mu}{1000} \cdot \omega^2 \right) \text{ and } E_2 = \gamma \cdot \left(\frac{\Delta F}{\Delta z} \cdot \sin(\phi) \right) \quad (5.23)$$

ΔF......... Difference between maximum and minimum force [N]
Δz.......... Difference between maximum and minimum deformation of the specimen [mm]

Both values ΔF and Δz can be derived analytically from the received regression functions (2.2), (2.3) or (2.4). In a first step, the first derivation of the regression function for the force and one deformation sensor is set to zero. Then the time of these extremal values are calculated and these times are applied into the regression function. Thus, the extremal values and ΔF and Δz can be obtained.

With the four different Φ values, four different values for E_1 and E_2 can be derived and further by using (5.20) four different φ. The difference between the measured phase lags Φ and the material phase lags φ is the mass factor μ that takes into account inertia effects of the moveable parts of the test machine. As defined in (5.17) the influence of μ on E_1 increases to the second power with respect to frequency. The difference between Φ and φ gets larger with increasing dynamic masses M and m, test frequency, decreasing material stiffness and a higher measured phase lag (the ratio between the 1st and 2nd term in (5.17) increases). Since the stiffness of HMA decreases with increasing temperature the largest effect of the mass factor on the phase lag was found at test conditions of 50°C, 30 Hz and a standard, non-modified bitumen. The maximum difference between Φ and φ was around 0.4° at these extreme conditions. It can therefore be stated that the mass factor cannot be neglected, but the error made if left unconsidered is rather small, if the maximum difference of 0.4° is compared to material phase lags of e.g. an AC 11 70/100 at 50°C and 30 Hz of around 15° (= 2.67% difference).

V.1.3 Advanced Viscoelastic Material Parameters

Since CCTs result in relevant strain also in radial direction, the test results can be used to describe the quasi-3-d behavior of HMA under compressive loading. For this reason further material parameters, i.e. the Poisson's Ratio and the shear modulus are introduced.

For a material under uniaxial tension the **Poisson's Ratio** ν characterizes the ratio between the strain transverse to the tension direction ε_{tr} and the strain in direction of the loading ε_{ax}:

$$\varepsilon_{tr} = -\nu \cdot \varepsilon_{ax} \quad (5.24)$$

The upper equation (5.24) is valid if the strain is uniform over the specimen's length. For isotropic, linear elastic materials the Poisson's Ratio range between 0 and 0.5.

When the Poisson's Ratio is 0, no transversal strain occurs, whereas a value of 0.5 shows that no change in volume can be expected. Anisotropic, porous materials can exhibit Poisson's Ratios higher than 0.5 due to changes in their void content. (Mang, et al., 2000)

One important precondition for the determination of v is, that the transversal strain is uniform over the height of the specimen since the Poisson's Ratio is related to the change in volume. In the CCT friction occurs at the contact area between the load plates and the specimen. Although the friction is reduced by using a lubricant on the upper and lower end planes of the specimens, the deformation in radial direction is still not uniform over the height of the specimen.

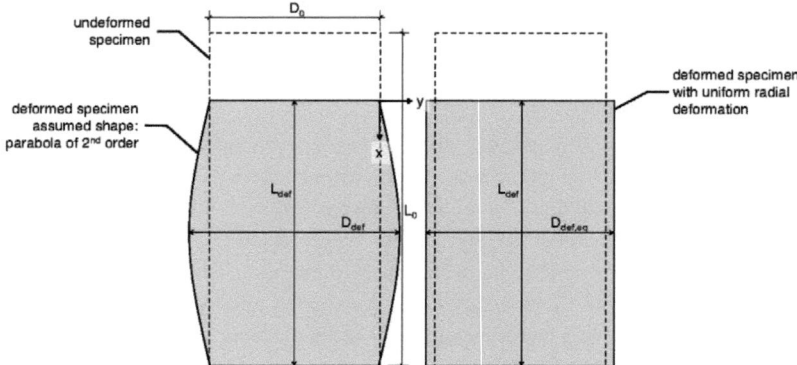

Figure V-4. Deformed specimen after CCT with actual shape (left) and adapted shape with uniform radial deformation (right).

As shown in the left sketch in Figure V-4, it can be assumed from experience that there is hardly any radial deformation at the end planes and the maximum deformation occurs in the middle of the specimen. This is the position where the radial deformation is obtained from SG data. If the measured value is used for calculating the Poisson's Ratio, it leads to incorrect results. The measured deformation must be transferred to an equivalent mean radial deformation in terms of specimen volume. For the calculation of a mean equivalent radial deformation, the shape of the deformed specimen is assumed to be a parabola of 2nd order (Figure V-4). This assumption corresponds to measurements of actual HMA specimens after TCCTs. If a coordinate system is set originating on top of the specimen and the x-direction running along the specimen's height and the y-direction transverse to it, the following functional values can be derived if the deformed shape is a parabola of 2nd order:

$$f(x=0) = f\left(x = L_{def}\right) = 0$$
$$f\left(x = \frac{L_{def}}{2}\right) = D_{def} - D_0 = \Delta D \quad (5.25)$$

These three functional relationships define a parabola of 2nd order of the following form:

$$f(x) = -\frac{4 \cdot \Delta D}{L_{def}^2} \cdot x^2 + \frac{4 \cdot \Delta D}{L_{def}} \cdot x \qquad (5.26)$$

If an equal uniform radial deformation (ΔD_{eq}) should be derived with the requirement that the deformed height of the specimen and its volume are equal, then the area outside the dotted lines in the left sketch of Figure V-4 must be equal to the area outside the dotted lines in the right sketch of this figure:

$$\int_{x=0}^{L_{def}} f(x)dx = L_{def} \cdot \Delta D_{eq}$$

$$\frac{1}{3} \cdot \frac{4 \cdot \Delta D}{L_{def}^2} \cdot L_{def}^3 + \frac{1}{2} \cdot \frac{4 \cdot \Delta D}{L_{def}} \cdot L_{def}^2 = L_{def} \cdot \Delta D_{eq}$$

$$\frac{2}{3} \cdot \Delta D \cdot L_{def} = L_{def} \cdot \Delta D_{eq} \qquad (5.27)$$

$$\frac{2}{3} \cdot \Delta D = \Delta D_{eq}$$

From the derivation made above, it was shown that the obtained radial deformation ΔD must be reduced by 2/3 to obtain an equal deformation ΔD_{eq} that is uniform along the height of the specimen and can be used for obtaining the Poisson's Ratio.

In case of cyclic dynamic tests v can be calculated by the ratio between the amplitude of radial strain $\Delta \varepsilon_{rad}$ and axial strain $\Delta \varepsilon_{ax}$. Analogue to the complex modulus E^* discussed earlier in this chapter, a **complex Poisson's Ratio** v^* and its elastic (v_1) and viscous (v_2) part are introduced if a phase lag between axial and radial deformation $\delta_{ax,rad}$ can be measured:

$$v^* = -\frac{\Delta \varepsilon_{rad}}{\Delta \varepsilon_{ax}} \cdot (\cos \delta_{ax,rad} + i \cdot \sin \delta_{ax,rad}) = v_1 + i \cdot v_2$$

$$v_1 = -\frac{\Delta \varepsilon_{rad}}{\Delta \varepsilon_{ax}} \cdot \cos \delta_{ax,rad} \qquad (5.28)$$

$$v_2 = -\frac{\Delta \varepsilon_{rad}}{\Delta \varepsilon_{ax}} \cdot \sin \delta_{ax,rad}$$

$$|v^*| = \sqrt{v_1^2 + v_2^2}$$

By combining the complex modulus E^* and the complex Poisson's Ratio v^* a **complex shear modulus G^*** and its elastic and viscous moduli can be given. The phase lag $\varphi_{ax,ax}$ shall be measured between axial loading and axial deformation and the phase lag $\delta_{ax,rad}$ between axial deformation and radial deformation:

$$G^* = \frac{E^*}{2\cdot(1+v^*)} \cdot [\cos(\varphi_{ax,ax} + \delta_{ax,rad}) + i\cdot\sin(\varphi_{ax,ax} + \delta_{ax,rad})] = G_1 + i\cdot G_2$$

$$G_1 = \frac{|E^*|}{2\cdot(1+|v^*|)} \cdot \cos(\varphi_{ax,ax} + \delta_{ax,rad}) \quad | \quad v^* \in \mathbb{R}$$

$$G_2 = \frac{|E^*|}{2\cdot(1+|v^*|)} \cdot \sin(\varphi_{ax,ax} + \delta_{ax,rad}) \quad | \quad v^* \in \mathbb{R} \tag{5.29}$$

$$|G^*| = \sqrt{G_1^2 + G_2^2} = \frac{|E^*|}{2\cdot(1+|v^*|)} \quad | \quad v^* \in \mathbb{R}$$

V.2 Test Program

To investigate the impact of time (loading frequency) and temperature on the viscoelastic parameters of HMA specimens in the compressive cyclic dynamic domain, CCTs were carried out at different test temperatures within a range of frequencies.

Carrying out cyclic dynamic tests with confining pressure (=TCCT) with one or two SGs attached to the specimen is more time-consuming than carrying out UCCTs, and it contains more risks of failure due to leakage. A special device was constructed to pass the cables of the SG within the membrane through the lower load plate outside the triaxial cell. But still, this system is more prone to leakage and thus failure of the test than UCCTs. Whenever possible, the cyclic dynamic tests were carried out as uniaxial tests (UCCTs) without confining pressure. This was possible for 10°C and 30°C tests. At 50°C all tests were run as TCCTs because the stability of the HMA specimens' structure was too low to test them without confining pressure. If the tests are carried out within the linear viscoelastic domain of the material, the stress deviator should not have an influence on the viscoelastic parameters. Results from TCCTs and UCCTs are therefore comparable in the linear viscoelastic domain. This fact will be checked and findings discussed in section V.4.

The following parameters are standard test conditions for the majority of tests. To deal with a couple of special research questions, tests were also run with parameters different from the standard conditions. In this case, it will be stated within the regarding section. The applied stresses for each test temperature and the standard tests are listed in Table V-2. The last row, highlighted in grey, gives the TCCT conditions according to (EN 12697-25, 2005).

To investigate the influence of time on the viscoelastic parameters in the compressive domain, each standard test was run with a range of frequencies from 0.1 Hz to 30 Hz. Before the actual test started, a 120 s preloading phase with 2 % of the maximum stress $(\sigma_{m,ax} + \sigma_{a,ax})$ is applied. After the loading phase, a stress free recovery phase of 10 min is allowed for most specimens. This test procedure is shown in Table V-3. The first tests for this investigation did not include the 0.5 Hz frequency. For the range from 0.1 Hz to

10 Hz, 200 samples were recorded for each load cycle. For higher frequencies, the sampling rate had to be decreased due to the limited sampling capacity of the record unit to 100 samples per load cycle. In the following chapters that deal with the analysis of test data, the 30 Hz frequency was usually neglected because the quality of the test data was not sufficient to produce sound results. In many cases also the 20 Hz packet did not bring sufficient quality for further analysis.

Table V-2. Standard test conditions for the determination of viscoelastic properties.

Temperature [°C]	Test Type	Mean Axial Compressive Stress $\sigma_{m,ax}$ [N/mm²]	Axial Stress Amplitude $\sigma_{a,ax}$ [N/mm²]	Radial Confining Pressure $\sigma_{c,rad}$ [N/mm²]
10	UCCT	0.60	0.50	0.00
30	UCCT	0.25	0.15	0.00
50	TCCT	0.45	0.30	0.15

Table V-3. Standard test characteristics for the determination of viscoelastic properties.

Test Phase	Frequency [Hz]	Number of Load Cycles per Frequency [-]	Recorded Samples per Load Cycle
Preload	-	-	-
Consolidation Phase	0.1	25	200
	0.5	50	200
	1.0	200	200
Main Phase	3.0	600	200
	5.0	1000	200
	10.0	1000	200
	20.0	1000	100
	30.0	1000	100
	0.1	25	200
	0.5	25	200
	1.0	100	200
Recovery	-	-	-

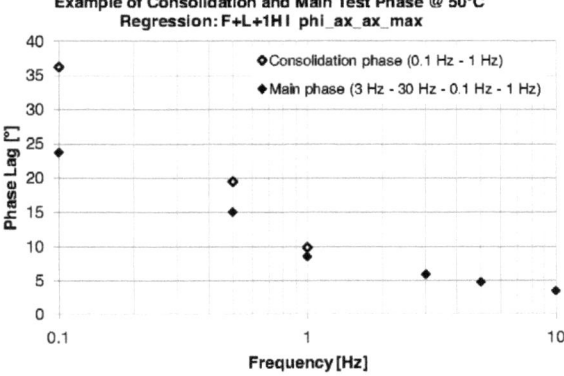

Figure V-5. Example of test results from consolidation and main test phase at 50°C, MV of $\varphi_{ax,ax,max}$.

Since the tests are carried out in the compressive domain a consolidation phase with frequencies from 0.1 Hz to 1 Hz is followed by the main test phase. Within the consolidation phase plastic deformations are dominant. Therefore the results from this part of the test are not taken for further analysis. As Figure V-5 shows the phase lags within the consolidation phase are much higher than in the main phase. At 1 Hz, the results from the first phase and the main phase become comparable.

V.3   Influences of the Connection between Specimen and Load Plates

When cyclic dynamic material tests are carried out in the compressive domain only, the specimen is usually not glued to the load plates as it is the case in tests with both tension and compression. Besides the TCCT and UCCT, the indirect tensile test (ITT) is a prominent example for a test type without a firm connection between specimen and load plates (in the case of CCT) or load strips (in the case of ITT). It is assumed that load plate and specimen are connected force-fit at all times, since no tensile load is applied. This assumption is especially important if viscoelastic material parameters (e.g. phase lags, dynamic modulus) are obtained from the test results. Material parameters are obtained from the applied loading (action) and the reaction of the material (i.e. deformation). Cyclic dynamic tests can be force- and displacement-controlled respectively. When a test is performed force-controlled, the control unit drives the hydraulic aggregate in a way that the force signal resembles a defined function (e.g. sinusoidal), whereas in a displacement-controlled test, the control unit actuates the hydraulic circuit so that the signal of the deformation sensor follows a certain function. The problem with unglued cyclic dynamic material tests in the compressive domain is that the test actually represents a mix of both control types even when it is carried out in a force-controlled way. Figure V-6 shows the principle: While the loading phase from the point of minimum displacement to its maximum is force-controlled and the test machine control is the active part producing an oscillation (e.g. sinusoidal), the situation in the unloading phase is displacement-controlled. From the point of maximum displacement to its minimum the specimen's recovery deformation controls the force given by the test machine. If the recovery deformation is slower than the sinusoidal force would be (which is the case for viscoelastic materials), the test machine can only react to the specimen's deformation since the specimen is not glued to the load plate.

A mix of force- and displacement-controlled testing has no impact when purely elastic materials are taken into consideration as there is no phase lag between action and reaction. However, in case of viscoelastic material testing, force- and displacement-controlled testing could mean that the specimen will never reach a steady state in stress or strain mode. This may not be critical when the main interest of the test is not the determination of material parameters but of e.g. creep characteristics. The main purpose of the standard TCCT according to (EN 12697-25, 2005), for example, is to produce creep curves representing the permanent deformation behavior under traffic loading at elevated temperatures (> 40°C). Thus, the unglued configuration is perfectly correct as it represents the situation under a wheel passing the pavement.

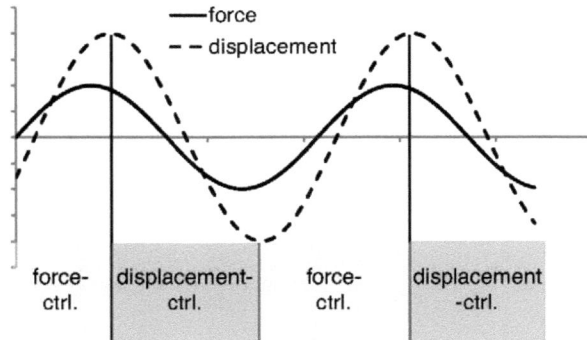

Figure V-6. Scheme of the test control in case of the specimen not being glued to the load plates.

Yet, if material parameters should be obtained from this test type it has to be verified that the mix-controlled test setup leads to similar results as the force-controlled setup. The question is especially delicate for unglued compressive tests where European Standards demand to obtain material parameters. One example is the ITT for stiffness (EN 12697-26, 2004) and fatigue resistance (EN 12697-24, 2007) testing of HMA. The main results of this test type are dynamic moduli and phase lags.

V.3.1   Test Setup

By taking into account the situation described above that the unglued cyclic dynamic compressive tests provide data from a mix-controlled test where the specimen may never reach a steady state, the question must be raised if it is correct to implement an approximation function $(F+L+1H)$ to fit the test data more accurately than a standard sine as suggested by (Kappl, 2007). If the test setup did not provide data for a correct assessment of viscoelastic material properties, a functional approximation that provides higher quality in mathematical means would not lead to more reliable results because of the incorrect test setup, strictly speaking.

To investigate the shortcome described above, two different test setups and the regarding results were compared. Figure V-7 shows a sketch of the two setups. The left setup (a) is a standard UCCT without a firm connection between load plate and specimen. In the right setup (b) the specimen is glued to both load plates with a two-component adhesive, thus making sure that the loading as well as the unloading phase is force-controlled and the specimen is in a steady state.

For each setup four specimens were tested. Details of the HMA specimens of the AC 11 mix are given in Table V-4. The tests were carried out without confining pressure as UCCTs at 30°C, a mean axial stress of 0.25 N/mm² and a stress amplitude of

0.15 N/mm². A frequency sweep was carried out with frequencies ranging from 0.1 Hz to 30 Hz.

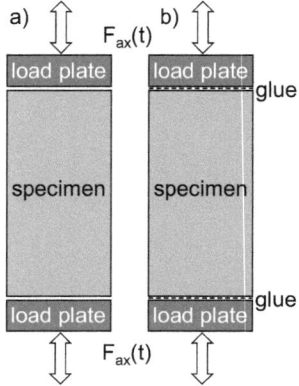

Figure V-7. Two test setups to compare results of mix-controlled (a) and force-controlled (b) CTTs.

The relevance of this investigation would be even higher if the tests were carried out at higher temperatures of 40°C or 50°C which are the standard test temperatures for TCCTs according to the EN standard. Since the stability of the material is too low at these temperatures to carry out a UCCT, the glued and unglued tests would have had to be run as TCCTs with confining pressure. Unfortunately, the present test equipment does not allow glued TCCTs. Thus, the investigation was carried out at 30°C. Since stiffness tests according (EN 12697-26, 2004), usually used to derive viscoelastic material parameters, are carried out only up to 35°C, the analysis presented in this chapter is still significant.

The test data was evaluated with the standard function $F+L$ as well as the advanced function $F+L+1H$. Test results for all four tested specimens were merged and statistically analyzed. This analysis was carried out separately for each test frequency. The diagrams in the following section always present the 5% and 95% quantiles as well as the median value (50% quantile) of the results for the glued and the unglued setup.

Table V-4. AC 11 specimen characteristics used for tests to compare results of glued and unglued setups.

Specimen	Binder	Binder Content [% (w/w)]	Binder:Fines (1:x)	Volume of air voids [% (v/v)]	glued/unglued
T404C				6.4	
T406E				5.5	
T406G				5.0	unglued
T406H	PmB	5.3	1.4	7.0	
T404A	25/55-65			3.8	
T405C				3.5	
T406A				5.2	glued
T406B				3.5	

V.3.2 Analysis of Signal Data from Force Sensor

If the assumption is correct that the unglued setup is rather mix-controlled than purely force-controlled, the regression quality of the standard $F+L$ function for an approximation of the force data should be lower than for the glued setup, since the shape of the sinusoidal force should be distorted. A distorted shape of the force oscillation is presumed in the unglued case because the unloading phase is controlled by the recovery deformation of the specimen and the test machine cannot provide a perfect sinusoidal unloading since it is influenced by the (delayed viscoelastic) reaction of the specimen. When the specimen is glued to the load plates, the test machine can enforce a sinusoidal function of the force for the complete load cycle. The top diagram in Figure V-8 provides a comparison of the coefficients of determination for the approximation of the force sensor with the standard $F+L$ function for both setups with respect to the test frequency. It is obvious that the fit quality is practically identical for both setups, and that the quality of the regression is at a very high level above 0.9999 until a frequency of 10 Hz is reached. From this test frequency on, the fit quality decreases and the scattering increases indicating that the test machine loses the ability to control a perfect sine the higher the frequency gets. This effect is more dominant for unglued specimens.

Since both setups result in similar fit quality this can be taken as one piece of evidence that the assumption made above was incorrect. Obviously it has no impact on the fit quality of the force sensor data whether the specimen is glued to the load plates or not, and thus the shapes of the force oscillation do not seem to differ so much.

The lower diagram in Figure V-8 gives the analogue information for the advanced $F+L+1H$ approximation. Again both setups are shown in the diagram. The situation until 10 Hz is similar to the quality of fit for the standard approximation. What can be proved by the diagram is that at higher test frequencies, when the test machine has more and more difficulty to provide a perfect sinusoidal force, the advanced sine is able to compensate this imperfect shape. The coefficient of determination for 10 Hz and 20 Hz is clearly higher when the $F+L+1H$ function is used for regression. And, again, both setups result in similar fit qualities.

Section II.3.4 presented an analysis of how the amplitude ration AR between 1st harmonic and fundamental oscillation of the approximation together with the shift factor γ influence the shape of the advanced function $F+L+1H$. It was shown that $F+L+1H$-analysis is a proper tool to check the shape of functions in a quick way without having to look at each individual oscillation graphically. It can therefore also be used to find out whether the shape of the sinusoidal force differs between the two setups. All this is said keeping in mind that great care has to be taken on the quality of the fit itself. If the coefficient of determination is lower than 0.95 the shape of the approximation function may differ considerable from the test data.

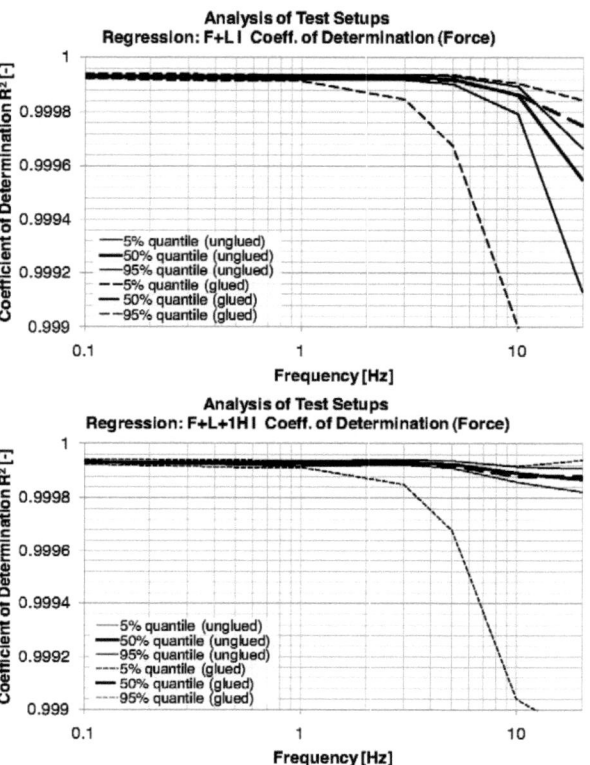

Figure V-8. 5%, 50% and 95% quantiles of the coefficient of determination R^2 for the force sensor for glued vs. unglued test setup; F+L approximation (top) and F+L+1H approximation (bottom).

If it was true that the unglued test setup results in distorted shape of the oscillating force due to the mix-controlled constellation, the degree of distortion could be described by means of *AR* and *γ*. In the case of a distortion of the sine, there should be clear differences between both setups. Figure V-9 presents the results. By taking a look at the amplitude ratio in the top diagram it becomes clear that there is no distinguishable distortion since *AR* is below 1‰ for lower test frequencies. *AR* increases with increasing frequency but still stays at very low level. Even at 10 Hz is the mean *AR* only around 5‰. At 20 Hz the unglued setup shows a mean *AR* of 1.8% vs. 1.1% for the glued setup (median values). This shows that the test machine can keep the sinusoidal load function in a near to perfect state more easily when the specimen is glued to the load plates and the complete load cycle is force-controlled.

Data of the shift factor are presented in the lower diagram. Again both setups produce the same *γ*. The scatter is rather large when testing at low frequencies. Due to the small *AR* this has no impact on the shape of the force approximation. The phase lag at 20 Hz is around -150°. Still, the amplitude ratio is so small that this has no noteworthy effect on the shape of the sinusoidal force.

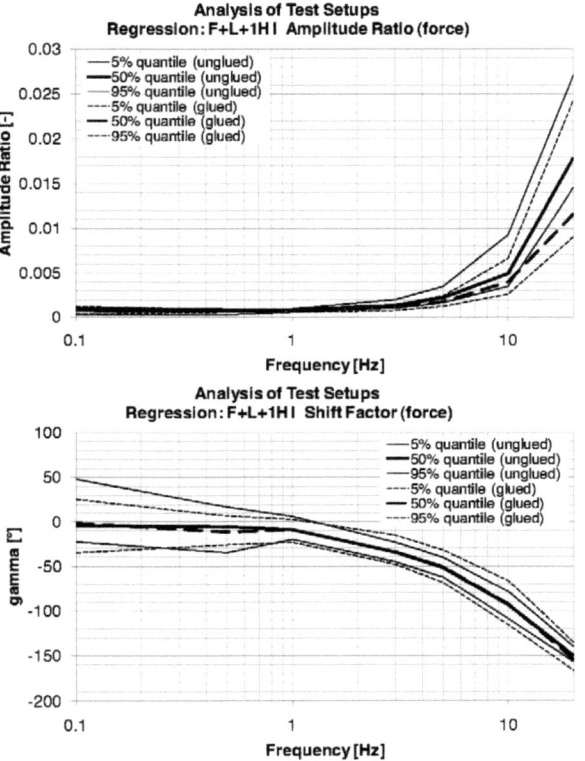

Figure V-9. 5%, 50% and 95% quantiles of AR (top) and γ (bottom) for the force sensor for glued vs. unglued test setup at 30°C.

This section proves clearly that the test setup, i.e. whether the specimen is glued to the load plates or not, has no relevant impact on the shape of the sinusoidal force signal. The force- as well as the mix-controlled test, both result in a near to perfect sine of the load. Only at high frequencies (above 10 Hz) the quality of the fit for the standard sine starts to decrease. This happens for both setups and seems to be caused by the test machine which loses its ability to drive a sinusoidal force oscillation the higher the test frequency gets.

V.3.3   Analysis of Signal Data from Deformation Sensor

Data from the force signal was analyzed in the previous section. This section shows analogue considerations for the axial deformation sensor. Axial deformation data are mean values from the two axial deformation sensors (LVDTs) (see Figure II-6).

Figure V-10 presents the coefficient of determination of both setups. The top diagram gives information about the regression with the standard approximation. Clearly, both setups are fitted with a high quality ($R^2 > 0.995$). Still, the glued setup can be described with the standard sine in a slightly better way. Scatter for both cases is similar. Although there is a small difference between both setups in the quality of the fit, it can be assumed that both setups produce similar deformations regarding the shape of the oscillating part.

The data show that the standard sine approximates the deformation data better with higher frequency. From 3 Hz on, the coefficient of determination R^2 is around or above 0.998. At low frequencies when the viscous part of the material behavior is still dominant, the standard approximation function describes the material behavior only at a lower, although good quality level.

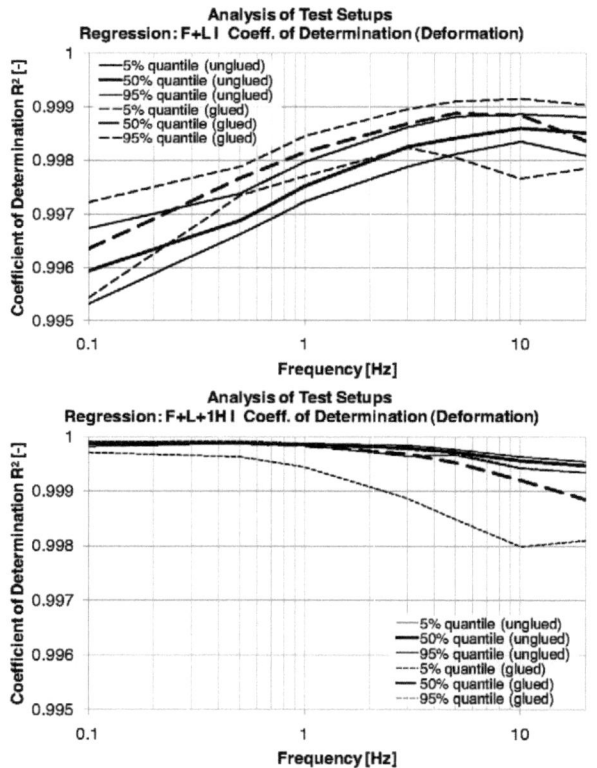

Figure V-10. 5%, 50% and 95% quantiles of the coefficient of determination R^2 for the deformation sensor for glued vs. unglued test setup at 30°C; F+L approximation (top) and F+L+1H approximation (bottom).

The lower diagram in Figure V-10 shows the results for the coefficient of determination R^2 for both setups with the advanced $F+L+1H$ approach. Compared to the standard regression the quality of fit of both cases is considerably higher. For the unglued specimens the quality of fit R^2 is above 0.999 for the complete range of test frequencies with a low scattering. Especially at low frequencies where the viscous part of the material behavior is more dominant, the difference to the standard approximation is clear. Above 3 Hz the quality of fit decreases slightly. Due to increasing stiffness, the amplitude of the deformation decreases as the amplitude of loading is kept constant. As the deformation is recorded by LVDTs, a smaller amplitude means that the noise of the signal gets more dominant and reflects on the quality of fit in a negative way.

Interestingly enough, the unglued specimens produce better fit quality than the glued setup when evaluated with the $F+L+1H$ approach. Especially the scatter of results is by far larger for glued specimens. A plausible explanation is an impact of the glued setup. Although both end planes of the specimens are grinded to ensure an optimum in terms of co-planarity there is still a certain amount of tilting. This means that the end planes are not perfectly perpendicular to the vertical axis but exhibit a deviation of maximal 1°. The other problem is that although the load plates are glued to the specimen's end planes with high precision, small eccentricities occur. Thus, when the axial loading is applied an additional momentum is induced by the two factors described above. This influences the axial deformation and leads to a lower quality of fit or rather a larger scattering of the quality.

Figure V-11 presents the amplitude ratio AR and the shift factor γ for the axial deformation data. There is a clear relation between AR and the test frequency. The higher the frequency, and thus the more dominant the elastic part of the behavior gets, the lower becomes the share of the 1st harmonic term. Data from tests with both setups start around the same values, the unglued setup at a range of 6.2% to 7.1%, and the glued setup ranges from 5.5% to 7%. The ratio decreases more quickly with increasing frequency for the glued setup reaching 1.5% to 2.5% at 20 Hz. The unglued setup results in a ratio of 2.4% to 3.8% at this frequency. In terms of the median values the unglued setup shows a 7% higher amplitude ratio at 0.1 Hz increasing to a 64% higher ratio at 20 Hz compared to the glued setup. It can be stated that the impact of the 1st harmonic declines with increasing test frequency. This goes along with the data presented in Figure V-10 where the coefficient of determination was shown for the standard approximation function. The quality of fit of the standard function increases with increasing frequency. This is a logic result as the share of the 1st harmonic gets less dominant at these test conditions. For the whole frequency range, the ratio is smaller for the glued setup, showing that the deformation produced by the glued specimens is more related to a sinusoidal shape than for the unglued specimens.

The lower diagram in Figure V-11 presents results of the shift factor γ. This value starts at -17° for the unglued and -19° for the glued setup (median values) at 0.1 Hz. As shown in section II.3.4 (Figure II-25) it produces a steeper incline of the approximation function in the loading and a flatter decline in the unloading phase. Up to 5 Hz, the phase lag stays beyond -45°. Simultaneously, AR decreases indicating that the difference in the gradient between loading and unloading phase gets smaller. When the test frequency is increased to 10 Hz and 20 Hz, γ becomes even larger reaching -70° for the glued and -100° for the unglued setup respectively. Together with a still decreasing amplitude ratio, the distortion of the deformation oscillation is less and less dominant.

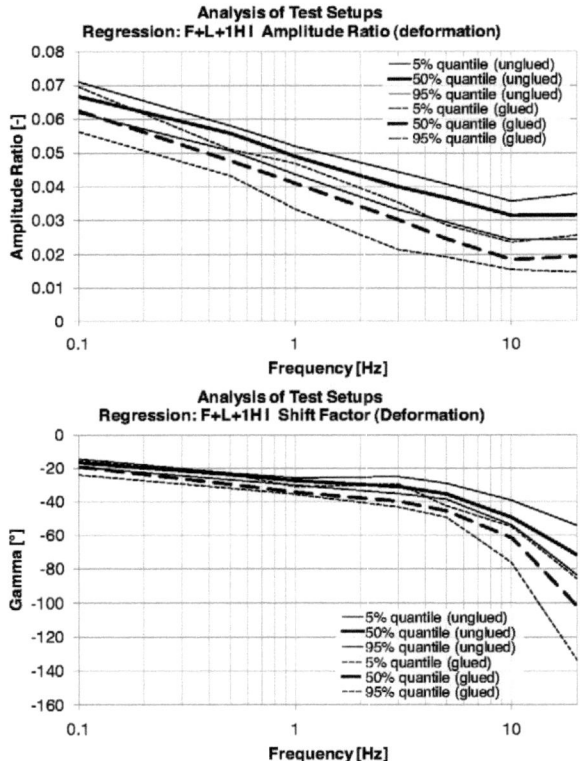

Figure V-11. 5%, 50% and 95% quantiles of AR (top) and γ (bottom) for the deformation sensor for glued vs. unglued test setup at 30°C.

It was shown in this section that the shape of the deformation cannot be described as perfectly with the standard $F+L$ regression as with the advanced $F+L+1H$ function. This is especially true for low frequencies below 3 Hz. Glued specimens produce slightly better qualities of the fit with the standard approximation function. It is therefore assumed that glued specimens react with a less distorted sine to sinusoidal load in terms of deformation. This thesis is confirmed by the amplitude ratio between 1st harmonic and fundamental. It is also higher for the glued setup throughout the frequencies.

V.3.4 Analysis of Material Parameters

So far the analysis was carried out for data from the force and deformation sensor. An interesting question is, if and how the setup influences on the derived viscoelastic material parameters. In Figure V-12 the top diagram shows the axial phase lag at the maximum loading ($\varphi_{ax,ax,max}$) between force and axial deformation when the regression is done with the $F+L$ function.

The difference between the two analyzed setups in terms of the axial phase lag is not significant. The lower diagram in Figure V-12 shows results of the dynamic modulus $|E^*|$ for both setups. The higher the test frequencies the larger gets the difference between the two setups, the glued setup always being stiffer. The median value of $|E^*|$ of the unglued setup starts at a value of 88% of the glued setup at 0.1 Hz, and this ratio decreases constantly to 82 % at 20 Hz.

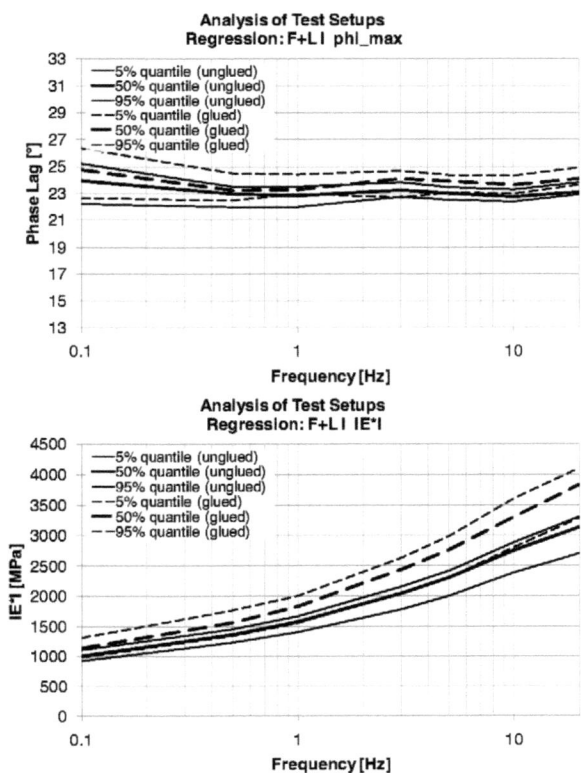

Figure V-12. 5%, 50% and 95% quantiles of $\varphi_{ax,ax,max}$ (top) and $|E^*|$ (bottom) for glued vs. unglued test setup at 30°C; F+L approximation.

Since the difference between the phase lags is not significant, the results in terms of elastic and viscous part of the dynamic modulus E_1 and E_2 can be omitted from this analysis. They would reflect the situation of $|E^*|$ in Figure V-12.

When both setups are evaluated with the advanced $F+L+1H$ function and the results are compared in terms of $|E^*|$, the stiffness values are nearly identical with the situation of the standard function in Figure V-12. Neither the absolute values nor the ranking of the two setups are different. The glued, force-controlled setup results in a stiffer behavior especially at higher frequencies. Unlike the fit quality in terms of coefficient of determination, the material stiffness is influenced by the way the test is controlled.

Figure V-13 shows the four phase lags obtained for the advanced approximation function. The upper two diagrams give information about the axial phase lag at the extrema of the loading indicated as φ_{min} and φ_{max}. Compared to the results of the standard regression no notable differences can be found. Again the glued setup results in a slightly more viscous behavior than the unglued setup. It is interesting to look at the results for the phase angle in the loading (φ_{MV+}) and unloading (φ_{MV-}) phase. While the phase lag is increasing with increasing test frequency in the loading phase for both setups, the lag is decreasing with increasing frequency in the unloading phase. Thus the unloading phase represents the material behavior that would be expected if it followed the theory of viscoelasticity that states that an increase in frequency or a decrease in temperature leads to a stronger influence of the elastic parts of the material behavior. All four phase lags start at different levels (15° to 30°, $\Delta = 15°$) at low frequencies, but the higher the test frequency gets the smaller the difference becomes between the different phase lags (19° to 27°, $\Delta = 8°$).

From this data it appears that the advanced $F+L+1H$ function accounts for the varying material reaction in frequency domain. At low frequencies the viscous material behavior is more dominant. The material reaction depends more on whether it is in the loading or in the unloading phase, at least in the compressive domain. The higher the test frequency and thus the elastic part of the behavior, the smaller becomes the influence of the loading state. As the results coincide for the glued as well as the unglued setup, a crucial influence of the setup on the resulting material behavior cannot be found.

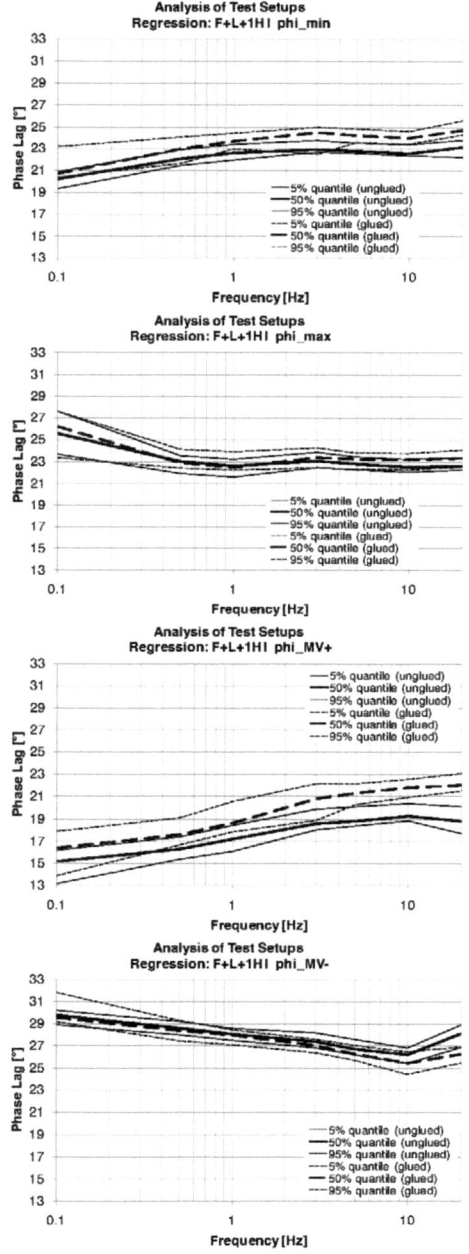

Figure V-13. 5%, 50% and 95% quantiles of $\varphi_{ax,ax,min}$ (top), $\varphi_{ax,ax,max}$ (2nd from above), $\varphi_{ax,ax,MV+}$ (3rd from above) and $\varphi_{ax,ax,MV-}$ (bottom) for glued vs. unglued test setup at 30°C; F+L+1H approximation.

V.3.5 Conclusions

From the analysis of two test setups – one with specimens glued to the load plates and thus guaranteeing a force-controlled test for loading and unloading phase and one standard unglued setup – the following conclusions can be drawn:

- The advanced $F+L+1H$ approximation function is a proper tool to quickly check the shape of sinusoidal functions. If the sine gets distorted, the amplitude ratio AR and the shift factor γ are able to describe the shape and magnitude of the distortion.
- Regarding the data of the force sensor, no significant difference between both test setups was found. Both setups result in a high quality of fit with the standard and the advanced regression approach. Since the amplitude ratio AR is very low (<5‰) for frequencies up to 10 Hz, there is no noteworthy distortion of the sinusoidal force.
- The analysis of data of the deformation sensor revealed that the deformation cannot be approximated as well with the standard sinus as the force data. The fit quality of the glued setup is better when the $F+L$ function is used for regression analysis. The advanced approach results in coefficients of determination of 0.999 and higher for both setups. By considering the amplitude ratio AR it was found that especially at lower frequencies (with a more dominant viscous material behavior) the shape of the deformation is distorted with a steeper incline of the loading phase and a flatter decline in the unloading phase. The effect is stronger for unglued specimens.
- In terms of mechanical material parameters it was found that there is no difference in phase lags between both setups. Glued specimens react stiffer with an increasing difference between both setups at higher the frequencies.
- The advanced approximation function accounts for the changing material response within one single load cycle. This was shown by the four phase lags at different amplitude values. Especially at low frequencies the difference between the phase lags is obvious. The effect gets less dominant with increasing frequency.
- $F+L+1H$ can give valuable information about the shape of oscillating test data. Problems with the test machine control as well as shape and magnitude of the deformation oscillation can be found and described easily. In addition, the advanced approach is capable of describing varying material reaction at different stages of a load cycle.
- When it comes to material stiffness, it makes a difference whether the specimen is glued to the load plate prior to testing or not. Unglued specimens result in a lower stiffness (80% to 90% of the glued setup depending on the test frequency).
- The results of this investigation are based on tests carried out at 30°C and one specific HMA used for the investigation. The present test setup does not allow a similar analysis at higher temperatures since TCCTs cannot be run out with

glued specimens at the moment. But the results at 30°C show that the difference between both setups in terms of $|E^*|$ grows with increasing stiffness of the material. It is therefore assumed that the effect will become less important at lower material stiffness and thus at higher test temperatures. Still, a validation of this assumption by carrying out this investigation at 50°C will be an important future task.

V.4 Linear vs. Nonlinear Viscoelastic Behavior of HMA

According to the theory of viscoelasticity (Findley, et al., 1989) the viscoelastic material parameters are independent of the state of stress as long as the material is stressed within the linear viscoelastic domain. For HMAs a strain value of about 10^{-4} ensures that the material reacts in a linear viscoelastic way according (Airey, et al., 2003). For the binder this limit is much higher and depends on the temperature. (Airey, et al., 2003) found strain values from some 10^{-1} at high temperatures down to some 10^{-2} for lower temperatures.

The following investigation compares various deviatoric stress amplitudes at different temperatures to find out whether the load level chosen for the research stresses the material in a linear or non-linear viscoelastic way. If it is found that the material is within a non-linear domain, the effects of non-linear effects on the material parameters are analyzed.

Tests at 10°C

At 10°C UCCTs were carried out at three different deviatoric stress levels. Since there was no radial confining pressure, the axial stress equals the deviatoric stress. Table V-5 contains information about the tested specimens. The HMA is an AC 11 70/100 with a binder content of 5.3% (m/m). The right column in Table V-5 shows the deviatoric stress range as the lower ($\sigma_{dev,min}$) and upper ($\sigma_{dev,max}$) extrema of the oscillation. For the lowest stress level from 0.1 N/mm² to 0.7 N/mm² data from only one specimen is available. It is worth noting that the stress conditions chosen for this temperature are different from the stress levels for the standard TCCT. This is due to the fact that the stiffness of the material is at a much higher level at 10°C and thus the standard stress conditions according to (EN 12697-25, 2005) result in very small deformations. These deformations could not be recorded with sufficient quality by the deformation/strain sensors. The test frequencies range from 0.1 Hz to 30 Hz.

The test data was evaluated with the advanced function $F+L+1H$. Test results for all specimens tested at the same conditions were merged and statistically analyzed. This analysis was carried out separately for each test frequency. The diagrams in the following always present the 5% and 95% quantiles as well as the median value (50% quantile) of results for all three stress levels.

Table V-5. Specimen characteristics used for tests to compare results at different deviatoric stress levels at 10°C.

Specimen	Binder	Binder Content [% (w/w)]	Binder:fines (1:x)	Volume of air voids [% (v/v)]	$\sigma_{dev,min}$-$\sigma_{dev,max}$ [N/mm^2]
T335B	70/100	5.3	1.4	2.3	0.1-0.7
T328A				2.9	0.1-0.9
T328B				2.3	
T333B				2.3	0.1-1.1
T334B				2.2	

To check the quality of the fit of the approximation function to the test data, the coefficient of determination R^2 is presented in Figure V-14 for the radial and the axial deformation data. There is a correlation between the stress amplitude and the fit quality.

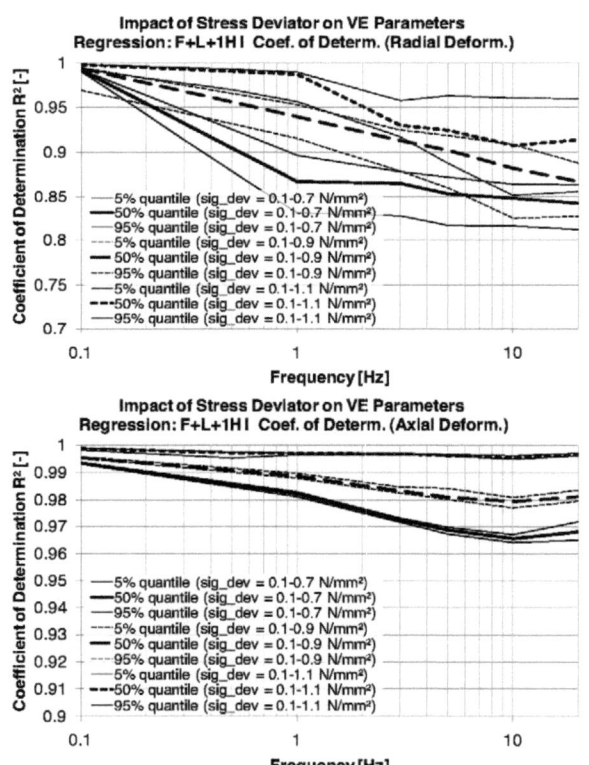

Figure V-14. 5%, 50% and 95% quantiles of the coefficient of determination R^2 for the radial (top) and axial (bottom) deformation sensor for three deviatoric stress levels at 10°C; F+L+1H approximation.

The lower the stress amplitude the lower is the quality of approximation. This effect is more dominant for the radial deformation data. The logical explanation is that with increasing stress amplitude, the deformation increases as well if it is assumed that the stiffness of the specimen is not crucially influenced by the stress level. An increase in the deformation amplitude means that the noise of the sensors gets less dominant and therefore the quality of fit rises. For the radial deformation R^2 is even below 0.8 for the lowest stress amplitude for frequencies higher than 10 Hz. But also for lower frequencies the quality of fit cannot be considered as high. A larger scattering of results is expected from these findings.

As stated above, the strain level – in the case of cyclic dynamic tests the strain amplitude – is an important benchmark to check whether an HMA specimen was tested in the linear viscoelastic domain or not. Figure V-15 shows the strain amplitude for the radial and axial reaction. The radial strain is below 10^{-4} for all stress levels from 0.5 Hz upwards. The axial strain is higher for all three stress levels and stays below $2·10^{-4}$ for frequencies of 0.5 Hz and higher. But even at 0.1 Hz the strain amplitude lies within a range in which linear viscoelasticity can be assumed.

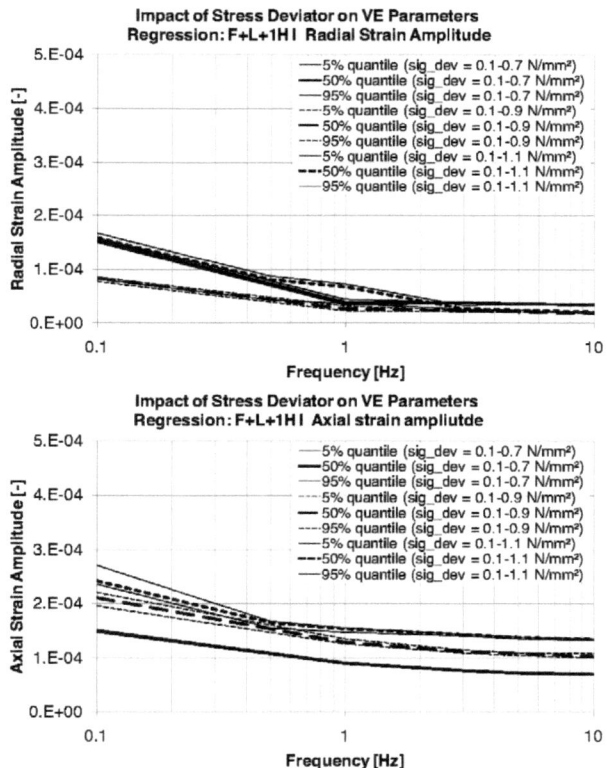

Figure V-15. 5%, 50% and 95% quantiles of the radial (top) and axial (bottom) strain amplitude for three deviatoric stress levels at 10°C; F+L+1H approximation.

Regarding the radial phase lag, Figure V-16 shows that the difference between the different stress levels is not significant taking into consideration the scattering of the results and the accuracy of the approximation. At the minimum of the loading and in the loading phase, the highest stress level produces slightly higher phase lags, whereas at the maximum loading and in the unloading phase, the lowest stress amplitude reacts slightly more elastically. Taking into consideration the radial strain amplitude (Figure V-15) these results were expectable since the strain level is low enough for linear viscoelastic behavior.

Analogue data for the axial phase lag is shown in Figure V-17. Interestingly enough the highest stress level results in an obviously more elastic behavior than the other two stress amplitudes. The median values are 1° to 5° lower depending on the amplitude value at which the phase lag is looked at. Clear differences occur at the point of minimum loading and in the loading phase. Part of this difference may be explained by the fact that the axial strain amplitudes are higher for the highest stress level and therefore some non-linear effects may have occurred at this test.

The stiffness of the specimens in terms of the dynamic modulus $|E^*|$ is depicted in Figure V-18. All three stress levels result in similar stiffness if the frequency stays below 1 Hz. Above this frequency, the material reacts clearly stiffer at the lowest stress amplitude than at the other two stress amplitudes. $|E^*|$ is about 14% higher for the low stress level (median value). The quality of the fit of the low stress amplitude is clearly below the other two deviatoric stresses. Part of the difference in $|E^*|$ can be correlated to this lower quality of the fit. Effects of nonlinearity cannot be found from this data because the strain amplitudes are highest at low frequencies. At these frequencies all three stress levels exhibit similar dynamic moduli.

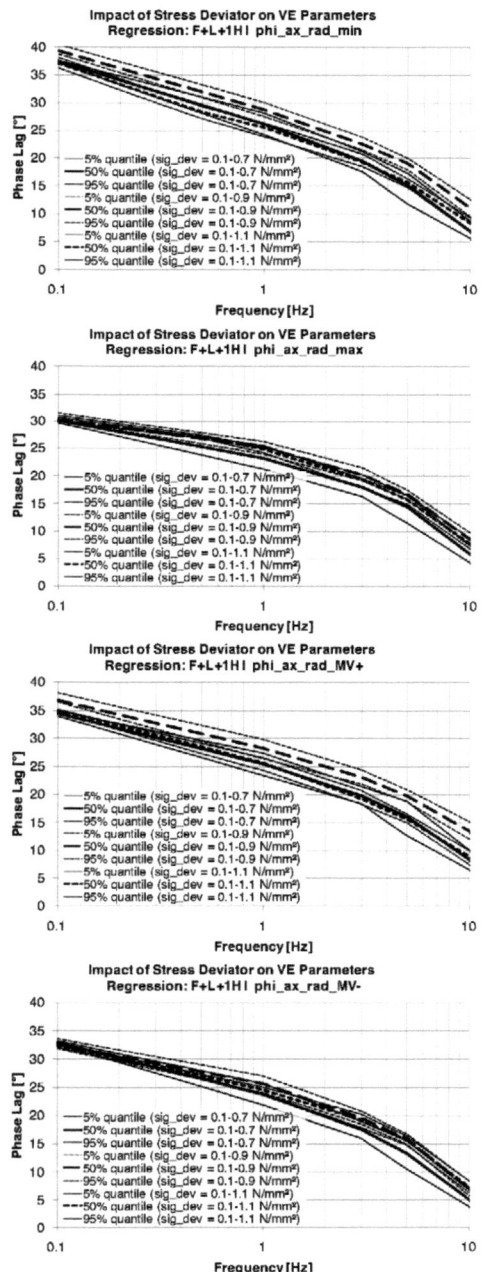

Figure V-16. 5%, 50% and 95% quantiles of $\varphi_{ax,rad,min}$ (top), $\varphi_{ax,rad,max}$ (2nd from above), $\varphi_{ax,rad,MV+}$ (3rd from above) and $\varphi_{ax,rad,MV-}$ (bottom) for three deviatoric stress levels at 10°C; F+L+1H approximation.

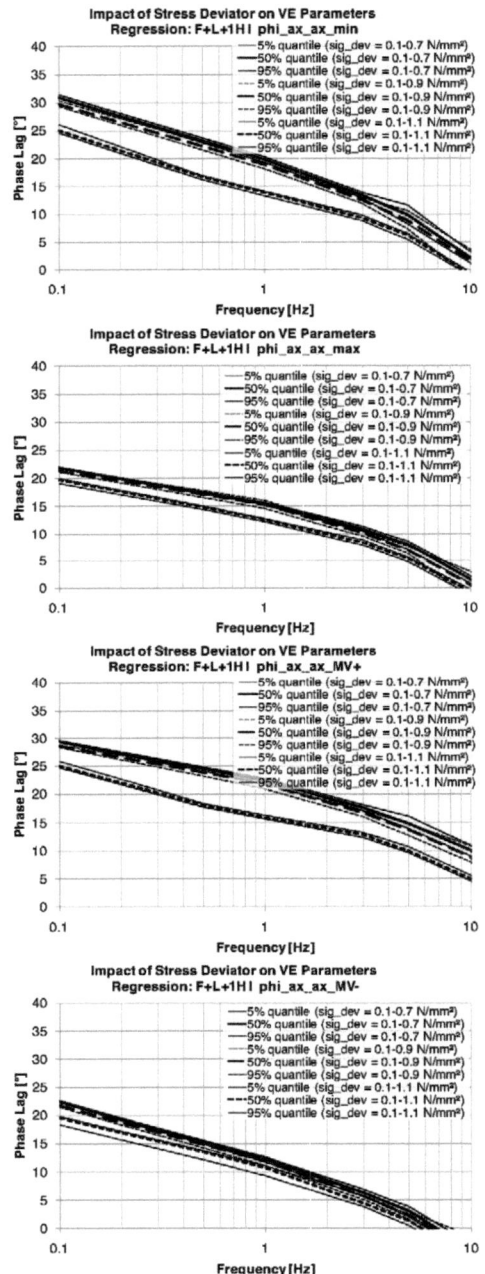

Figure V-17. 5%, 50% and 95% quantiles of $\varphi_{ax,ax,min}$ (top), $\varphi_{ax,ax,max}$ (2nd from above), $\varphi_{ax,ax,MV+}$ (3rd from above) and $\varphi_{ax,ax,\,MV-}$ (bottom) for three deviatoric stress levels at 10°C; F+L+1H approximation.

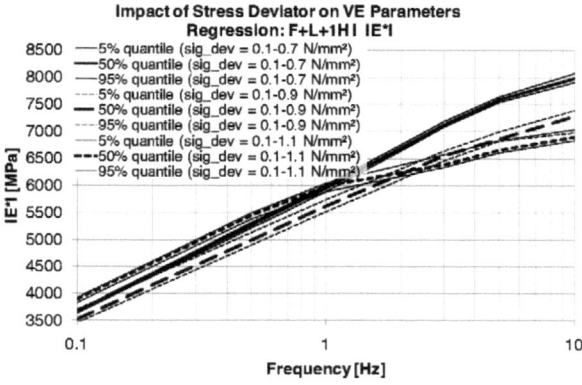

Figure V-18. 5%, 50% and 95% quantiles of |E*| for three deviatoric stress levels at 10°C; F+L+1H approximation.

Tests at 30 °C

Tests to analyze the impact of the deviatoric stress on the viscoelastic material parameters were also run at 30°C. Three different deviatoric stress amplitudes were investigated. Tests at two deviatoric stress levels were carried out with constant confining pressure as TCCTs and at the third stress level without radial pressure as UCCTs. Table V-6 gives an overview on the tested specimens. The mix is an AC 11 70/100 with a binder content of 5.3% (m/m). The range of the deviatoric stress is shown in the right column. For the first test condition with a range of σ_{dev} from 0.0 N/mm² to 0.4 N/mm² only one specimen was tested successfully. As usual the tests were carried out with a frequency sweep from 0.1 Hz to 30 Hz.

Table V-6. Specimen characteristics used for tests to compare results at different deviatoric stress levels at 30°C.

Specimen	Binder	Binder Content [% (w/w)]	Binder:fines (1:x)	Volume of air voids [% (v/v)]	$\sigma_{dev,min}$-$\sigma_{dev,max}$ [N/mm²]
T284A	70/100	5.3	1.3	2.1	0.0-0.4 (TCCT)
T284D				2.1	0.0-0.6 (TCCT)
T292C				2.0	
T333A			1.4	3.2	0.1-0.4 (UCCT)
T336B				2.5	

A look at the quality of the fit in terms of the coefficient of determination R² (Figure V-19) for the radial and axial deformation data shows that the radial deformation is approximated with an R² of 0.95 or above until 10 Hz. A higher deviatoric stress amplitude results in a higher quality of the fit. The effect has already been explained in the section about tests at 10°C. The axial deformation data are approximated much better than the radial deformation. The coefficient of determination never lies below 0.98.

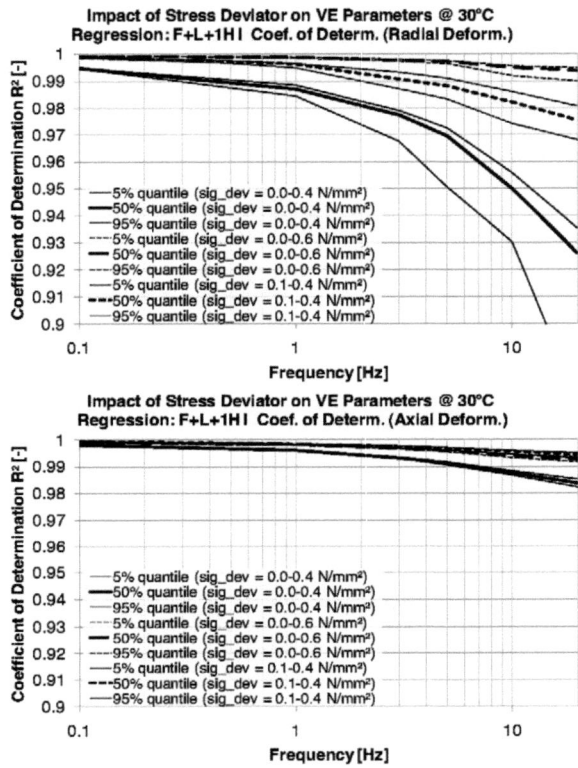

Figure V-19. 5%, 50% and 95% quantiles of the coefficient of determination R^2 for the radial (top) and axial (bottom) deformation sensor for three deviatoric stress levels at 30°C; F+L+1H approximation.

Figure V-20 presents the strain amplitudes in radial and axial direction for the three stress levels. Clearly, the TCCT with the higher stress amplitude (0.0 to 0.6 N/mm²) produces significantly higher strain amplitudes, especially in the low frequency range. The other two stress levels exhibit similar strain amplitudes. All strain amplitudes are above 10^{-4} for lower frequencies and approach the linear VE strain limit at 1 Hz according to literature given at the beginning of this section.

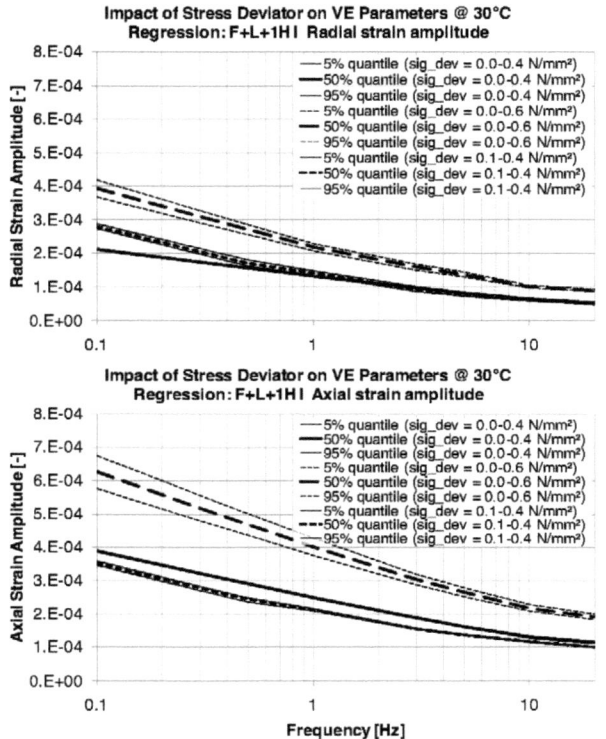

Figure V-20. 5%, 50% and 95% quantiles of the radial (top) and axial (bottom) strain amplitude for three deviatoric stress levels at 30°C; F+L+1H approximation.

The influence of the deviatoric stress on the radial phase lag is not consistent for all four different phase lags; the results are shown in Figure V-21. The phase lag at the maximum loading is quite similar for all three stress levels with the largest variety at 0.1 Hz. The TCCT with a stress deviator ranging from 0.0 N/mm² to 0.4 N/mm² has a radial phase lag of 23.2° at 0.1 Hz, the TCCT with a range from 0.0 N/mm² to 0.6 N/mm² 26.2° and the UCCT (0.1 N/mm² to 0.4 N/mm²) 28.8°. The phase lag at the minimum loading is similar for the two TCCTs, the UCCT produces far higher lags at lower frequencies. The higher the test frequency the more the results convert from all three test setups. The same can be stated about the phase lag in the unloading phase (MV-). In the loading phase (MV+) the UCCT and the TCCT with the higher deviatoric stress level show similar results, whereas the phase lags for the TCCT with lower deviatoric stresses produce much lower phase lags in the low frequency domain.

The results indicate that at least the phase lag at the maximum of the loading is hardly influences by the stress deviator. It is proven in section VI.3 that the material reacts according to the theory of linear viscoelasticity since master curves can be obtained from test data when the results evaluated at the maximum of the loading are used.

For the phase lags at the other three amplitude values, it seems that at the material reacts differently at low frequencies. The difference may be explained by the fact that the strain amplitudes are quite high at low frequencies – so there may be non-linear effects that lead to stress-dependent behavior.

An analogue illustration for the axial phase lag is depicted in Figure V-22. The relation between the different phase lags is the similar to the radial phase lags for three amplitude values (min, max, MV-). In the loading phase (MV+), the TCCT with the lower deviator stress and the UCCT exhibit similar behavior in terms of axial phase lags. This is different from the radial phase lag. Interestingly enough, the behavior at maximum of the loading is again similar for all three stress states. Again, largest differences between the three test conditions can be found at low test frequencies.

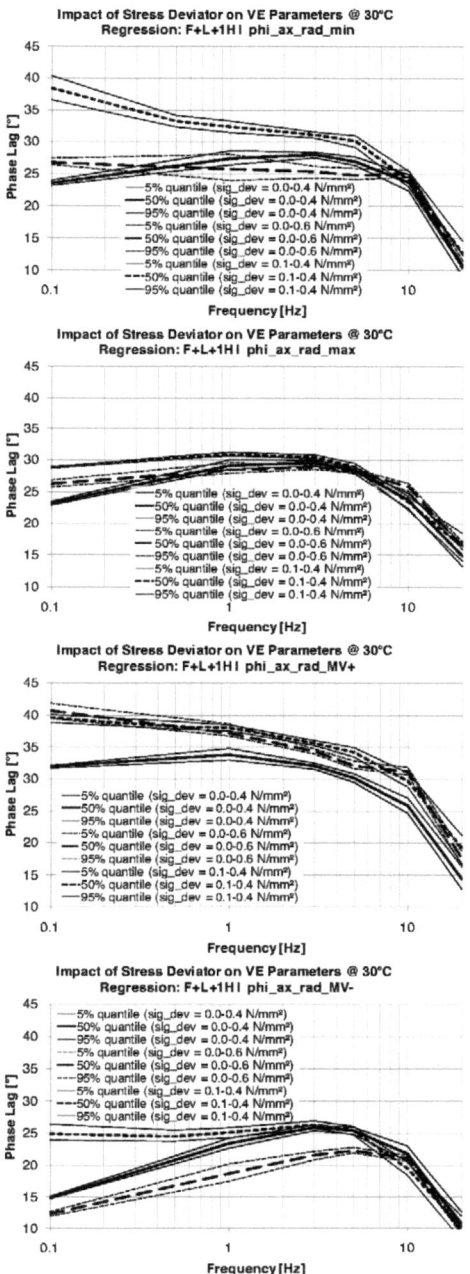

Figure V-21. 5%, 50% and 95% quantiles of $\varphi_{ax,rad,min}$ (top), $\varphi_{ax,rad,max}$ (2nd from above), $\varphi_{ax,rad,MV+}$ (3rd from above) and $\varphi_{ax,rad,MV-}$ (bottom) for three deviatoric stress levels at 30°C; F+L+1H approximation.

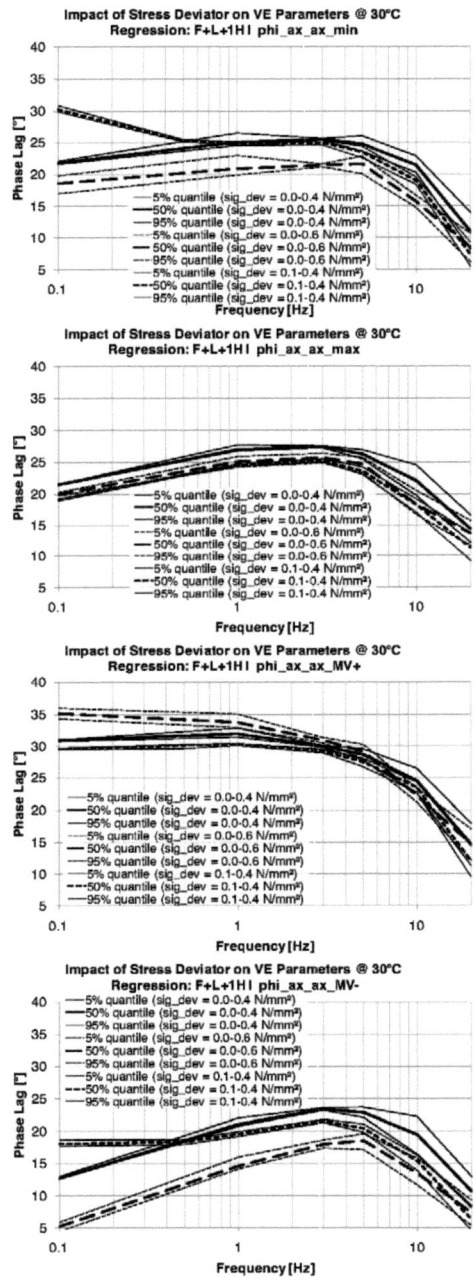

Figure V-22. 5%, 50% and 95% quantiles of $\varphi_{ax,ax,min}$ (top), $\varphi_{ax,ax,max}$ (2nd from above), $\varphi_{ax,ax,MV+}$ (3rd from above) and $\varphi_{ax,ax,MV-}$ (bottom) for three deviatoric stress levels at 30°C; F+L+1H approximation.

Concerning the stiffness of the mix (Figure V-23), the following can be stated:
- $|E^*|$ ranges from 770 MPa (UCCT) and 872 MPa (TCCT high deviatoric stress, +13%) to 943 MPa (TCCT low deviatoric stress, +22%) at 0.1 Hz.
- It increases to 2780 MPa (UCCT), 2890 MPa (TCCT high deviatoric stress, +4%) and 3310 MPa (TCCT low deviatoric stress, +19%) at 20 Hz (median values).

The difference between the various test conditions is significant. In the UCCT the material develops the lowest stiffness; in the TCCT the lower stress amplitude results in the highest stiffness. Again, an influence of the deviatoric stress amplitude is possible which would mean that the material is outside the linear viscoelastic domain.

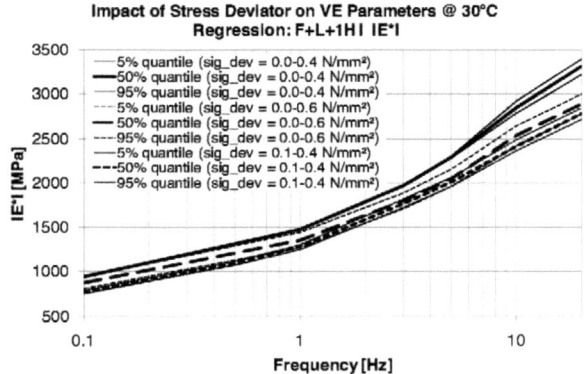

Figure V-23. 5%, 50% and 95% quantiles of $|E^*|$ for three deviatoric stress levels at 30°C; F+L+1H approximation.

Tests at 50 °C

To finalize this investigation, TCCTs were run at 50°C and two different deviatoric stress amplitudes. A third amplitude with a range from 0.0 N/mm² to 0.8 N/mm² was tried but the specimens failed at these high stresses. The mix was the same as for 30°C and 10°C. Two specimens were tested for each test condition. Table V-7 provides the most important specimen characteristics.

Table V-7. Specimen characteristics used for tests to compare results at different deviatoric stress levels at 50°C.

Specimen	Binder	Binder Content [% (w/w)]	Binder:fines (1:x)	Volume of air voids [% (v/v)]	$\sigma_{dev,min}$-$\sigma_{dev,max}$ [N/mm²]
T295C	70/100	5.3	1.4	2.0	0.0-0.4
T322A				3.3	(TCCT)
T325A				3.3	0.0-0.6
T334C				2.1	(TCCT)

Figure V-24 shows the quality of the fit for both test setups. The radial as well as the axial deformation can be fitted with high quality. The coefficient of determination never decreases below 0.997 (median values). The scatter for the radial deformation data and the lower stress level is significantly higher than for the higher stress level or the axial deformation sensor.

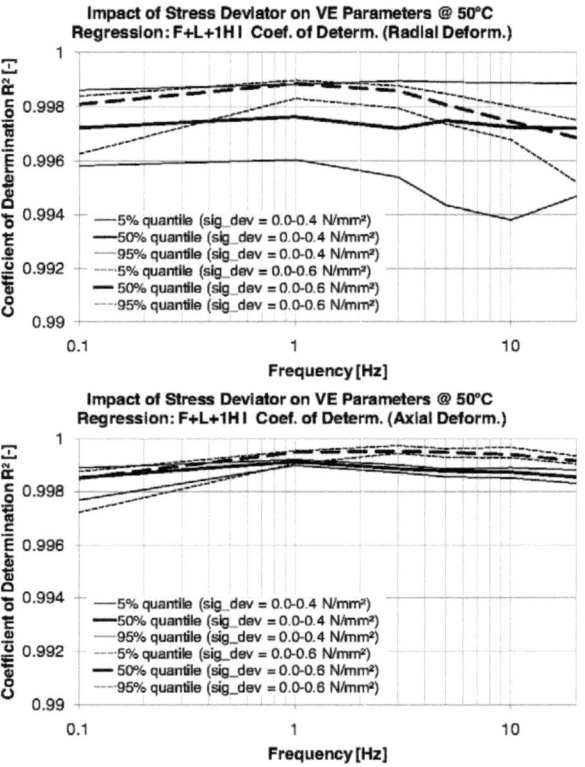

Figure V-24. 5%, 50% and 95% quantiles of the coefficient of determination R^2 for the radial (top) and axial (bottom) deformation sensor for three deviatoric stress levels at 50°C; F+L+1H approximation.

The strain amplitudes in radial and axial direction are depicted in Figure V-25. Both amplitude levels are significantly above the limit of linear viscoelastic behavior (10^{-4}) according to literature and the difference between both test setups is obvious. The higher deviatoric stress amplitude starts at around $9 \cdot 10^{-4}$ for the axial strain amplitude whereas the lower stress amplitude exhibits an axial strain amplitude of around $5.5 \cdot 10^{-4}$. The strain amplitude decreases slightly for the lower stress level and stronger for the higher stress level as the test frequencies are increased. According to the limits given in literature, both test conditions will activate a non-linear viscoelastic behavior of the specimen to some extent. The non-linear viscoelastic behavior will be more dominant for the higher stress amplitude.

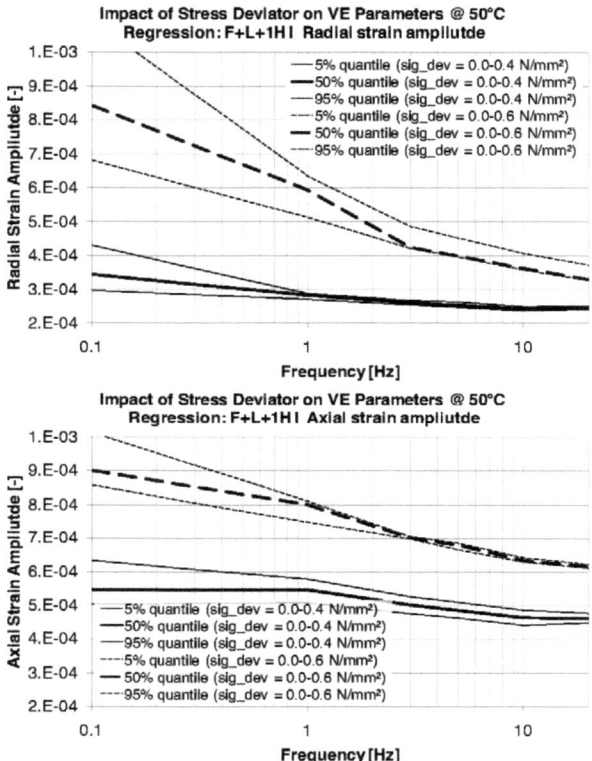

Figure V-25. 5%, 50% and 95% quantiles of the radial (top) and axial (bottom) strain amplitude for three deviatoric stress levels at 50°C; F+L+1H approximation.

The difference between the radial phase lags at different functional values is obviously significant, as shown in Figure V-26. Again, the difference is higher at lower frequencies and decreases as the frequency increases. Analogue to the results at 30°C, the difference is smallest at the maximum of loading. In the unloading phase (MV-), the time lag between radial deformation and axial loading is at a low level and even negative for 10 Hz and 20 Hz, which is physically impossible, since the outer force is the active component and the deformation as the reaction can only occur simultaneously (for purely elastic materials) or with a certain positive phase lag. The negative phase lags seem to have their origin in the inevitable inaccuracy (noise) of the test data.

The differences between results from tests with both stress amplitudes indicate that the material is stressed in a non-linear viscoelastic domain. Interestingly enough, the phase lag evaluated at the maximum loading is impacted by the stress level only to a minor extent.

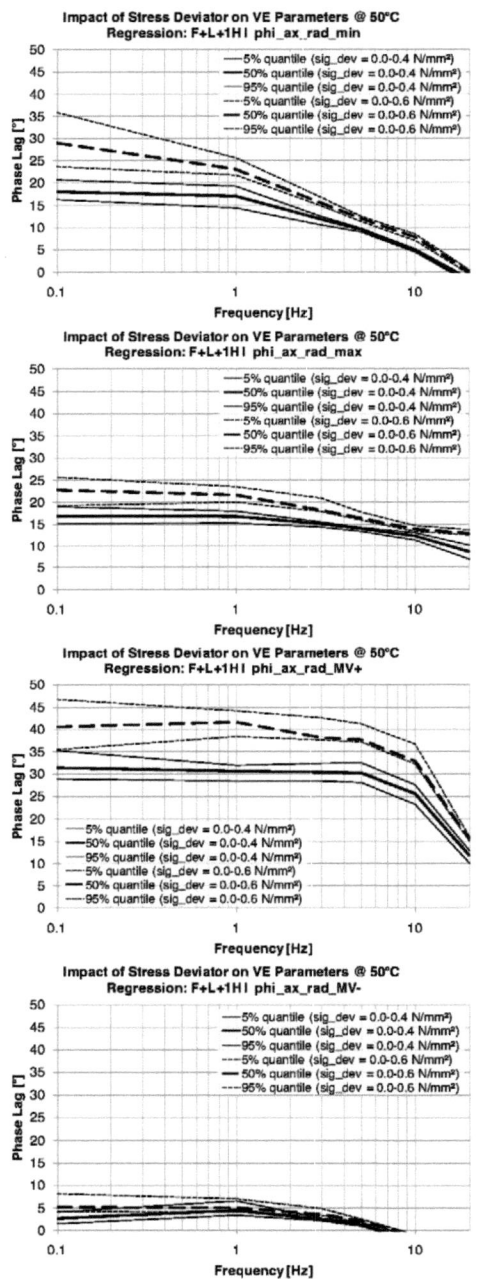

Figure V-26. 5%, 50% and 95% quantiles of $\varphi_{ax,rad,min}$ (top), $\varphi_{ax,rad,max}$ (2nd from above), $\varphi_{ax,rad,MV+}$ (3rd from above) and $\varphi_{ax,rad,MV-}$ (bottom) for three deviatoric stress levels at 50°C; F+L+1H approximation.

The difference between the axial phase lags between both deviatoric stress amplitudes is not as high as for the radial phase lag. In Figure V-27 data for the radial phase lag is presented. The phase lag at the maximum loading is again quite similar for both stress amplitudes. Larger differences than at the maximum loading occur in the loading phase (MV+) and at the minimum loading. In the unloading phase (MV-) again negative axial phase lags are derived from the test data.

The stiffness (Figure V-28) of the mix is also not comparable for both test conditions. At low frequencies, the higher deviatoric stress leads to lower results (582 MPa vs. 627 MPa; -7%) – although looking at the scatter, the difference is insignificant. At higher frequencies, the situation is reversed and the higher deviatoric stress amplitude leads to higher results (904 MPa vs. 786 MPa; +15%).

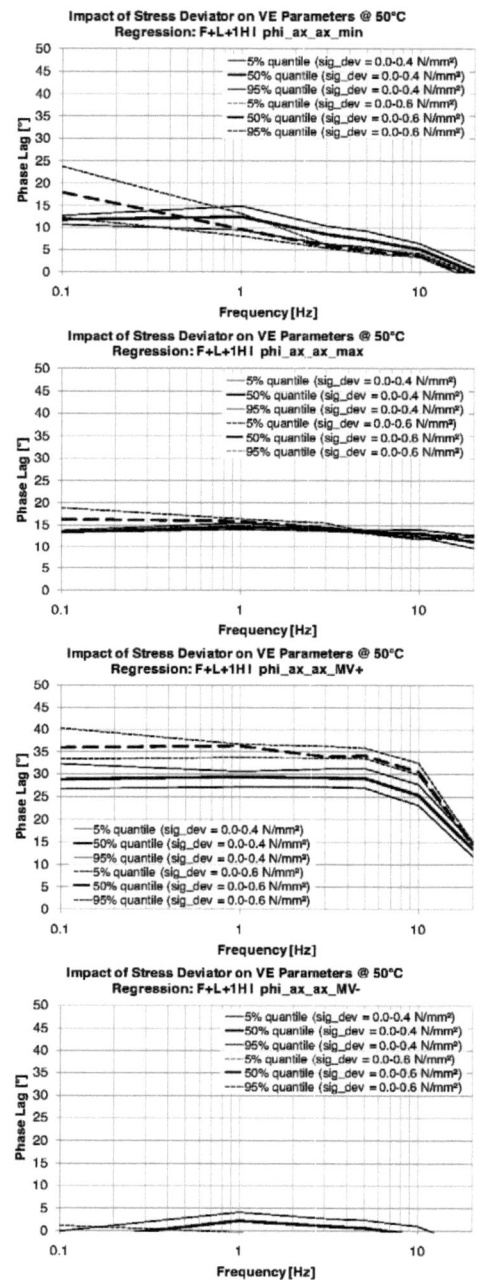

Figure V-27. 5%, 50% and 95% quantiles of $\varphi_{ax,ax,min}$ (top), $\varphi_{ax,ax,max}$ (2nd from above), $\varphi_{ax,ax,MV+}$ (3rd from above) and $\varphi_{ax,ax,MV-}$ (bottom) for three deviatoric stress levels at 50°C; F+L+1H approximation.

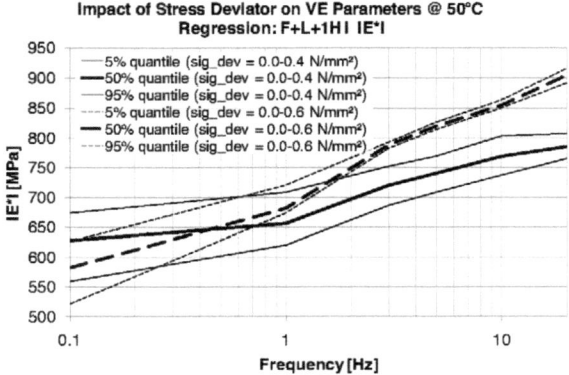

Figure V-28. 5%, 50% and 95% quantiles of |E*| for three deviatoric stress levels at 50°C; F+L+1H approximation.

V.4.2 Conclusions

From the tests at different temperatures and deviatoric stress amplitudes the following conclusions can be given:

At lower temperatures (10°C) it is possible to test the HMA specimens within the linear viscoelastic domain and still produce deformation signals that can be approximated with an adequate quality. Influences of the deviatoric stress level on the material parameters were not found. The stress level for the test routine is therefore set to 0.1 N/mm² to 1.1 N/mm² for further tests. It can be assumed that the material stays within the linear viscoelastic domain and produces deformation that can be approximated with satisfactory quality.

At 30°C and 50°C the stress amplitudes used for the investigations seem to activate non-linear effects to a higher extent at low frequencies and vice versa. The phase lags (to a higher degree) as well as material stiffness (to a lower degree) are influenced by the applied stress. Only if the phase lag at the maximum point of loading is considered, all phase lags are comparable. Thus, it can be concluded that if the $F+L+1H$ regression function is used for evaluation and results from the maximum of the load cycle are used for interpretation, the impact of non-linear viscoelastic effects can be minimized and results will be comparable.

For further material tests at 30°C, UCCTs with the stress amplitude from 0.1 N/mm² to 0.4 N/mm² will be carried out. At 50°C, TCCTs with the deviatoric stress amplitude from 0.0 N/mm² to 0.4 N/mm² will be run. The stress range at 50°C matches the standard test conditions according to (EN 12697-25, 2005).

V.5 Viscoelastic Material Parameters Derived from the Compressive Domain

In the following descriptive analysis, the viscoelastic parameters ($|E^*|$, $\varphi_{ax,ax}$, $\varphi_{ax,rad}$) of HMAs from CCTs are presented and impacts of time (frequency of loading), temperature and mix design parameters (binder content and air void content) are explained. This analysis is carried out for HMAs with both binders, the unmodified 70/100 and the PmB 25/55-65. Sections V.7.1 and V.7.2 will continue this investigation by expanding the field to the dynamic Poisson's Ratio $|v^*|$ and the dynamic shear modulus $|G^*|$.

The analysis of viscoelastic parameters presented in the following is based on limited data and thus the findings are not supported by statistical analysis. The intention of the investigation is to check for unambiguous connections between viscoelastic material behavior and time, temperature and mix design parameters.

The diagrams always contain results for three test frequencies (0.1 Hz, 1.0 Hz and 10.0 Hz) and three different HMAs. They either differ in binder content or content of air voids. The materials are named in the following way: *"specimen_bindercontent_airvoid-content_temperature"*, so a *"t371d_53_16_10g"* would indicate that the results were derived from a mix with 5.3% (m/m) binder content, 1.6% (v/v) air void content tested at 10°C. The name of the mix and binder is always given in the heading of the diagram.

V.5.1 Impact of Binder Content

For the AC 11 70/100 mixes with 5.0% (m/m), 5.3% (m/m) and 5.6% (m/m) binder content are compared in the following. For the mixes with modified PmB 25/55-65, the variation of the binder content was enlarged from 4.8% (m/m) to 5.3% (m/m) and 5.8% (m/m). All mixes were produced with a target content of air voids of 3.0% (v/v).

Six diagrams in Figure V-29 and Figure V-30 show the effect of the binder content on the dynamic modulus of the mixes at three different temperatures. The diagrams in Figure V-29 are from AC 11 70/100, those in Figure V-30 from AC 11 PmB 25/55-65. The top diagram presents results from tests at 10°C, the second from above at 30°C and the lower diagram at 50°C. In addition a table below each diagram gives the values shown in the diagram. Specimens from the mix with unmodified binder 70/100 and high binder content were not tested successfully at 50°C and are therefore not shown in the results.

At 10°C (top diagrams in Figure V-30 and Figure V-29) there seems to be an optimum at a binder content of 5.3% (m/m) which also happens to be the optimal content according to Marshall. This is true for the AC 11 70/100. No such optimum can be found for the mix with PmB 25/55-65. The dynamic modulus is rather constant at frequencies of 1 Hz and 10 Hz. At the lowest frequency $|E^*|$ drops strongest by 11% when the binder content is raised from 4.8% (m/m) to 5.8% (m/m) (+21%).

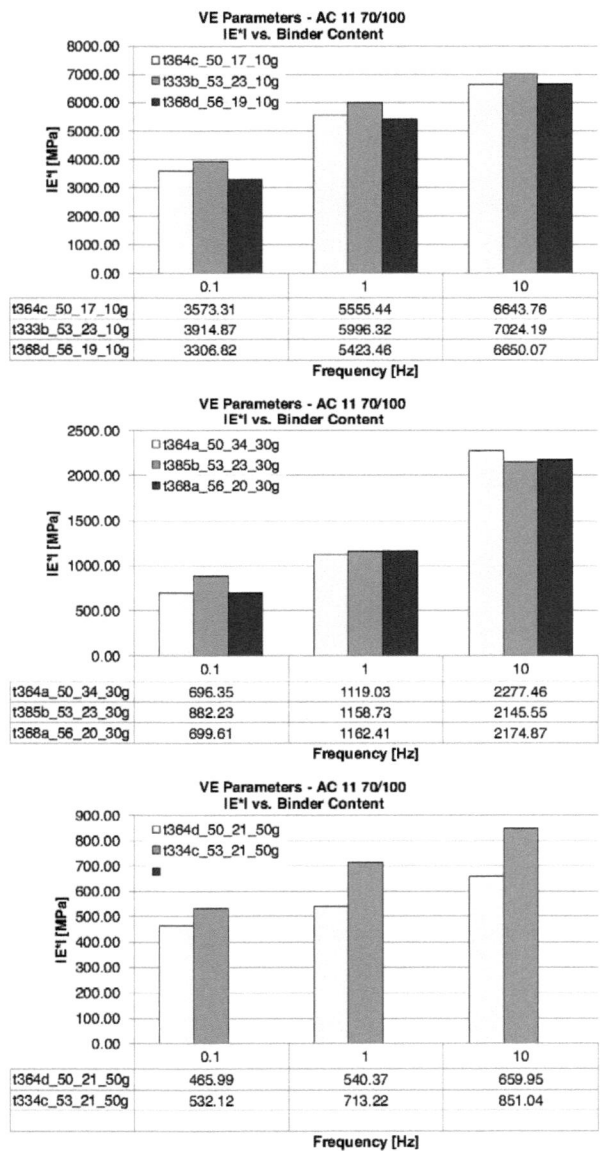

Figure V-29. Mean values of $|E^*|$ for AC 11 70/100 vs. binder content for 10°C (top), 30°C (middle) and 50°C (bottom) and 0.1 Hz, 1 Hz and 10 Hz.

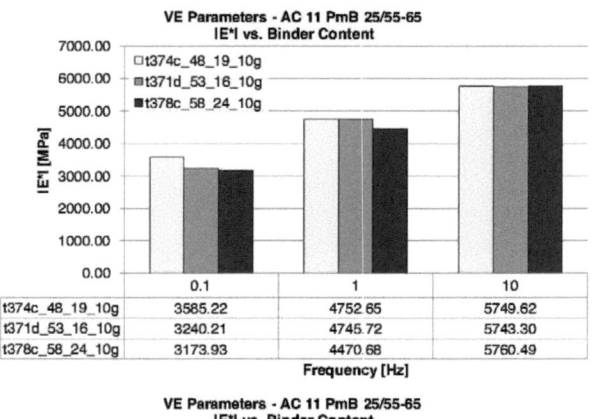
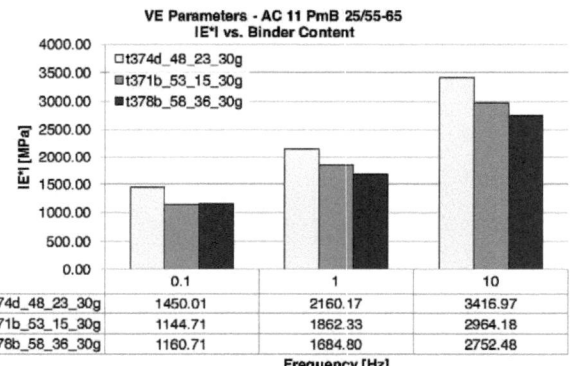
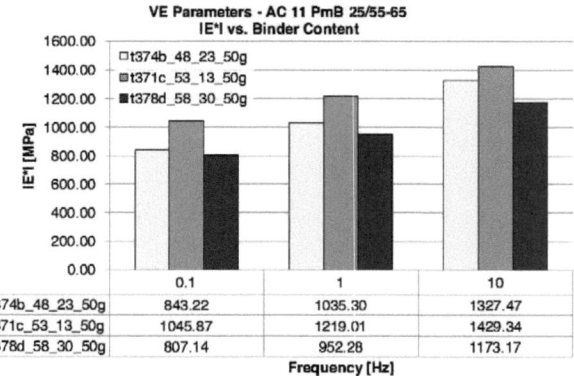

Figure V-30. Mean values of |E*| for AC 11 PmB 25/55-65 vs. binder content for 10°C (top), 30°C (middle) and 50°C (bottom) and 0.1 Hz, 1 Hz and 10 Hz.

At 30°C the decrease in stiffness with increasing binder content is more severe for the PmB 25/55-65 mixes. When the content is raised by 0.5% (m/m), the stiffness drops by 13% (10 Hz) and 21% (0.1 Hz) and another 7 to 10% for 1 Hz and 10 Hz when the binder content is increased by another 0.5% (m/m). The situation for the 70/100 mixes (left column of diagrams) is different again. An optimum is notable at the lowest frequency and 5.3% (m/m) but no significant difference occur at higher frequencies.

More conclusive findings can be found for the temperature effect on the dynamic modulus. For the 70/100 mix the stiffness decreases by around 80% for all binder contents and low frequencies when the temperature is raised from 10°C to 30°C. For high frequencies the decrease is also independent from the binder content but is only around 70%. When the temperature is raised for another 20 K to 50°C the loss in stiffness is now less severe for the low frequency range (-40%) than for the 10 Hz (-60 to -70%). The same tendency can be found for the PmB 25/55-65 mix:

- -60% stiffness for 0.1 Hz and 1 Hz when the temperature is changed from 10°C to 30°C compared to -40% to -50% at 10 Hz and
- -35% for 0.1 Hz and 1 Hz when the temperature is raised to 50°C compared to -50% to -60% at 10 Hz.

Figure V-31 and Figure V-32 contain results for the axial phase lag $\varphi_{ax,ax}$ at all three temperatures. Interestingly there is a minimum for the phase lag at all three temperatures for a binder content of 5.3% (m/m) for the PmB 25/55-65 mix (Figure V-32). Thus it can be concluded that there is a binder content where the mix exhibits an optimal material behavior in terms of elasticity. This is true for low and high frequencies. At intermediate loading durations (1 Hz) the difference is less significant.

For the unmodified 70/100 mix (Figure V-31) the situation is not so conclusive. At 10°C no strong effect of increasing binder content can be found. If there is any impact, the viscosity is slightly increasing with increasing binder content. At 10 Hz and 10°C, the axial phase angle appears to be negative, which is physically impossible. The negative results must be affected by the inaccuracy of the chain from measuring sensor, ADC and test data evaluation. The inaccuracy is mainly caused by noise that cannot be canceled out completely. A measuring sensor with a smaller measuring range and higher resolution is not affected so much by the noise. But when HMAs are tested with a large range of stiffness vs. temperature/frequency, there will always occur to problem that

- the measuring range of the sensor is too small for low stiffness behavior where the deformation is higher than for high stiffness behavior and
- noise affects your signal to a greater extent if a sensor is used with a larger measuring range in the high stiffness range of the material where deformation is smaller.

As an alternative, a sensor with a small measuring range could be used and the loading for low stiffness behavior be reduced to keep the deformation small enough to be measured by the sensor. But, in this case, the noise of the load cell will increase and affect the data negatively.

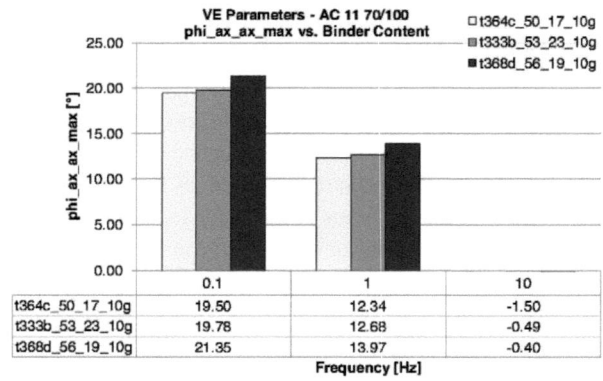
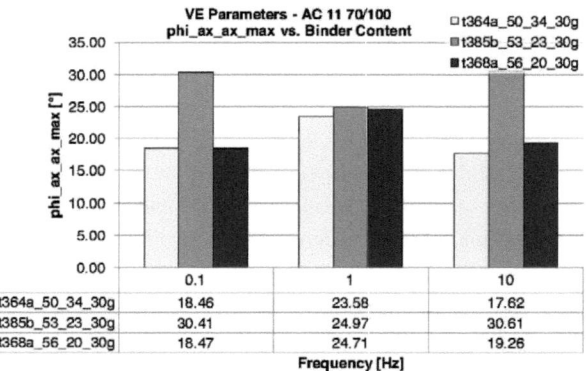
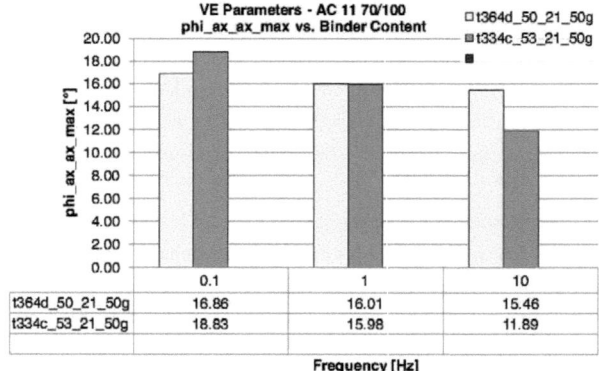

Figure V-31. Mean values of $\varphi_{ax,ax,max}$ for AC 11 70/100 vs. binder content for 10°C (top), 30°C (middle) and 50°C (bottom) and 0.1 Hz, 1 Hz and 10 Hz.

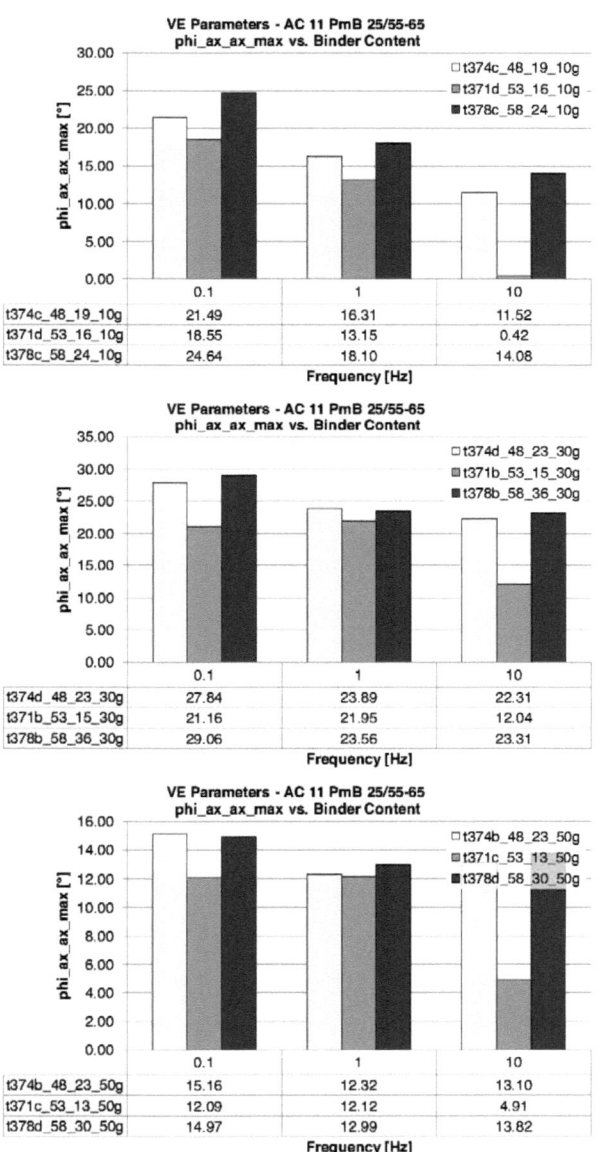

Figure V-32. Mean values of $\varphi_{ax,ax,max}$ for AC 11 PmB 25/55-65 vs. binder content for 10°C (top), 30°C (middle) and 50°C (bottom) and 0.1 Hz, 1 Hz and 10 Hz.

It is interesting that the mix seems to react nearly elastically at low temperatures and frequencies. At 30°C the axial phase lags show a maximum at a binder content of 5.3% (m/m). This is the inverted case compared to the PmB. The maximum is significant for low and high frequencies. Again the 50°C data have only limited informative value, since only two binder contents are available. Also the evolution of the axial phase lag is not conclusive, since it increases for low frequencies and decreases for high frequencies and increasing binder content.

The radial phase lag $\varphi_{ax,rad}$ in Figure V-33 and Figure V-34 shows an optimal binder content again, but only at 30°C and 50°C for the PmB mix (Figure V-34). At 10°C and the low frequencies of 0.1 Hz and 1 Hz, the phase lag increases with increasing binder content. A distinctive optimum in terms of binder content can be found at 10 Hz. An important finding is that the radial phase lag is not a constant value as it is stated in a number of publications, e.g. (von der Decken, 1997) and (Weise, et al., 2008).

For the 70/100 mixes (Figure V-33) the radial phase lag does not seem to be influenced by different binder contents at 10°C. At 30°C there is a maximum radial viscosity for the medium binder content, which is in contrary to the material behavior of the PmB mix.

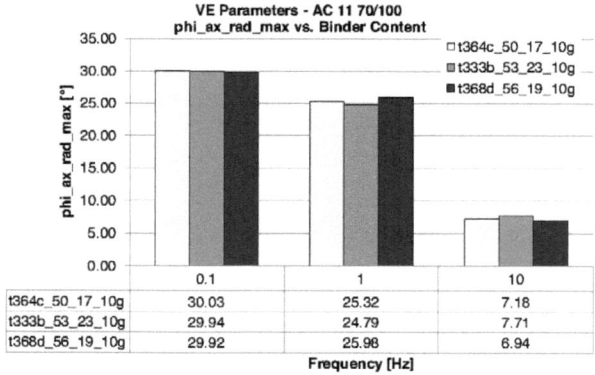
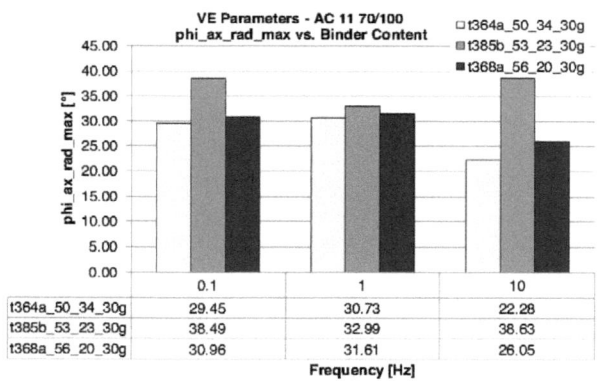
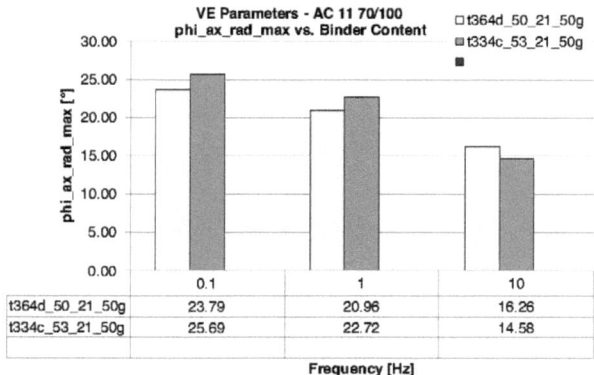

Figure V-33. Mean values of $\varphi_{ax,rad,max}$ for AC 11 70/100 vs. binder content for 10°C (top), 30°C (middle) and 50°C (bottom) and 0.1 Hz, 1 Hz and 10 Hz.

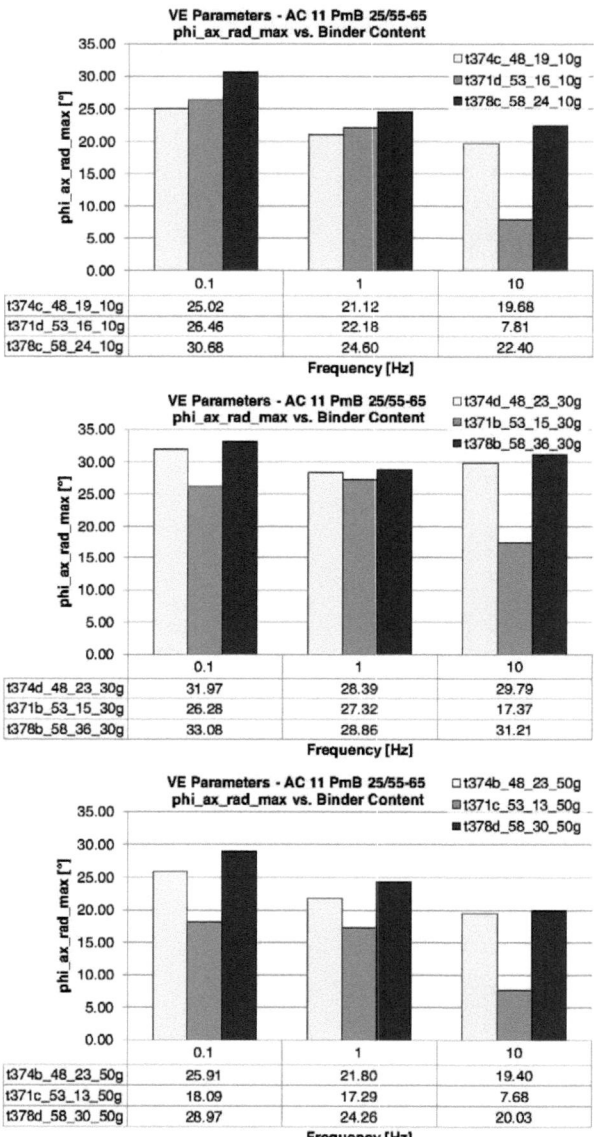

Figure V-34. Mean values of $\varphi_{ax,rad,max}$ for AC 11 PmB 25/55-65 vs. binder content for 10°C (top), 30°C (middle) and 50°C (bottom) and 0.1 Hz, 1 Hz and 10 Hz.

V.5.2 Impact of Air Void Content

In this section, the influence of a change in the volume of air voids is investigated. AC 11 70/100 and AC 11 PmB 25/55-65 mixes with 5.3% (m/m) binder content and three different target contents of air voids (3% (v/v), 5% (v/v) and 8% (v/v)) are compared. For the 70/100 mixes, tests at 10°C with medium content of air voids and at 30°C with high content of air voids failed and cannot be employed in this study.

Figure V-36 show the evolution of the dynamic modulus. The void content cannot be achieved with the same precision as the binder content in the process of specimen preparation. Thus, only qualitative levels of void content (low, medium, high) can be given. At 10°C, the air void content has no dominant effect on the stiffness of the PmB mix, neither for 0.1 Hz or 1 Hz, nor for 10 Hz. For the 70/100 mix at 10°C the high air void mix reacts significantly less stiff than the low air void mix. The decrease ranges between 22% for 0.1 Hz, 27% for 1 Hz and 19% for 10 Hz.

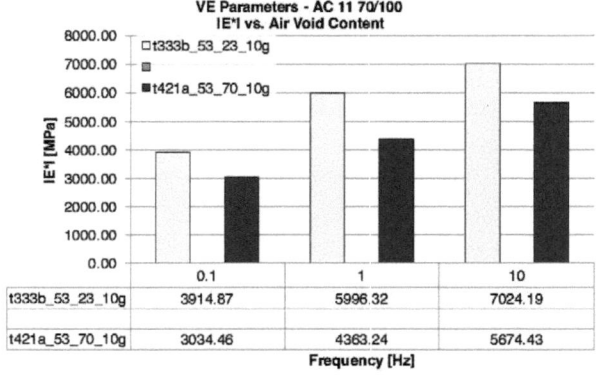
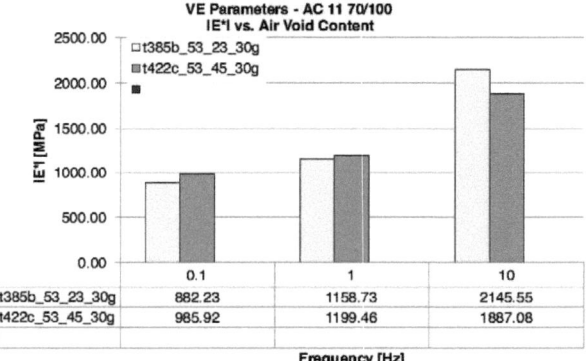
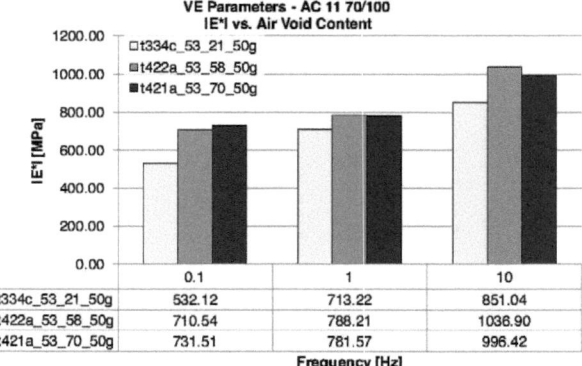

Figure V-35. Mean values of $|E^*|$ for AC 11 70/100 vs. air void content for 10°C (top), 30°C (middle) and 50°C (bottom) and 0.1 Hz, 1 Hz and 10 Hz.

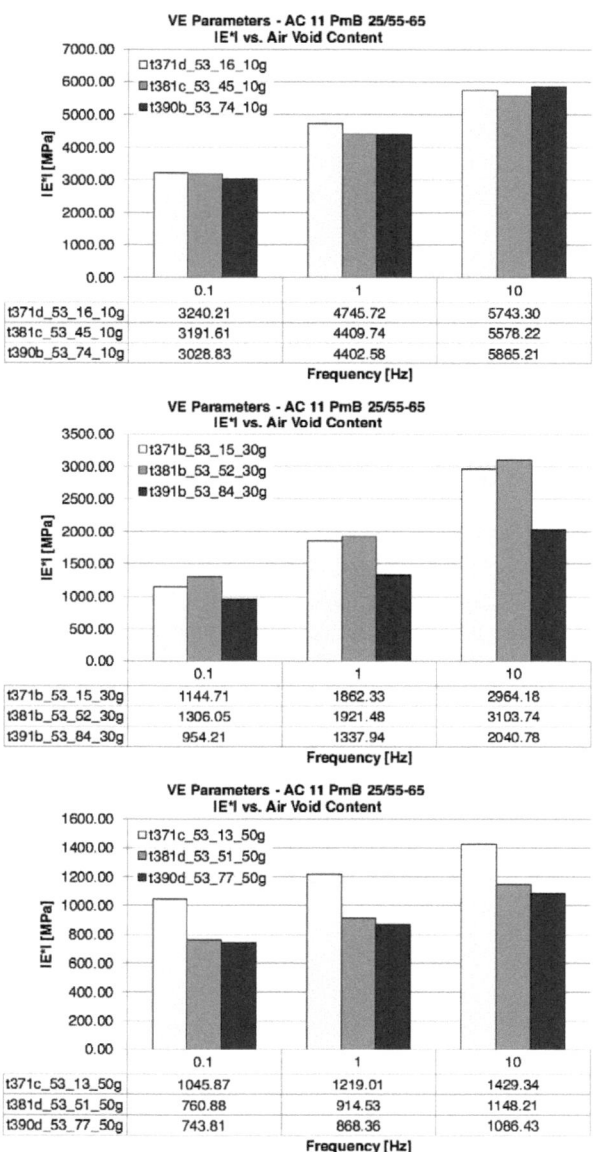

Figure V-36. Mean values of $|E^*|$ for AC 11 PmB 25/55-65 vs. air void content for 10°C (top), 30°C (middle) and 50°C (bottom) and 0.1 Hz, 1 Hz and 10 Hz.

At 30°C, the two lower air void contents result in similar stiffness values at all frequencies, the medium content showing a slightly higher dynamic modulus (+3% to +14%) for the PmB mix. A significant drop in stiffness can be found when the air void content is increased even higher from 5.2% (v/v) to 8.4% (v/v) in case of PmB. Compared to the medium content the dynamic modulus decreases by 27% to 34%. For the 70/100 mix the stiffness slightly increases for 0.1 Hz and 1 Hz (12% and 4%) and decreases for 10 Hz (12%) between the low and medium air void content at 30°C.

At 50°C, the dynamic modulus decreases strongly from the lowest to medium void content by 20% to 27% for the PmB mix. The difference between medium and high void content is not significant ranging between -5% and -2%. The situation is turned around for the unmodified 70/100 mix: The stiffness increases between 11% and 34% depending on the frequency when the void content changes from low to medium. When the void content is increased once more to a high air void content, the dynamic modulus does not change significantly anymore for this mix.

Figure V-37 and Figure V-38 show findings for the axial phase angle $\varphi_{ax,ax}$. At 10°C, the increase between low and high air void content leads to a strong increase of the viscosity for all frequencies by 7° to 13° for the PmB mix. The same is true for the 70/100 mix.

At 30°C, the incline of viscosity with higher content of air voids is strong for the 0.1 Hz and 1.0 Hz and not significant for higher frequencies for the PmB mix. For the 70/100 mix the phase lag also increases at this temperature for 0.1 Hz when the air void content is increased. No significant findings can be stated for higher frequencies.

At 50°C, on the other hand, viscosity is similar at low and medium void content (at least for 0.1 Hz and 1 Hz) and increases strongly at high air void contents for the PmB mix. This shows that viscoelastic behavior inhibit mechanisms where temperature, frequency and mix design parameters interact in different ways. Depending on these three conditions, there seem to be different threshold values that lead to a significant change in the viscoelastic behavior of the mix. For the 70/100 mix, increasing content of air voids results in decreasing phase lags for 0.1 Hz and 1 Hz with no significant change for 10 Hz at 50°C. This is contrary to the situation for the PmB mix and thus the evolution of the viscosity seems to be connected to the binder type as well as the mix composition.

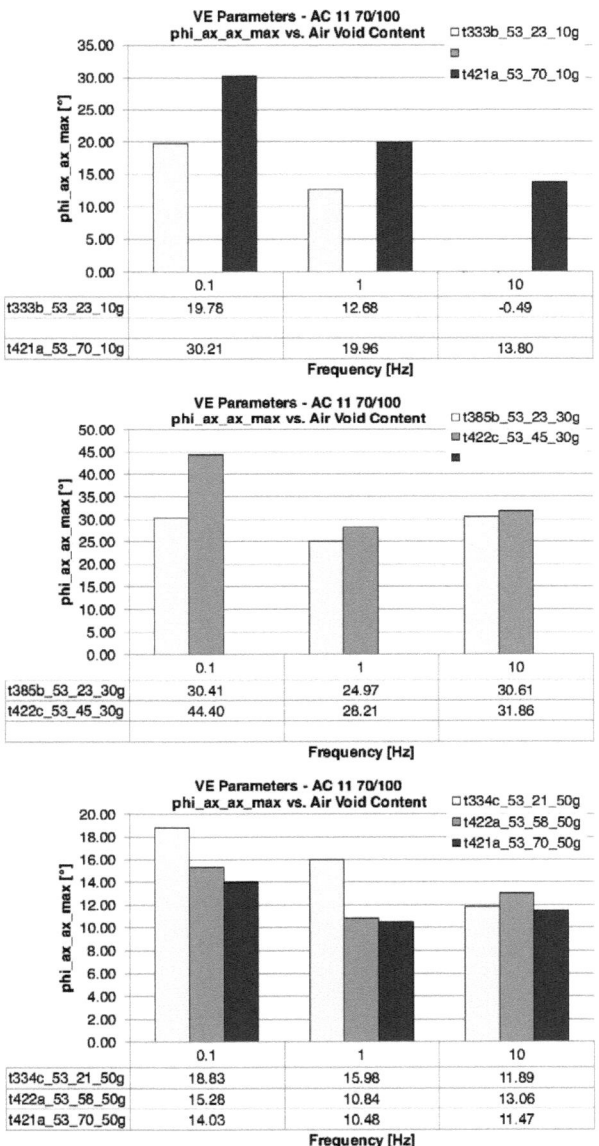

Figure V-37. Mean values of $\varphi_{ax,ax,max}$ for AC 11 70/100 vs. air void content for 10°C (top), 30°C (middle) and 50°C (bottom) and 0.1 Hz, 1 Hz and 10 Hz.

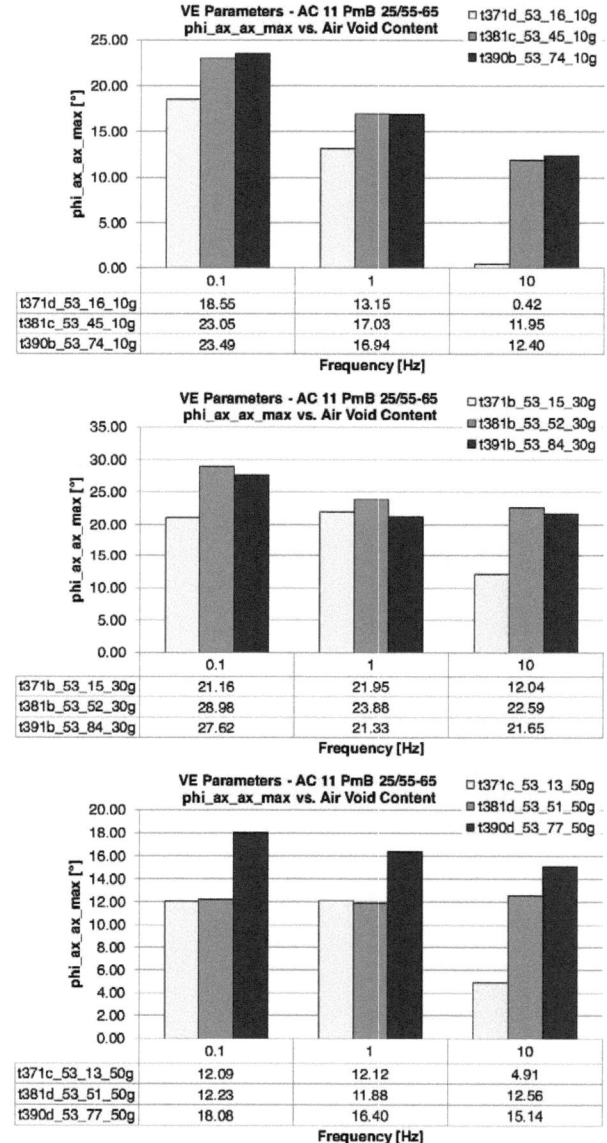

Figure V-38. Mean values of $\varphi_{ax,ax,max}$ for AC 11 PmB 25/55-65 vs. air void content for 10°C (top), 30°C (middle) and 50°C (bottom) and 0.1 Hz, 1 Hz and 10 Hz.

Figure V-39 and Figure V-40 deal with the radial phase lag at different air void contents. In general there seems to be an increasing trend with increasing air void content for all temperatures and both mix types, although the intensity of increase is different at different temperatures and frequencies.

For the PmB mix the effect is rather small with the exception of 10 Hz at 10°C. At 30°C the increase in radial viscosity is occurs from the low to medium air void content at 0.1 Hz and 10 Hz. At 1 Hz, changing air void contents do not lead to a significant change in radial phase lags. The most dominant effect of changing air void content on the radial phase angle can be found at 50°C. The higher the void content the higher is the time shift between axial loading and radial reaction.

For the unmodified 70/100 mix significant changes can be found at 10°C for all frequencies, strongest for 10 Hz, at 30°C and 0.1 Hz and at 50°C at 10 Hz.

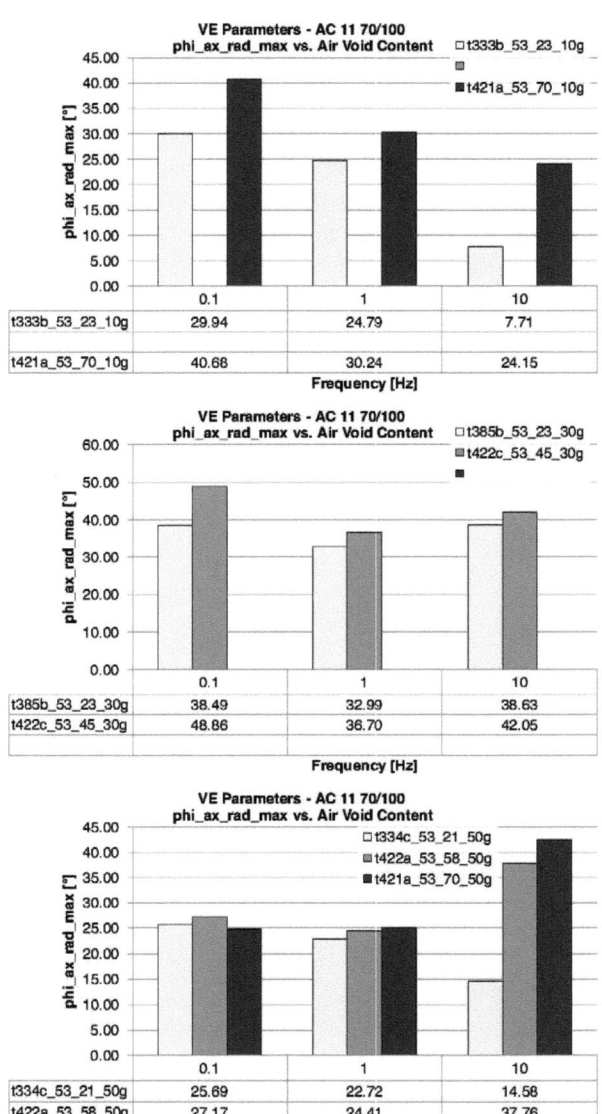

Figure V-39. Mean values of $\varphi_{ax,rad,max}$ for AC 11 70/100 vs. air void content for 10°C (top), 30°C (middle) and 50°C (bottom) and 0.1 Hz, 1 Hz and 10 Hz.

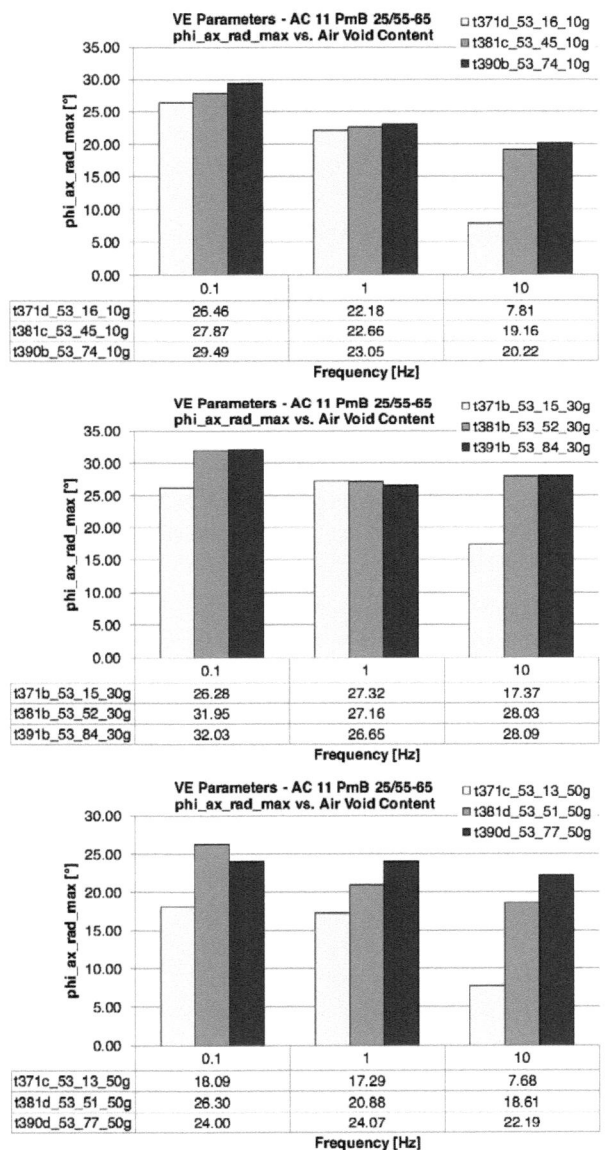

Figure V-40. Mean values of $\varphi_{ax,rad,max}$ for AC 11 PmB 25/55-65 vs. air void content for 10°C (top), 30°C (middle) and 50°C (bottom) and 0.1 Hz, 1 Hz and 10 Hz.

V.5.3 Conclusions

From the analysis of viscoelastic parameters in this chapter, the following conclusions can be drawn, taking into account that only limited data is available:

Temperature has a strong and conclusive impact on the dynamic modulus of HMA. For the PmB mix the decrease with temperature is dominant between 10°C and 30°C and less obvious between 30°C and 50°C. For the unmodified mix, the loss in stiffness is even higher between 10°C and 30°C and still significant between 30°C and 50°C. This can be explained by the different binders. While the stiffness of both binders is similar at low temperatures, the unmodified 70/100 exhibits a stronger loss in stiffness with increasing temperature.

The axial and radial phase lags show a maximum of viscosity not at the highest temperature of 50°C but at 30°C for both binder types. This seems to be in conflict with the behavior of the pure binder. When the stiffness ratio of binder to mix is taken into consideration (see Figure VI-11 in section VI.3) and especially its evolution with temperature, the effect is explainable. Since the stiffness ratio decreases dramatically between 30°C and 50°C the impact of the bitumen on the overall mix behavior also drops and the elastic behavior of the aggregate skeleton gains more influence. Thus, the viscosity decreases when the temperature is raised from 30°C to 50°C.

The **binder content** influences the viscoelastic behavior of the mix differently depending on the temperature. In accordance to the optimal binder content according to Marshall, the mixes show a maximum in stiffness at the optimal binder content at 50°C. At lower temperatures, the dynamic modulus either decreases slightly (30°C) or stays stable (10°C). This is true for the PmB mixes. For the unmodified HMAs, the stiffness shows a vague maximum at the medium binder content for 10°C and no conclusive changes at 30°C.

In terms of the axial viscosity expressed by the axial phase lag, there is a minimum in the phase angle for all three temperatures at the optimal Marshall binder content of 5.3% (m/m) for the PmB HMAs. The effect tends to be stronger for higher frequencies. This cannot be found for the unmodified mixes, where the phase lags are either stable (10°C, 50°C) or show even a maximum at 5.3% (m/m) (30°C). The radial phase lag also shows a minimum for the PmB HMAs at 5.3% (m/m) and, again, no (10°C, 50°C) or the inversed effect (30°C) for 70/100 mixes.

The **content of air voids** reflect on the dynamic modulus of the mix in different ways. For the PmB mixes, the material seems to be insensitive to changes in the void content at 10°C. At 30°C, the dynamic modulus drops from medium to high air void content and at 50°C from low to medium content of air voids. It seems that there is a certain threshold value in terms of the void content which brings a significant loss in stiffness. This threshold value tends to wander towards lower air void contents the higher the temperature is set. For the unmodified mix only limited data are available at 10°C and 30°C, but at least at 50°C where three mixes were successfully tested, the hypothesis raised above does not seem to be valid, since the stiffness increases slightly with increasing void content.

Analogue to the stiffness, the viscosity of the modified HMAs also increases strongly at the same level of air voids at 30°C and 50°C. At 10°C, there is also an increase in viscosity between low and medium air void content despite a loss in dynamic modulus at these conditions. Again, the situation is reversed for the 50°C data and the 70/100 mixes. Here, the mix reacts more elastically when the content of air voids is set to a higher level.

Summing up the findings, it can be confirmed that temperature and frequency have a clear and conclusive impact on stiffness and viscosity of mixes made from both binder types. Mix design parameters like the content of binder and air voids tend to reflect differently depending on the binder type of the HMA.

V.6 A Study on the Difference between Axial and Radial Phase Lag

During the analysis of the radial ($\varphi_{ax,rad}$) and axial phase lag ($\varphi_{ax,ax}$) for CCTs carried out at different conditions in terms of test temperature, binder type and content and stress deviator, it was found that there is a notable difference between these two phase angles. The axial phase lag was always smaller than the radial lag.

To find an answer to this phenomenon, a couple of assumptions are proposed. By taking into account the characteristics of HMA specimens, the effect may be explained by one of the following reasons:

- Anisotropy of the material. The composite material HMA is compacted leading to an anisotropic orientation of aggregates within the mix. Different orientation of specimens can lead to different material properties.
- Radial deformation is not uniform around the circumference. Since the standard setup of SGs was one 150 mm SG glued to the specimen's surface, a different setup with two SGs allocated on both sides of the specimen could lead to different results.
- The measuring system which is used to obtain radial strain is the problem. SGs are glued to the specimen's surface with a two-component adhesive which develops a high stiffness (Young's modulus around 14,000 MPa). The higher stiffness of the adhesive combined with the SG could work like a restraint that prevents or rather delays deformation resulting in higher phase lags – at least in the loading phase. In the unloading phase of course, this effect should be reversed since the stiff measuring system would increase the speed of recovery deformation. If this was true, the effect would have to vary for different stages of the load cycle and show a correlation between the stiffness of the specimen (i.e. the test temperature and frequency). A measuring system with a much lower stiffness would also not produce the delayed radial deformation.

If none of the assumption above is capable of explaining the existence of a phase lag between axial and radial deformation, the reason for this must be related to the
- rheological behavior of the material. This means that the transfer from axial load to radial deformation takes more time than from axial load to axial deformation due to viscoelastic mechanisms.

The following chapters deal with each of these hypotheses, discussing the assumptions made by means of material testing and analysis of respective results.

V.6.1 Influence of Anisotropy of HMA

HMA layers used in road construction are usually compacted by roller compaction on site. To achieve consistency in compaction, both in the laboratory and on site, it is necessary to obtain reliable correlation between HMA laboratory performance and the observed in-service behavior. As presented in section III.2, specimens for this thesis were cored and cut from roller-compacted slabs according to (EN 12697-33, 2007). (Airey, et al., 2005) found that the roller compaction used to produce HMA-slabs in the lab provide the best correlation with field specimens in terms of aggregate structure and mechanical properties. As shown in (Hofko, et al., 2011b) the mechanical characteristics of roller-compacted specimens depend on the orientation of testing. It could be assumed that the anisotropy also affects the viscoelastic properties in terms of phase lags.

Test Setup

To find relevant impacts of the testing orientation on the material phase lag in axial and radial direction, two different specimen orientations were compared. The orientation of the coordinate system was set according to the compaction process presented in Figure V-41. Starting with the z-direction, the coordinate system is oriented in direction of the compaction force. The path of the roller is represented by the y-axis, the x-direction is orthogonal to the other two directions. In the standard procedure, specimens for CCTs are cored such that the principal dimension of the cylinder is in x-direction. The relevant (axial) loading and (axial) reaction is oriented in the same direction. Therefore this test setup is abbreviated *H-X-X*. It stands for high temperature behavior with the principle orientation of the loading and the reaction in x-direction.

Table V-8. Specimen characteristics used for tests to compare specimen orientation.

Specimen	Binder	Binder Content [% (w/w)]	Binder:fines (1:x)	Volume of air voids [% (v/v)]	Pattern
T406E				5.5	H Z Z
T406G				5.0	
T406H	PmB 25/55-65	5.3	1.4	7.0	
T404B				3.2	H X X
T404C				6.4	
T405D				4.0	

The other pattern is an *H-Z-Z* orientation. The specimen is cored from the slab in the direction of the compaction force. The principal loading and reaction are in z-direction. *H-Z-Z* represents the compaction and loading situation on the road. Table V-8 shows both patterns. Three specimens were tested for each orientation. Details on the specimens are also given in Table V-8. The tests were carried out as UCCTs at 30°C, a mean axial stress of 0.25 N/mm² and a stress amplitude of 0.15 N/mm². Frequency sweep ranged from 0.1 Hz to 30 Hz.

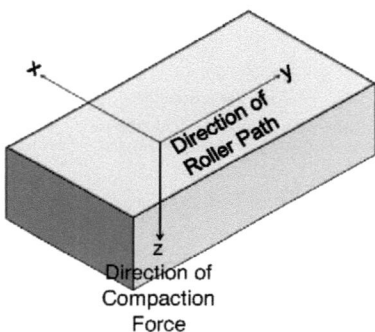

Figure V-41. Coordinate system of compaction.

The test data was evaluated with the advanced function $F+L+1H$. Test results for all three tested specimens were merged and statistically analyzed. This analysis was carried out separately for each test frequency. The diagrams in the following always present the 5% and 95% quantiles as well as the median value (50% quantile) of results for both orientations.

Results

The first step in each analysis was a check, whether the approximation function fits the test data with sufficient quality. Therefore, the coefficient of determination R^2 is presented for both deformations in Figure V-42. The top diagram deals with the radial deformation, the lower diagram with the axial counterpart. It is obvious that both deformations are described with a high quality. R^2 is clearly above 0.9995 for lower frequencies. Especially the *H-X-X* tests produce lower R^2 from 5 Hz on and a far larger scattering than the *H-Z-Z* specimens. It goes along with the results of (Hofko, et al., 2011b) where it was shown that *H-Z-Z* specimens always ended up in a distinctively smaller scattering of results. This is due to the fact that *H-Z-Z* specimens are tested in direction of compaction force.

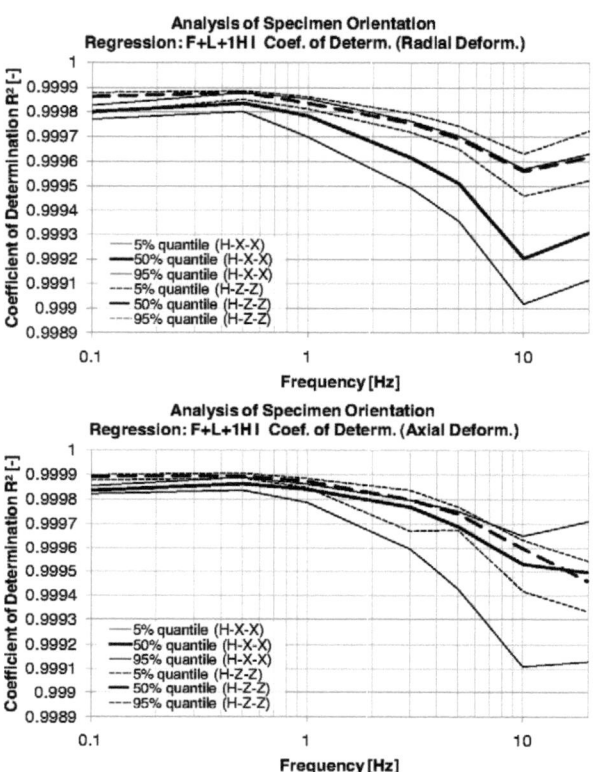

Figure V-42. 5%, 50% and 95% quantiles of R^2 for regression of the radial (top) and the axial deformation (bottom) for H-X-X vs. H-Z-Z orientation; F+L+1H approximation.

More information on the shape of the deformation function is given by the amplitude ratio between the 1st harmonic and the fundamental oscillation as presented in section II.3.4. Figure V-43 contains AR for the radial (top diagram) and axial (lower diagram) deformation. In both cases the ratio decreases with increasing frequency and starts off at 0.1 Hz at 6% to 7%, the *H-Z-Z* orientation producing a higher ratio. The difference between both orientations is leveled out towards higher test frequency (f > 1 Hz). For the axial deformation, AR is slightly higher at high frequencies than for the radial deformation. The radial deformation results in an AR of around 2% at 10 Hz, whereas the axial deformation shows a value of around 3% at this frequency.

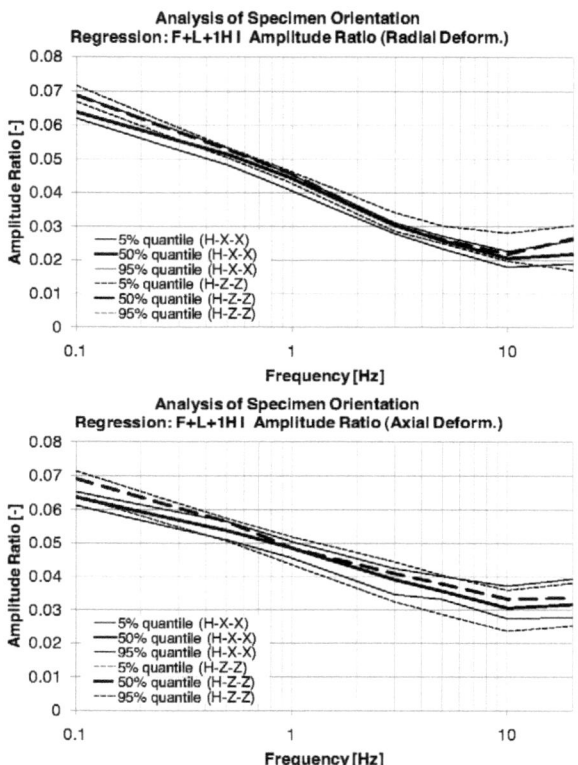

Figure V-43. 5%, 50% and 95% quantiles of *AR* of the radial (top) and the axial deformation (bottom) for H-X-X vs. H-Z-Z orientation; F+L+1H approximation.

The phase lags are analyzed separately in a first step. Figure V-44 displays the material phase lag $\varphi_{ax,rad}$ between axial loading and radial deformation at the maximum of loading, and Figure V-45 presents $\varphi_{ax,ax}$ for axial loading to axial deformation.

Regarding the radial phase lag (Figure V-44), it is interesting to observe that the *H-Z-Z* orientation results in significantly lower phase lags at frequencies from 0.1 Hz to 1 Hz. The material reacts more elastically when the specimens are stressed in the direction of the compaction force for low frequencies.

Figure V-45 presents results for the axial phase lags. The differences between the two specimen orientations *H-X-X* and *H-Z-Z* are not significant.

By comparing Figure V-44 and Figure V-45, it becomes clear, that the radial deformation lags behind the axial loading more than the axial deformation. The difference between axial and radial phase lag is presented in Figure V-46. The phase angle between two deformation components are described by δ in order not to mix them up with phase angles between deformation and force φ. To ensure a comparable scale in y-direction, all diagrams dealing with phase angles in this chapter are scaled to show a range of 20°. For the standard *H-X-X* orientation, the difference is quite constant until 5 Hz and increases from this frequency on. The *H-Z-Z* specimens produce smaller deformation phase lags. Only at 10 Hz and 20 Hz both orientations show similar differences.

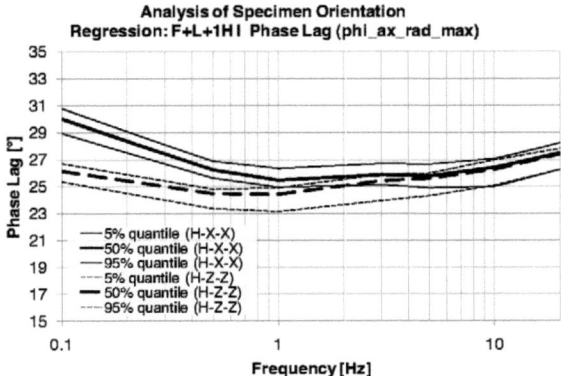

Figure V-44. 5%, 50% and 95% quantiles of $\varphi_{ax,rad,max}$ for H-X-X vs. H-Z-Z orientation; F+L+1H approximation.

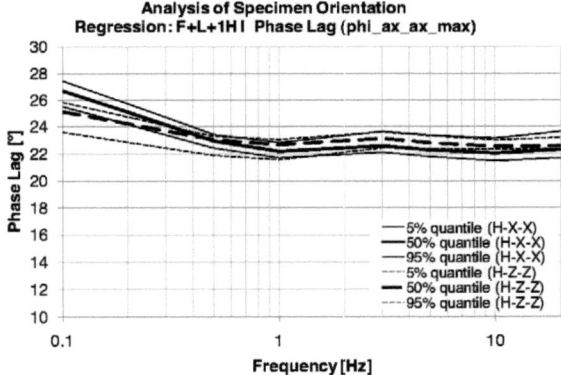

Figure V-45. 5%, 50% and 95% quantiles of $\varphi_{ax,ax,max}$ for H-X-X vs. H-Z-Z orientation; F+L+1H approximation.

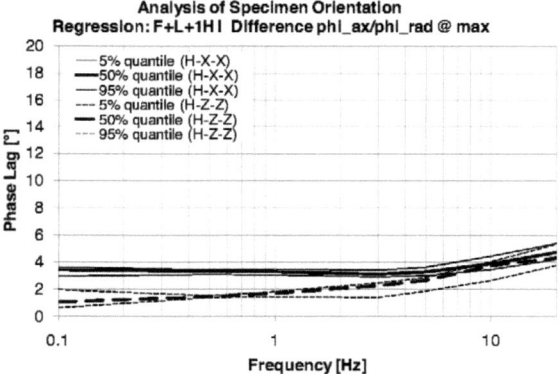

Figure V-46. 5%, 50% and 95% quantiles of $\delta_{ax,rad,max}$ between radial and axial deformation for H-X-X vs. H-Z-Z orientation; F+L+1H approximation.

It is revealed by the analysis that the specimen orientation does have an impact on the material parameters in terms of phase lags. This is an unambiguous sign that anisotropy influences these material characteristics. Especially at frequencies below 1 Hz H-Z-Z produces smaller radial phase lags than the standard H-X-X direction. The axial phase lag is similar for both cases. Thus the material reacts in an anisotropic way in terms of the deformation phase lag δ. Again, in the low frequency domain, the H-Z-Z orientation produces smaller values. Yet the radial phase lag $\varphi_{ax,rad}$ is still higher than the axial phase lag $\varphi_{ax,ax}$ also for the H-Z-Z specimens. The effect of material anisotropy provides interesting information but gives no satisfying answer why the radial phase lag is larger than the axial phase lag or, in other words, why the deformation phase lag $\delta_{ax,rad}$ is always positive.

V.6.2   Uniform Radial Deformation of Cylindrical Specimen

The axial deformation in the CCTs is recorded by two LVDTs to compensate any uneven deformation during the test by taking the mean value of both sensors for evaluation. For the radial deformation one SG with a grid length of 150 mm is attached to the specimen's surface in the standard case. There is a possibility that the radial deformation is not uniform around the circumference of the specimen as discussed at the International Conference on Asphalt Pavement 2010 in Nagoya, Japan after the presentation of (Hofko, et al., 2010). Therefore, a test series was carried out with different SG setups to deal with this assumption.

Test Setup

Three different SG setups were compared within the investigation. Figure V-47 depicts a top view on the specimens (cylinders with Ø 100 mm and height of 200 mm) with different SGs attached to them. The upper left picture shows the standard setup with one SG with a grid length of 150 mm attached to the specimen at both ends of the SG according to section II.2.3. The upper right picture represents the situation where two SGs with a grid length of 150 mm each are directly glued together and not firmly attached to the specimen's surface. This is possible since the carrier of the measuring grid and thus the overall length of the SG is 165.6 mm. So the SGs overlap 8.5 mm on both ends if the specimen's diameter is assumed to be 100 mm. The setup also prevents a concentrated stress induction compared to the standard setup at the points where the SG is glued to the specimen's surface. The lower picture shows a third alternative with two SGs with a grid length of 100 mm. Analogue to the standard setup, the ends of the SGs are glued to the specimen's surface.

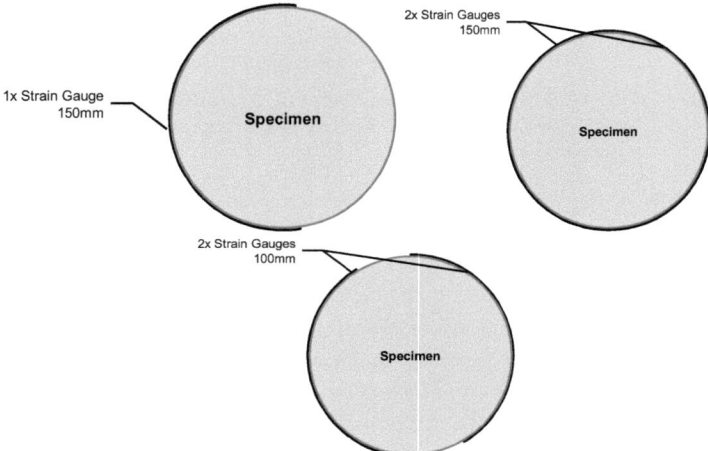

Figure V-47. Different SG setups to analyze the uniformity of radial deformation around the circumference; 1x SG 150 mm (top left = standard), 2x SG 150mm (top right) and 2x SG 100 mm (bottom).

The tests were carried out without confining pressure as UCCTs at a temperature of 30°C, a mean axial stress of 0.25 N/mm² and a stress amplitude of 0.15 N/mm². A frequency sweep was performed with a range from 0.1 Hz to 30 Hz. For each SG setup, two specimens were tested. Table V-9 shows characteristics of the six tested specimens.

The test data was evaluated with the advanced function $F+L+1H$. Test results for two tested specimens with the same setup were merged and statistically analyzed. This analysis was carried out separately for each test frequency. The diagrams in the following present the median value (50% quantile) of results for the three different setups. The 5% and 95% quantiles are not shown in the diagrams because three different data sets are compared, and the two additional quantile lines for each setup would confuse more than they could explain. Also, the analysis showed that all three setups resulted in similar scatterings.

Table V-9. Specimen characteristics used for tests to compare results with different SG setups.

Specimen	Binder	Binder Content [% (w/w)]	Binder:fines (1:x)	Volume of air voids [% (v/v)]	SG Setup
T423B	PmB 25/55-65	5.3	1.3	3.8	1x SG 150 mm
T423D				4.3	
T423C				2.8	2x SG 150 mm
T424C				2.5	
T424B				3.3	2x SG 100 mm
T424D				3.2	

Results

As an introduction to the analysis of results, the regression quality of the radial and axial deformation data is illustrated in two diagrams in Figure V-48. The radial deformation data have a quasi constant R^2 between 0.9996 and 0.9998 until 10 Hz. The standard SG setup with one SG 150 mm shows the lowest of the three R^2. The 20 Hz frequency packet results in significantly lower qualities of the fit.

The fit quality of the axial deformation sensors is even higher. All three setups have R^2 values around 0.9998 until 10 Hz. Since the axial deformation sensors were not changed between the three setups this also shows the reproducibility of test data at least in terms of quality of the fit.

Figure V-49 contains information on the radial phase lag to compare the three SG setups. The difference between the three cases can be described as follows: At the maximum of the load cycle the specimens with the standard setup result in higher radial phase lags for all frequencies. The difference is about 1.5° to 2°. At the maximum loading and low frequency, the two setups with two SGs also produce different values for the radial phase lag. The case with two SGs 100 mm produces lowest phase lags in this case.

It could therefore be assumed that the standard setup, where only one SG is glued to the specimen's surface leads to higher phase lags in the radial direction. If the explanation was raised that a one-sided SG with a high stiffness delays the radial deformation on the side of the specimen with the attached SG within one load cycle and thus produces higher phase lags then the whole picture was not taken into account. If the explanation above was true, the phase lag in the unloading phase should be significantly lower than the other two setups because it would enforce a faster recovery deformation due to the high stiffness. This is not the case as shown in the lower diagram in Figure V-49. In addition Figure V-50 shows the analogue diagram for axial phase lags. The data shows the same tendency as the results of the radial phase lag.

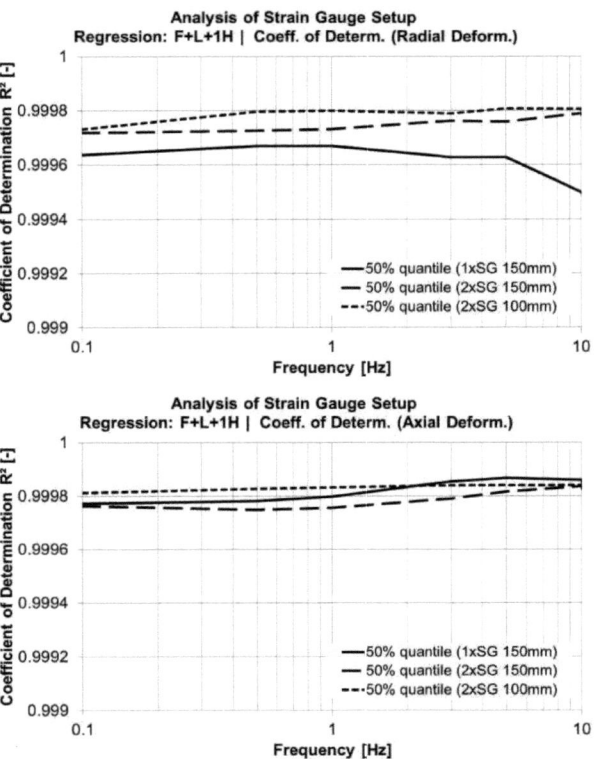

Figure V-48. 50% quantiles of R^2 for regression of the radial (top) and the axial deformation (bottom) for three different SG setups; F+L+1H approximation.

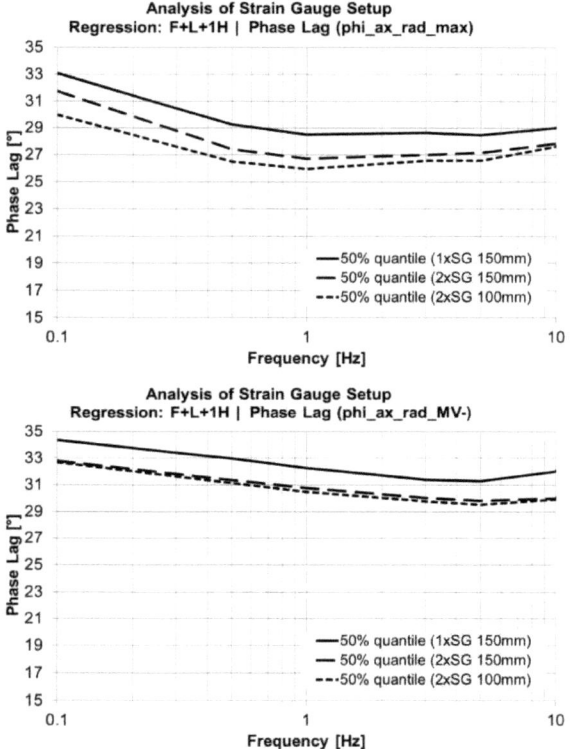

Figure V-49. 50% quantiles of $\varphi_{ax,rad,max}$ (top) and $\varphi_{ax,rad,MV\text{-}}$ (bottom) for three different SG setups; F+L+1H approximation.

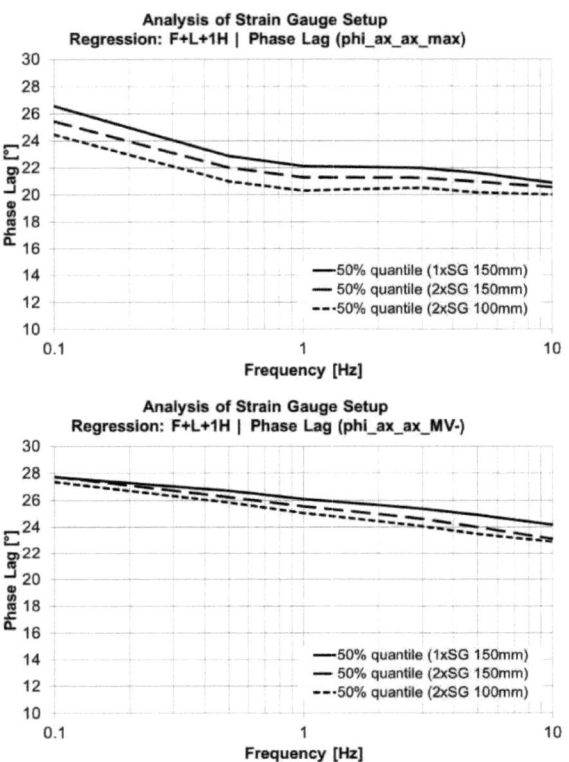

Figure V-50. 50% quantiles $\varphi_{ax,ax,max}$ (top) and $\varphi_{ax,ax,MV\text{-}}$ (bottom) for three different SG setups; F+L+1H approximation.

Thus, the reason for the difference in the radial phase lags must be related to a different phenomenon rather connected with the mix design parameters than with the SG setup.

The question of the analysis is, whether the SG setup has an influence on the deformation phase lag δ. Figure V-51 contains information about this parameter. Interesting enough, the values for different setups are not significant for a frequency higher than 0.1 Hz. It is assumed from the results of this analysis that the SG setup and thus the non-uniformity circumferential deformation is not a factor that influences either the radial phase lag or deformation phase lag in a crucial way.

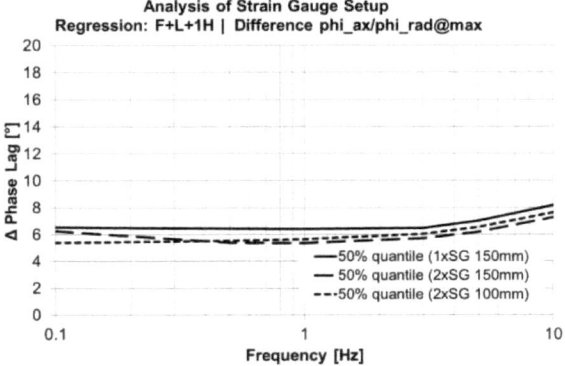

Figure V-51. 50% quantiles of $\delta_{ax,rad,max}$ between radial and axial phase lag for three different SG setups; F+L+1H approximation.

V.6.3 Impact of the Measuring System

While working on (Hofko, et al., 2011a), the question came up if the difference between the two force-deformation phase lags in axial and radial direction could be influenced by the measuring system. The assumption arose that the high stiffness of the measuring system consisting of the SG and the two-component adhesive could prevent and delay radial deformation and thus produce a higher radial phase lag. This phase lag would not be due to the material alone but increased by restraining effects of the measuring system.

The assumption stated above was based on a diagram in (Hofko, et al., 2011a) that is shown in Figure V-52. It is a comparison between the axial and radial phase lag at the point of maximum loading for an AC 11 70/100 derived from an UCCT at 30°C. Since the difference between the axial and radial phase lag decreases with increasing frequency, the hypothesis was that the increasing stiffness of the specimen with increasing frequency brought both phase lags closer together because the difference in stiffness between the measuring system and specimen got smaller. The choice of the data presented in the diagram was unfortunate, because it is one of only few examples, where the difference between axial and radial phase angle does actually get smaller with increasing frequency. Usually, and the data presented in section V.6.1 and V.6.2 confirm this, the difference between the two respective phase lags stays constant or even increases with increasing frequency.

Still, to verify the assumption about the correlation between stiffness and deformation phase lag δ, two analyses were carried out. First of all, the material stiffness in terms of $|E^*|$ was compared to the δ-values to see if any correlation can be found. Secondly tests with the standard measuring setup for radial deformation (SGs) was compared to another measuring system where the radial deformation is recorded by an LVDT-based device which is described in section II.2.1.

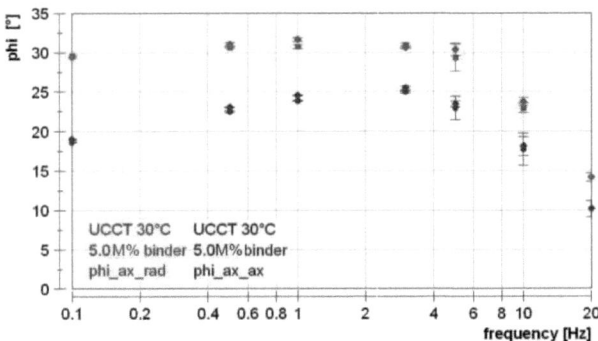

Figure V-52. MV and SD of $\varphi_{ax,ax,max}$ and $\varphi_{ax,rad,max}$ for an AC 11 70/100 and UCCT at 30°C. (Hofko, et al., 2011a)

Results of Investigations about Correlations between |E*| and δ

For the analysis of correlations between $|E^*|$ and δ at different functional values, test data from the two different standard mixes for this thesis were taken: the AC 11 PmB 25/55-65 and the AC 11 70/100. To analyze a wide spectrum of material stiffness, test data from three different temperatures (10°C, 30°C and 50°C) and from two specimens for each test temperature were used. UCCTs were carried out with a mean axial stress of 0.25 N/mm² and 0.60 N/mm² at 30°C and 10°C respectively. The stress amplitude was 0.15 N/mm² and 0.50 N/mm² at 30°C and 10°C respectively. TCCTs at 50°C were carried out with confining pressure. The mean axial stress was set to 0.45 N/mm² with an amplitude of 0.30 N/mm² and a radial confining pressure of 0.15 N/mm². The test frequencies ranged from 0.1 Hz to 30 Hz. Table V-10 shows characteristics of the six tested specimens for the AC 11 PmB 25/55-65 and

Table V-11 for the AC 11 70/100. It might be noticed that specimen T421A is given for the 10°C UCCT and the 50°C TCCT; this is correct. No noticeable damage occurs in the UCCTs at 10°C. Therefore T421A was re-used for one 50°C TCCT.

Table V-10. Specimen characteristics (AC 11 PmB 25/55-65) used for tests to investigate correlations between $|E^*|$ and δ.

Specimen	Test Temperature [°C] and Test Type	Binder Content [% (w/w)]	Binder:fines (1:x)	Volume of air voids [% (v/v)]
T381C	10 / UCCT			4.5
T394D				3.8
T393B	30 / UCCT	5.3	1.7	3.2
T394C				3.2
T373A	50 / TCCT			3.0
T393C				3.0

Table V-11. Specimen characteristics (AC 11 70/100) used for tests to investigate correlations between $|E^*|$ and δ.

Specimen	Test Temperature [°C] and Test Type	Binder Content [% (w/w)]	Binder:fines (1:x)	Volume of air voids [% (v/v)]
T421A	10 / UCCT		1.3	7.0
T421D				6.3
T333A	30 / UCCT	5.3	1.4	3.2
T336B				2.5
T421A	50 / TCCT		1.3	7.0
T422A				5.8

Test data were evaluated with the advanced function $F+L+1H$. Test results for two tested specimens at the same temperature were merged and statistically analyzed. This analysis was carried out separately for each test frequency. The diagrams in the following present the median value (50% quantile) of results in terms of δ vs. the median value of $|E^*|$. To find any relevant correlations, it was decided to carry out the analysis for all four points of the oscillation, the extrema of the loading cycle as well as the mean values of the loading and unloading cycle.

The first set of diagrams in Figure V-53 presents the data for the mix with the modified binder. In the following diagrams, the rightmost lines belong to the 50°C TCCT, the middle lines to the UCCT at 30°C and the leftmost lines with the highest stiffness to the 10°C UCCT. In some cases there seems to be a correlation between $|E^*|$ and δ over a wider range of $|E^*|$. The upper left diagram shows $\delta_{ax,rad,min}$. The 10°C results start off at the value where the 30°C results end. Still, this is not true for the start of the 30°C and the end of the 50°C results.

If the assumption was true that the deformation phase lag is due to the too high stiffness of the measuring system of SG and adhesive, a higher material stiffness should produce smaller differences. This is the case for some single test conditions. For example $\delta_{ax,rad,max}$ at 50°C is sharply decreasing with increasing material stiffness. The same parameter is inclining with increasing stiffness at 30°C, and nearly constant for 10°C.

As already mentioned above, the situation should be inverted if the deformation phase lag was considered in the loading (MV+) and unloading (MV-) phase. For the loading phase, the radial phase lag should be larger than the axial phase lag because the SG delays radial deformation. For the unloading phase on the contrary the radial recovery deformation should occur faster than the axial deformation because the high stiffness of the SG increases the recovery deformation. The diagrams do not reflect the assumption made above. In fact, the tests produce reversed results. The $\delta_{ax,rad,MV+}$ increases with increasing stiffness and the $\delta_{ax,rad,MV+}$ decreases with increasing stiffness at least for 10°C.

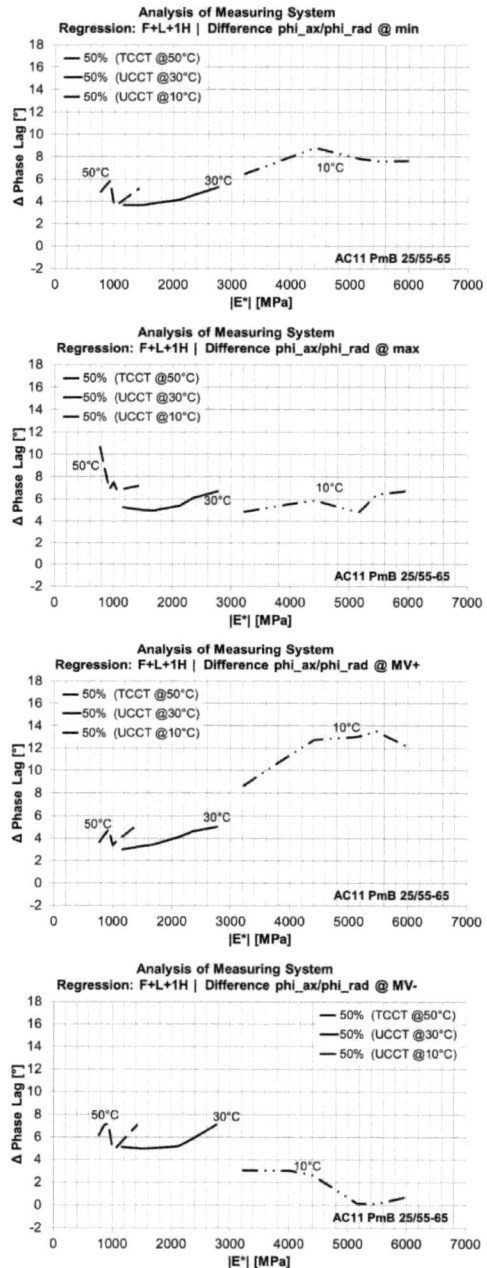

Figure V-53. 50% quantiles of $\delta_{ax,rad,\min}$ (top), $\delta_{ax,rad,\max}$ (2nd from above), $\delta_{ax,rad,\text{MV+}}$ (3rd from above) and $\delta_{ax,rad,\text{MV-}}$ (bottom) for 3 different test temperatures and AC 11 PmB 25-55/65; F+L+1H approximation.

With analogue considerations, the same conclusions can be drawn for the mix with the standard bitumen in Figure V-54. The results are not conclusive in terms of correlations between material stiffness and deformation phase angles. At least for 30°C all four δ decrease with increasing stiffness. Yet, if the assumption about the stiffness of the measuring system was true, it has to be said that the difference in phase lag in the unloading phase should be negative at low stiffness and increase towards zero with increasing material stiffness.

This analysis did not bring evidence for the proposed hypothesis. The results are not coherent, it seems that there are no clear relationships between the material stiffness and the $\delta_{ax,rad,}$ when the measuring system based on SGs is used.

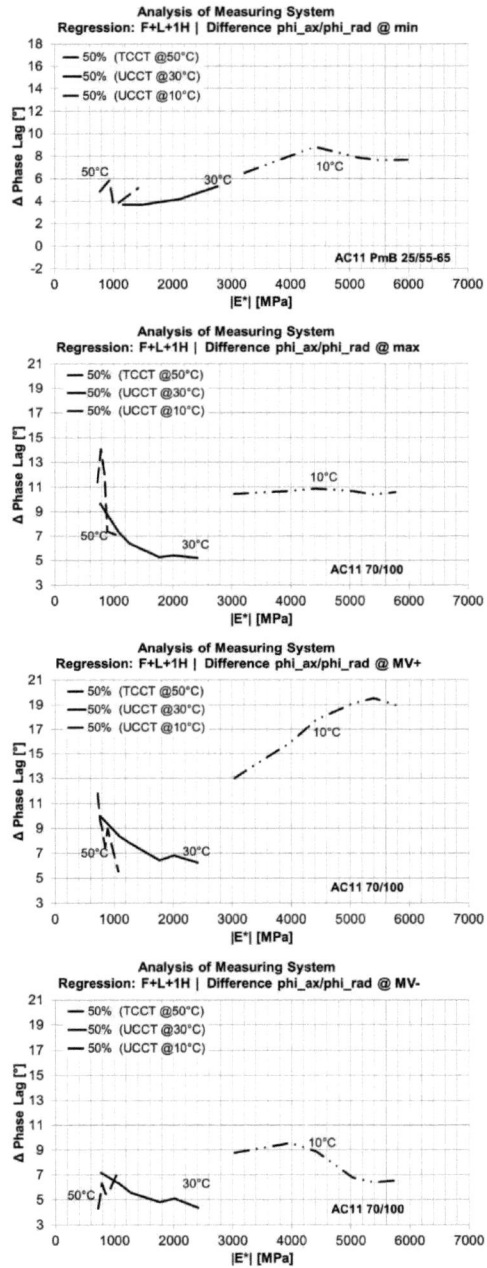

Figure V-54. 50% quantiles of $\delta_{ax,rad,\min}$ (top), $\delta_{ax,rad,\max}$ (2nd from above), $\delta_{ax,rad,MV+}$ (3rd from above) and $\delta_{ax,rad,MV-}$ (bottom) for 3 different test temperatures and AC 11 70/100; F+L+1H approximation.

Results of Tests with an Alternative Measuring System for Radial Deformation

The measuring device based on LVDTs was already described in section II.2.1; also the problems with the device and its inability to record radial deformation on the level of oscillations. After the hypothesis had been raised from data presented in (Hofko, et al., 2011a) that the measuring device consisting of SG and adhesive is too stiff for correct measurements on HMA, an alternative measuring system with a significantly lower stiffness was needed to prove that the measuring with SGs is a valid recording method. Therefore the LVDT-based measuring device was employed, thoroughly cleaned and lubricated to make it as smooth-moving as possible.

It was attached to an AC 11 70/100 specimen in the standard way, and a UCCT at 30°C with a mean axial stress of 0.25 N/mm² and an amplitude of 0.15 N/mm² was carried out. The frequency ranged from 0.1 Hz to 30 Hz. No SG was attached to the specimen to prevent any influence of this measuring device on the radial deformation. The test data were evaluated using the $F+L+1H$ approximation function. The quality of fit of the approximation to the recorded radial deformation was astonishingly high, ranging above 0.99. From earlier experience with the LVDT-based device, it was known that the accumulated radial deformation is recorded perfectly by the device but it failed to record the deformation on the oscillation level with satisfactory quality. From the recorded data (see Figure V-55) it is obvious that a perfectly clean and lubricated LVDT-based device is able to measure even the oscillating part of the radial deformation.

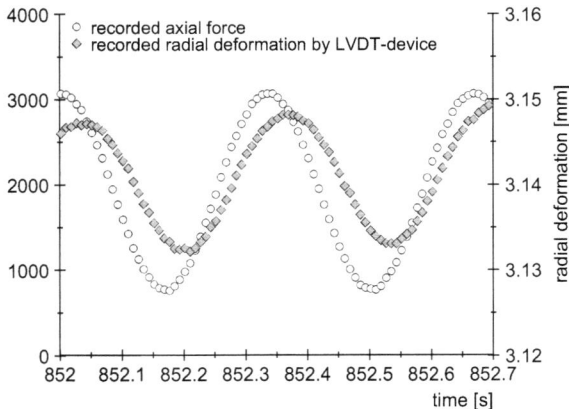

Figure V-55. Recorded axial force and radial deformation with the LVDT-based measuring device at AC 11 70/100 specimen tested in UCCT at 30°C.

For further analysis only those approximation blocks with a coefficient of determination for the radial deformation higher than 0.999 were used to ensure correct results. Figure V-56 presents the median values of δ between axial and radial deformation for the recordings with the LVDT based device vs. SG. The LVDT-based device results in even higher phase lags between the two deformation components.

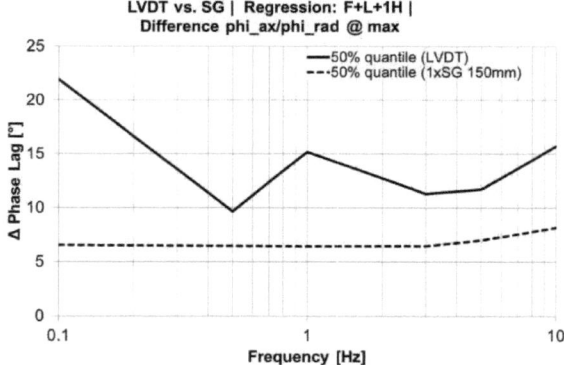

Figure V-56. 50% quantile of $\delta_{ax,rad,max}$ for LVDT-based radial deformation device vs. SG; F+L+1H approximation; AC 11 70/100 specimen tested in UCCT at 30°C.

V.6.4 Conclusions

After a comprehensive investigation and analysis of possible explanation for the existence of a phase lag between the axial and radial deformation $\delta_{ax,rad}$, it was conclusively found that this parameter is material-inherent. It cannot be explained by the material's anisotropy due to the method of compaction, neither to characteristics of the measuring device for radial deformation. It can be stated that $\delta_{ax,rad}$

- is related to material anisotropy to a minor extent,
- is not related to any uniformity of the radial deformation and
- is not related to the measuring system consisting of the SG and the adhesive.

The deformation phase lag seems to be material inherent and therefore a material parameter. Thus, the following section introduces another dynamic viscoelastic material characteristic, the dynamic Poisson's Ratio based on the deformation phase lag $\delta_{ax,rad}$.

V.7 Advanced Viscoelastic Material Parameters

V.7.1 Dynamic Poisson's Ratio

As a consequence of the measured time lag between axial loading and axial deformation for viscoelastic materials the complex/dynamic material stiffness parameters, i.e. E^*, E_1, E_2 and $|E^*|$ were introduced. For the same reasons, the so called complex Poisson's ratio v^* and the respective elastic and viscous part as well as its magnitude have been introduced in the rheological description of viscoelastic materials. For HMA (Di Benedetto, et al., 2007) used the complex Poisson's Ratio. It is an important characteristic not only for a better understanding of HMA behavior in three dimensions but also for modeling of pavement structures in 3-d. The theory has already been described in section V.1.3. In the following, the dynamic Poisson's Ratio $|v^*|$ and its elastic and viscous part is derived from test data for the same specimens used for analysis in section V.5.

Impact of Temperature and Frequency

Figure V-57 contains three diagrams which show $|v^*|$ (top), v_1 (middle) and v_2 (bottom) for the AC 11 PmB 25/55-65 mix at temperatures of 10°C, 30°C and 50°C. The material parameters were derived from the phase lag $\delta_{ax,rad,max}$ obtained at the maximum of loading since section V.4 showed that this phase lag is largely independent of the state of stress applied. The values for v ranging between 0.30 and 0.35 which is often found in literature and used for modeling and simulation is obviously only valid for intermediate temperatures at low frequencies or high temperatures at high frequencies. At 30°C the dynamic Poisson's Ratio ranges from 0.26 at 20 Hz to 0.35 at 0.1 Hz (median values). At 50°C this value runs from 0.37 to 0.50 from high to low frequencies. And at the lowest tested temperature of 10°C $|v^*|$ ranges from 0.10 to 0.16. It can also be stated that the parameter is strongly dependent on the frequency at high temperatures. At 50°C, the dynamic Poisson's Ratio drops from 0.1 Hz to 20 Hz a total of 0.13. At 30°C, the difference is around 0.09 and at 10°C, the difference is around 0.05. Therefore, especially at intermediate and high temperatures, not only the temperature but also the frequency of loading has to be taken into consideration. Analogue statements can be given for the elastic part of the dynamic Poisson's Ratio v_1.

v_2 is at a very low level for low and intermediate temperatures. At 10°C, v_2 is around 0.01 and at 30°C, v_2 is around 0.03. For both temperatures v_2 is independent of the frequency. The largest viscous Poisson's Ratio can be found at high temperatures which is logical since the viscous material behavior gets more dominant the higher the temperature is. v_2 starts at 0.14 at 0.1 Hz and decreases to 0.05 at 20 Hz (median values). Thus a clear dependency on the frequency can be distinguished at high temperatures.

The same analysis was carried out for the AC 11 70/100. Figure V-58 shows $|v^*|$, v_1 and v_2 for test at 10°C, 30°C and 50°C and a frequency sweep from 0.1 Hz to 20 Hz. Different to the situation for the mix with the polymer modified binder in the figure above, in this case, no significant difference can be found between results from 30°C and 50°C. Both are at the same level for the complete range of frequencies at least in terms of the dynamic Poisson's Ratio and its elastic part. The results show that the PmB (Figure V-57) enables higher mix stability at 30°C in terms of radial deformation than the 70/100 bitumen (Figure V-58). This higher stability compared to the 70/100 is lost when the temperature is raised from 30°C to 50°C. At 50°C both mixes, the PmB and the 70/100 mix, show similar dynamic Poisson's Ratios. The mix with the unmodified bitumen reaches an upper limit at a temperature of 30°C in terms of the Poisson's Ratio. In terms of the viscous part of v^* the material reacts more viscous at 50°C than at 30°C.

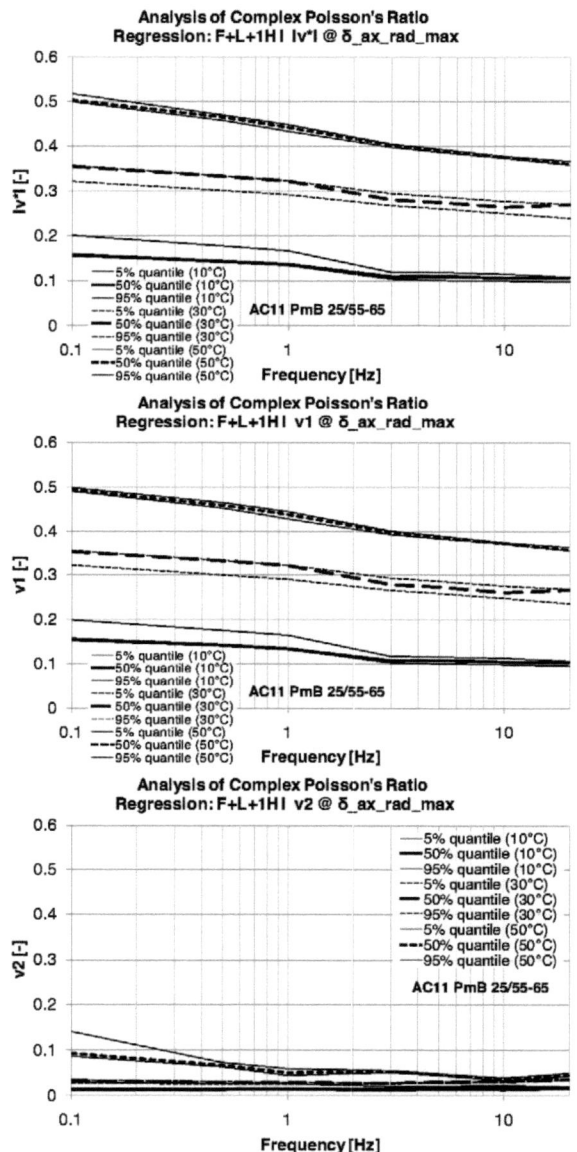

Figure V-57. 5%, 50% and 95% quantiles of $|v^*|$ (top), v_1 (middle), v_2 (bottom) for the AC 11 PmB 25/55-65 at $\delta_{ax,rad,max}$; F+L+1H approximation.

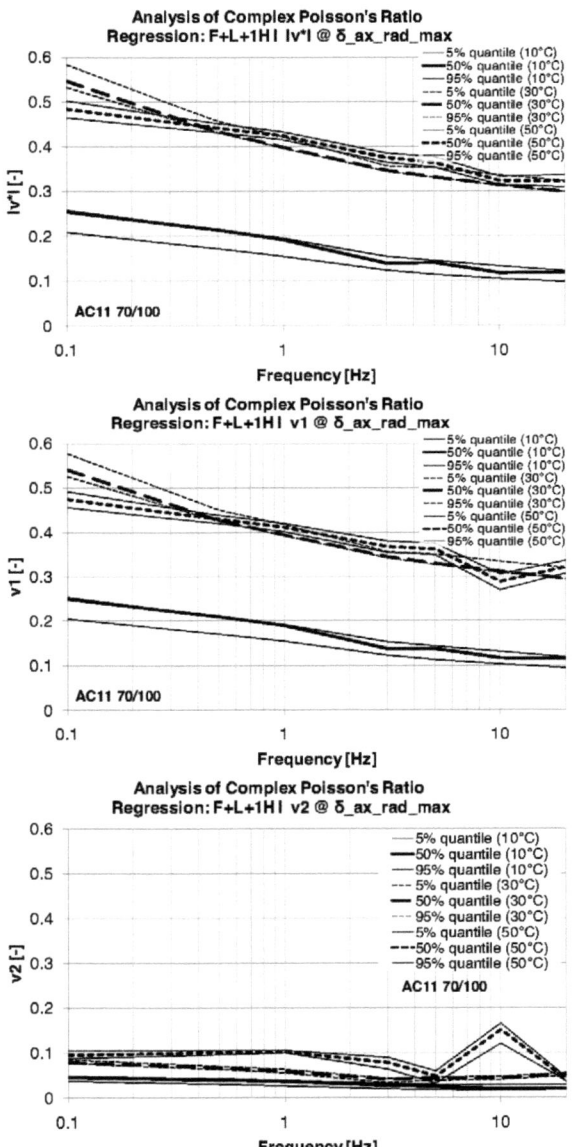

Figure V-58. 5%, 50% and 95% quantiles of $|v^*|$ (top), v_1 (middle), v_2 (bottom) for the AC 11 70/100 at $\delta_{ax,rad,max}$; F+L+1H approximation.

Impact of Binder Content

Analogous to the investigations carried out for the standard dynamic material parameters in section V.5.1 and V.5.2, the same mixes are used in the following to describes impacts of mix design parameters on the dynamic Poisson's Ratio and on the dynamic shear modulus.

Figure V-59 contains diagrams that show the impact of the binder content on the dynamic Poisson's Ratio of the PmB mix. At 10°C and 30°C, there are clearly increasing trends of this parameters with increasing binder content. It seems that a larger content of binder in the mix activates radial deformation to a higher extent. There is also a decreasing trend with increasing frequency and decreasing temperature.

At 50°C, it seems that the Poisson's Ratio is at a maximum value and a change in the binder content does not lead to a relevant change. $|v^*|$ ranges between 0.50 and 0.58 for 0.1 Hz and 0.36 and 0.43 at 10 Hz. As stated at the beginning of section V.1.3 dense materials cannot reach Poisson's Ratios higher than 0.5. In the case of porous materials, like HMA, higher values than 0.5 are possible. For Poisson's Ratios higher than 0.5, the volume of a specimen is increased when it is compressed due to increasing air void content. This happens for the AC 11 70/100 mix at a frequency of 0.1 Hz and 50°C. At these conditions the material shows a low stiffness and together with the long loading time, the composite structure of HMA can be altered by loading so that the volume increases due to additional air voids.

For the unmodified bitumen 70/100, the impact of a change in binder content is depicted in three diagrams in Figure V-60 for the three temperatures. Different from the situation for the PmB mixes, an increase of the binder content does not lead to increasing Poisson's Ratios at 10°C. There is a clear maximum at a binder content of 5.3% (m/m). The maximum is more dominant at low frequencies (0.1 Hz), where the difference between low and medium content is 0.17 (from 0.27 to 0.44). At 1 Hz the difference is 0.13 and at 10 Hz 0.05. The impact of binder content seems to decrease with increasing frequency of loading.

At 30°C and 0.1 Hz, an increase in binder leads to increasing Poisson's Ratios. On the other hand, at 1 Hz and 10 Hz there is hardly any change between the three bitumen contents. At 50°C data from only two mixes are available. But also here, an increasing binder content leads to higher ratios. The difference becomes smaller with higher frequencies. This could be shown for all three temperatures.

Compared to the PmB mixes, $|v^*|$ is higher for the 70/100 HMAs. This also indicates that the stiffness perpendicular to the cylinder axis is lower for the unmodified binder.

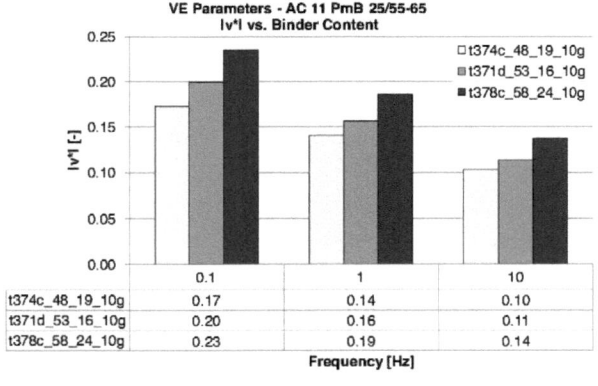
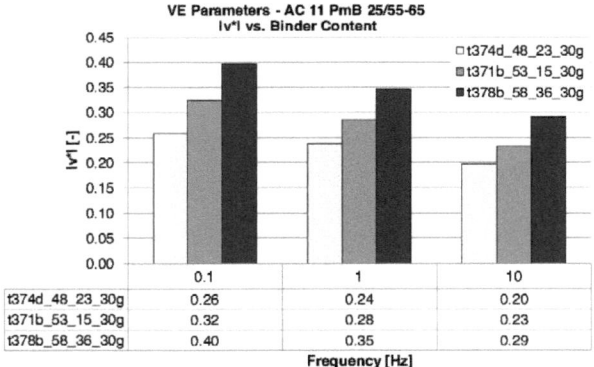
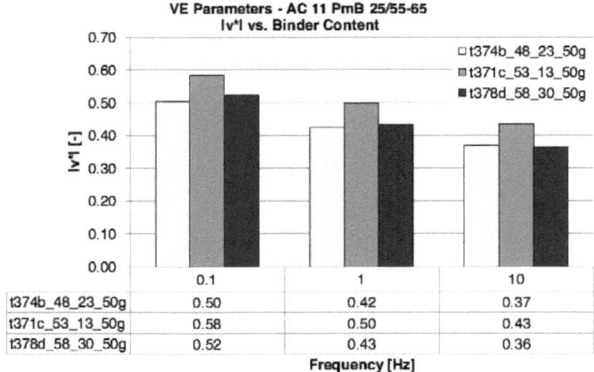

Figure V-59. Mean values of |v*| for AC 11 PmB 25/55-65 vs. binder content for 10°C (top), 30°C (middle) and 50°C (bottom) and 0.1 Hz, 1 Hz and 10 Hz.

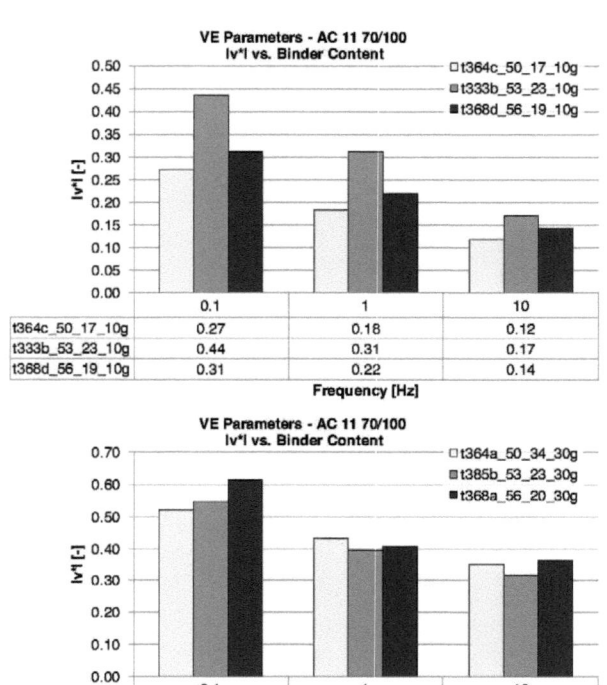
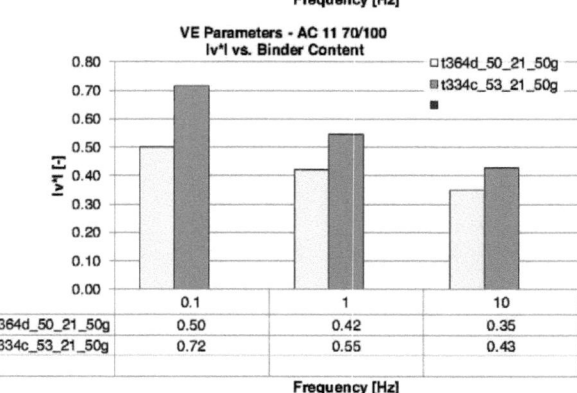

Figure V-60. Mean values of $|v^*|$ for AC 11 70/100 vs. binder content for 10°C (top), 30°C (middle) and 50°C (bottom) and 0.1 Hz, 1 Hz and 10 Hz.

Impact of Air Void Content

A variation of the air void content leads to different changes at different temperatures. Figure V-61 shows the results for the AC 11 PmB 25/55-65 mixes. At 10°C, there is a minimum of the Poisson's Ratio at a medium air void content. The difference to the other two mixes is not large. At 30°C, the two mixes with low and medium air void content develop similar Poisson's Ratios. At a high content of air voids, this parameter increases clearly from around 0.30 at 0.1 Hz to 0.46 and from 0.23 at 10 Hz to 0.32. When the temperature is set to 50°C, an increasing content of air voids leads to decreasing Poisson's Ratios. At 0.1 Hz, $/v^*/$ drops from 0.58 to 0.43, at 1 Hz from 0.50 to 0.39 and at 10 Hz from 0.43 to 0.33. The dynamic Poisson's Ratio and thus the radial strain compared to axial strain decreases with an increasing content of air voids. Thus, the hypothesis is raised that at high temperatures compressive loading leads rather to a reduction of air voids than to radial deformation. This seems logic since at 50°C the soft binder enables mineral aggregates to slide past each other.

A comparison of the dynamic Poisson's Ratio for the AC 11 70/100 mixes at different air void contents and temperatures is shown in Figure V-62. At 10°C, only two mixes with low and high volume of air voids are available. At this temperature the air void content seems to have a strong effect on the Poisson's Ratio. It decreases from 0.44 to 0.21 at 0.1 Hz, 0.31 to 0.16 at 1 Hz and 0.17 to 0.10 at 10 Hz. Again, as already shown for the binder content, the effect gets less dominant when the frequency is increased. At 30°C on the other hand, no significant impact of the air void content on $/v^*/$ can be found and at 50°C a decreasing trend occurs with increasing air void contents. The Poisson's Ratio drops strongly between low and medium air void content, e.g. from 0.72 to 0.50 at 0.1 Hz and less from medium to high air void content, e.g. from 0.50 to 0.47 at 0.1 Hz. And yet again higher frequencies damp this effect. The effect at 50°C can be explained by means analogue to the PmB mix results.

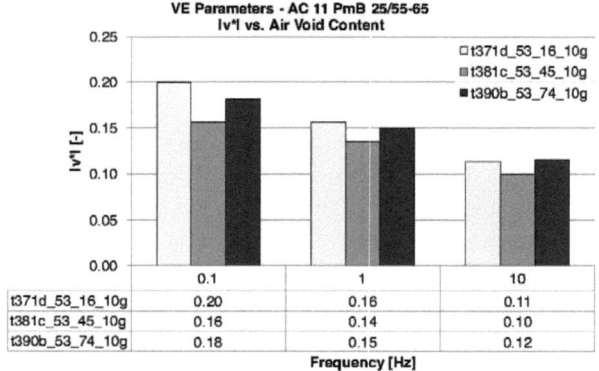
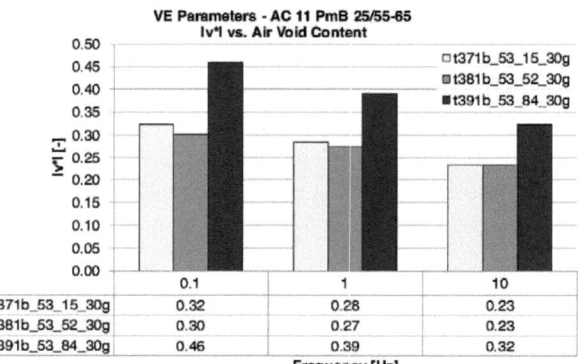
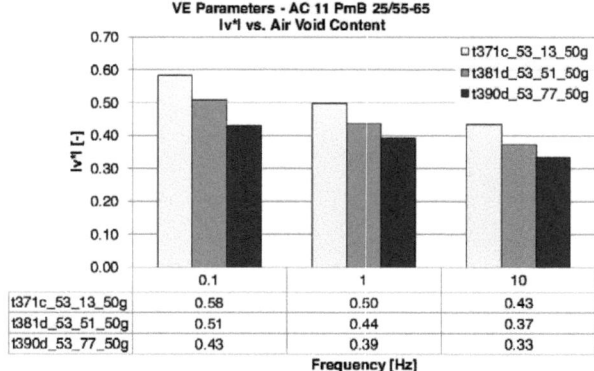

Figure V-61. Mean values of $|v^*|$ for AC 11 PmB 25/55-65 vs. air void content for 10°C (top), 30°C (middle) and 50°C (bottom) and 0.1 Hz, 1 Hz and 10 Hz.

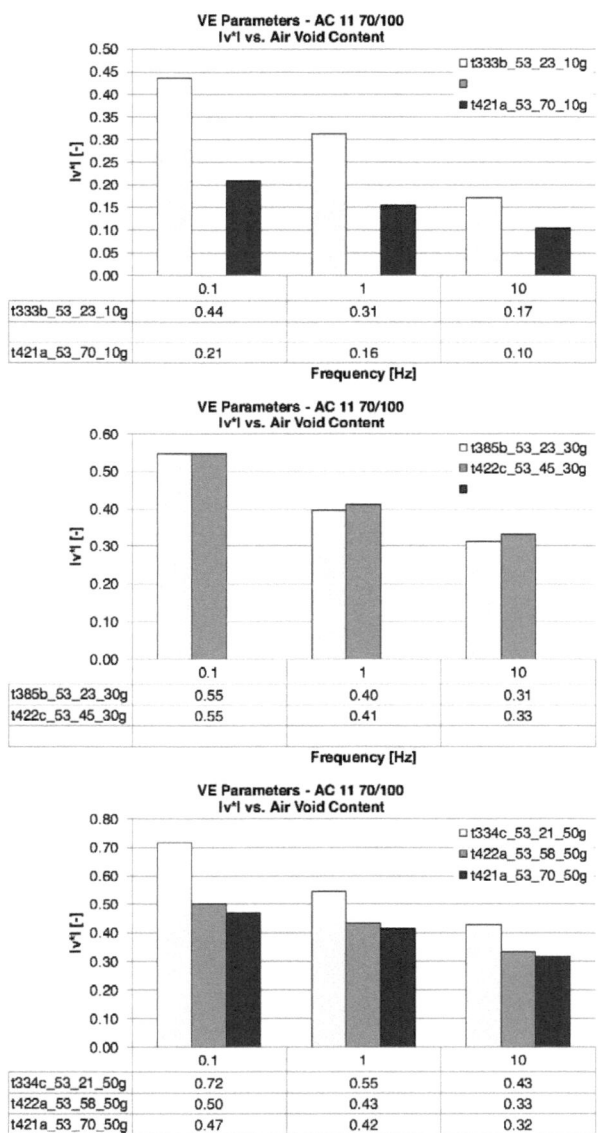

Figure V-62. Mean values of $|v^*|$ for AC 11 70/100 vs. air void content for 10°C (top), 30°C (middle) and 50°C (bottom) and 0.1 Hz, 1 Hz and 10 Hz.

V.7.2 Dynamic Shear Modulus

By combining the test results of E^* and v^* the complex shear modulus G^* and the magnitude of this complex number, the dynamic shear modulus $|G^*|$ and its elastic and viscous part can be derived by using formula (5.29). The dynamic shear modulus describes the stiffness behavior of a material perpendicular to the direction of main loading. In the following, the impact of temperature, frequency of loading, binder content and content of air voids on $|G^*|$ is described for the modified and unmodified HMAs.

Impact of Temperature

Temperature has a strong effect on the dynamic shear modulus (see Figure V-63) of the PmB mix. In addition to the information about the shear modulus, the diagrams in the follow also contain a percentage value that gives the ratio between $|G^*|$ and $|E^*|$.

At 0.1 Hz, $|G^*|$ drops from 1351 MPa at 10°C to 433 MPa at 30°C by 68% and to 330 MPa at 50°C (24%). The difference in stiffness between 10°C and 30°C gets smaller, the higher the frequency gets. It decreases by 65% at 1 Hz and 54% at 10 Hz. From 30°C to 50°C on the other hand, the difference increases to -44% at 1 Hz and -59% at 10 Hz. This is due to the fact that the dynamic shear modulus shows a stronger incline with frequency at 10°C than for 30°C and hardly any incline for 50°C. It seems that the material has reached the lower threshold value at 50°C. If the values of $|G^*|$ are compared to the values of $|E^*|$ it can be stated that the loss in stiffness is stronger for the dynamic shear modulus due to temperature. At 10°C, $|G^*|$ is still around 45% of $|E^*|$. This ratio drops to around 40% at 30°C and to around 35% at 50°C. Also, the increase in shear stiffness due to an increase in the test frequency is higher than for $|E^*|$. For example at 10°C the ratio starts with 42% at 0.1 Hz and increases to 43% and 45% at 1 Hz and 10 Hz.

For the unmodified mix, the change of $|G^*|$ with temperature is also presented in Figure V-63 (lower diagram). The impact is even higher than for the modified HMAs. When the temperature is increased from 10°C to 30°C, the mix loses 80% in shear stiffness at 0.1 Hz, 82% at 1 Hz and 73% at 10 Hz. The dynamic shear modulus decreases by another 46% when the temperature is put from 30°C to 50°C at 0.1 Hz, 45% at 1 Hz and 64% at 10 Hz. Compared to $|E^*|$ the same tendency occurs as for the modified mix. The shear stiffness decreases more strongly than $|E^*|$ with temperatures and the increase is higher with increasing frequencies.

It can be stated that the unmodified mix is clearly more sensitive to a change in temperature in terms of shear stiffness.

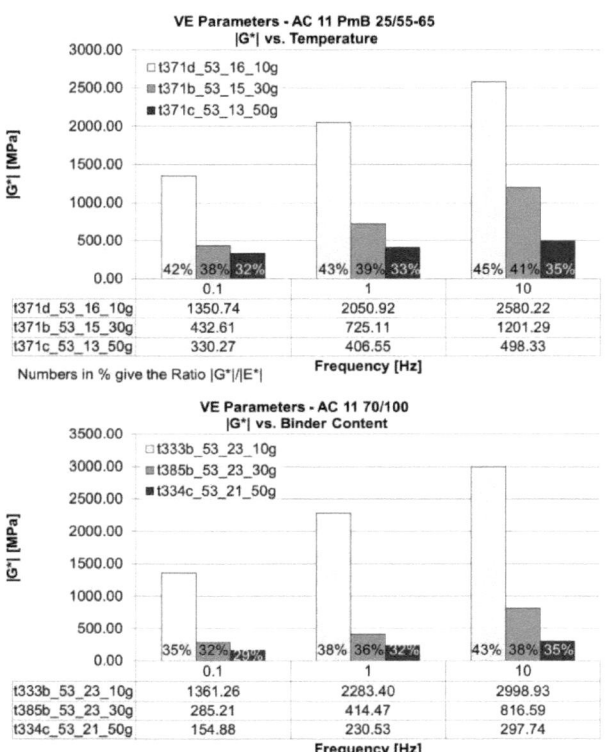

Figure V-63. Mean values of $|G^*|$ for AC 11 PmB 25/55-65 (top) and AC 11 70/100 (bottom) vs. temperature for 0.1 Hz, 1 Hz and 10 Hz.

Impact of Binder Content

As shown in Figure V-64, a change in binder content has different effects on the dynamic shear modulus for the PmB mix. At 10°C and 30°C, increasing the binder content means that the material reacts less stiff, although the decrease is small for 10°C. At 50°C the optimal binder content according to Marshall (5.3% (m/m)) prodcues a maximum in shear stiffness at all frequencies. The binder content seems to have the same effect for both stiffness parameters since the ratio between the two dynamic moduli $|G^*|/|E^*|$ does not change significantly with increasing binder content.

Rising binder contents reflect differently on the dynamic shear modulus at different temperatures for the 70/100 mixes as well. In Figure V-65 it can be seen that at 10°C and 30°C there is hardly any change between different binder contents neither in temperature, nor in frequency domain. It must be kept in mind that the variation in binder content for the unmodified mix was only ± 0.3% (m/m). The variation of the binder content could be too small to see the impact of this mix design parameter. Thus, the difference may not be significant. At 50°C and higher frequencies, a higher binder content increases the shear stiffness by 21% at 1 Hz and 10 Hz. Again, the binder content does not have an effect on the ratio between $|G^*|$ and $|E^*|$.

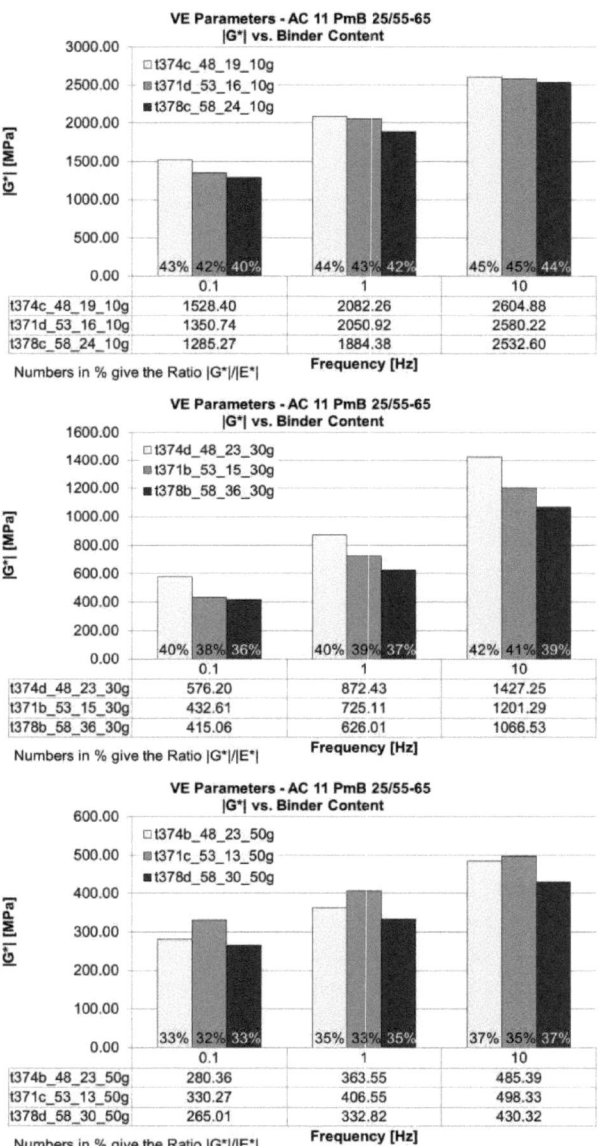

Figure V-64. Mean values of $|G^*|$ for AC 11 PmB 25/55-65 vs. binder content for 10°C (top), 30°C (middle) and 50°C (bottom) and 0.1 Hz, 1 Hz and 10 Hz.

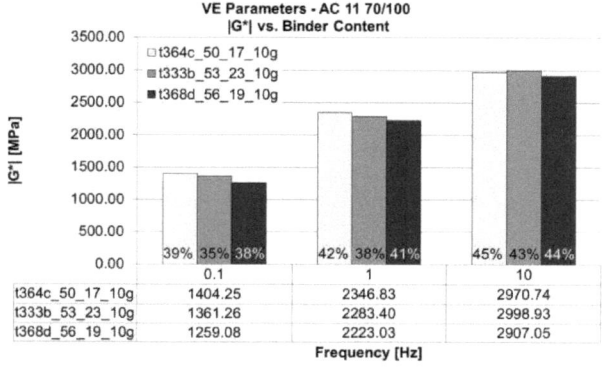
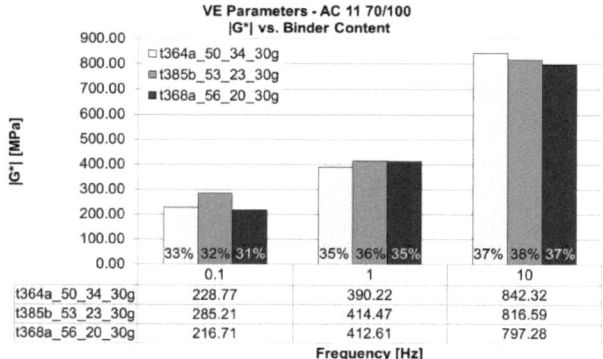
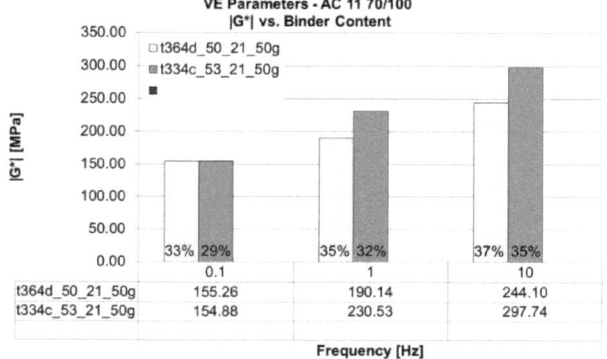

Figure V-65. Mean values of $|G^*|$ for AC 11 70/100 vs. binder content for 10°C (top), 30°C (middle) and 50°C (bottom) and 0.1 Hz, 1 Hz and 10 Hz.

Impact of Air Void Content

Impacts of the air void content on $|G^*|$ of AC 11 PmB 25/55-65 can be seen in Figure V-66. At 10°C, a change in the volume of voids in the mix does not reflect on the dynamic shear modulus in a significant way. At 30°C, there is a considerable drop in $|G^*|$ between the medium and high air void content, from 35% at 0.1 Hz to 38% at 10 Hz. When the temperature is set to 50°C, the drop in dynamic shear modulus occurs between low and medium air void content and ranges from 24% at 0.1 Hz to 16% at 10 Hz. $|G^*|$ seems to be sensitive to changes in the void content at higher temperatures. The higher the temperature gets, the more the drop in stiffness shifts towards the lower side of the air void content. Again, the results are analogue to those obtained from $|E^*|$.

The effect of the air void content $|G^*|$ for the 70/100 mix is depicted in the three diagrams in Figure V-67. At 10°C, only the low and high air void content is available. The dynamic shear modulus drops slightly at 0.1 Hz by 8% and stronger for 1 Hz (17%) and 10 Hz (14%). At 30°C, the dynamic shear modulus is practically stable at 0.1 Hz and 1 Hz and drops with increasing void content at 10 Hz. At 50°C, it can be shown that an increasing air void content also increases the dynamic shear modulus, when the content is changed from low to medium. $|G^*|$ increases by 50% at 0.1 Hz, by 20% at 1 Hz and 30% at 10 Hz. This is interesting, since the situation is inverse to the case for PmB mixes. Again, changing void contents do not change the ratio between the two dynamic moduli $|E^*|$ and $|G^*|$.

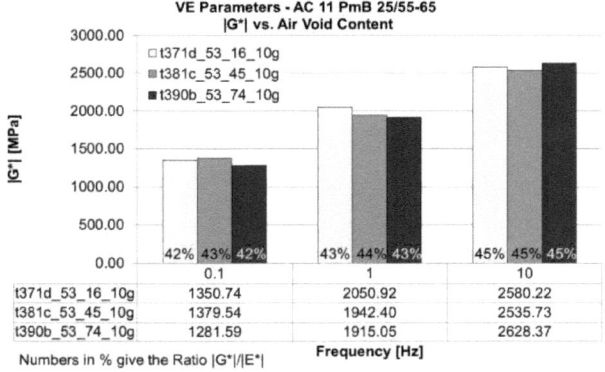
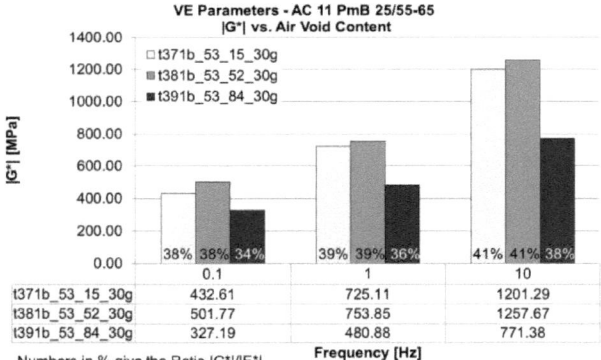
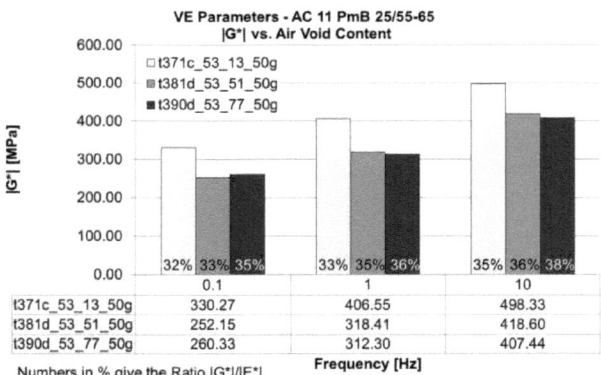

Figure V-66. Mean values of $|G^*|$ for AC 11 PmB 25/55-65 vs. air void content for 10°C (top), 30°C (middle) and 50°C (bottom) and 0.1 Hz, 1 Hz and 10 Hz.

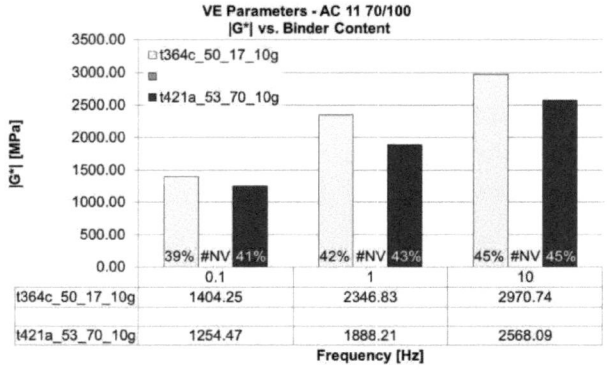
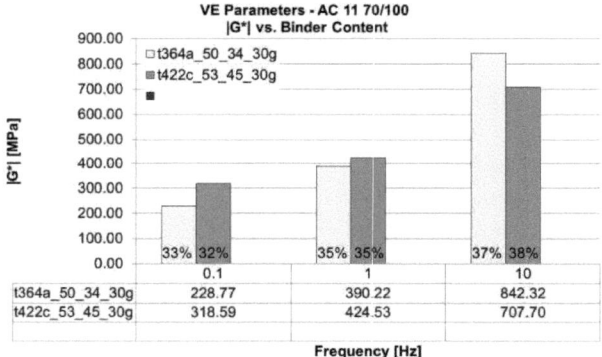
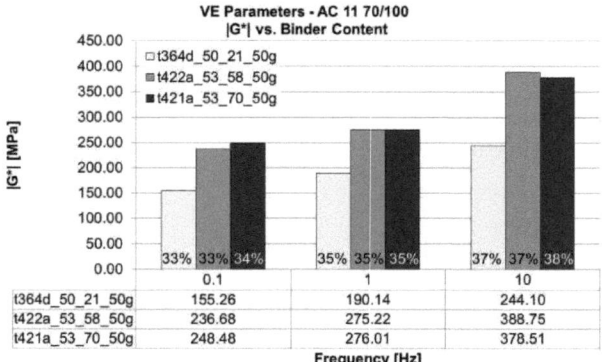

Figure V-67. Mean values of $|G^*|$ for AC 11 70/100 vs. air void content for 10°C (top), 30°C (middle) and 50°C (bottom) and 0.1 Hz, 1 Hz and 10 Hz.

Connection with Viscoelastic Material Parameters

An interesting interrelation was found between the viscoelastic and permanent deformation behavior of HMA. Therefore, the dynamic shear modulus $|G^*|$ is evaluated for the standard TCCTs according to (EN 12697-25, 2005) on AC 11 70/100 specimens with three different binder contents and on AC 11 PmB 25/55-65 specimens with one binder content. The results of these standard TCCTs in terms of creep rates have already been presented in section IV.2.2.

When the linear or the logarithmic creep rate is depicted vs. the dynamic shear modulus of the mixes at the standard TCCT conditions (50°C, 3 Hz) (see Figure V-68), a simple power function can be employed to describe the relation between both parameters with acceptable quality. Each dot in the diagram represents one test result.

The correlations show that an increasing dynamic shear modulus leads to a better resistance to permanent deformation, if the permanent deformation is represented by the linear or logarithmic creep rate. The power function that links the dynamic shear modulus to the creep rate indicates that the creep rate approaches zero with increasing $|G^*|$. A creep rate of 0 describes a material behavior where no permanent deformation occurs when the material is loaded in the compressive range. So, the stiffer the material gets, the smaller becomes the permanent deformation. On the other hand when the dynamic shear modulus decreases, the creep rate decreases strongly due to the power function. In the low dynamic shear modulus range a small loss in $|G^*|$ leads to a strong increase of the permanent deformation.

Figure V-69 shows the quality of the fit of the diagrams in Figure V-68 by means of quantiles. In 95% of all cases, the creep parameter that is predicted from the dynamic shear modulus by the power function lies between 87% and 122% of the creep rate derived from the TCCT. This is true for the linear case. For the logarithmic case, the 95% confidence interval ranges from 89% to 125%. Thus, both approaches exhibit equal fit qualities.

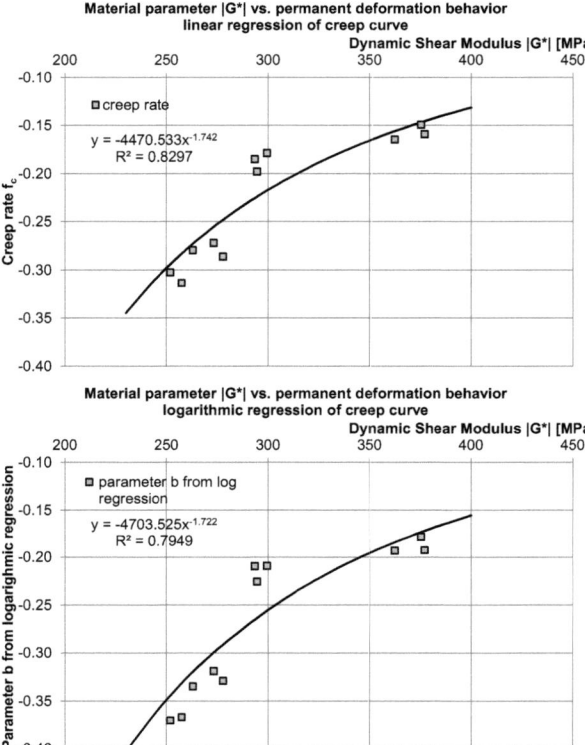

Figure V-68. Linking viscoelastic material behavior to the permanent deformation behavior for the linear creep rate (top) and the logarithmic creep rate (bottom) for standard TCCTs.

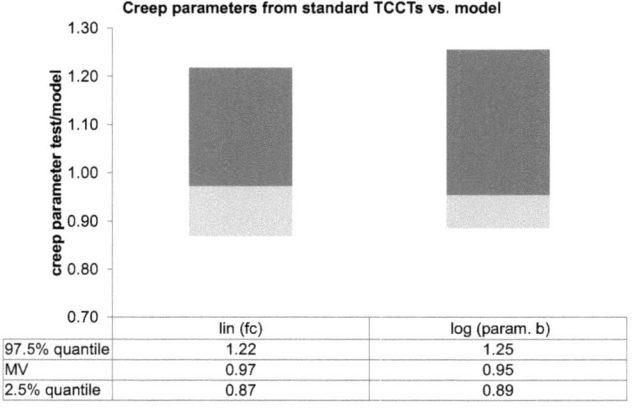

	lin (fc)	log (param. b)
97.5% quantile	1.22	1.25
MV	0.97	0.95
2.5% quantile	0.87	0.89

Figure V-69. Fit quality of the link between $|G^*|$ and the creep parameter.

V.7.3 Conclusions

An in-depth study on the complex Poisson's Ratio and its elastic and viscous parts was carried out in this section. It is defined analogous to the complex modulus and has been described in (Di Benedetto, et al., 2007). The mentioned paper deals with the magnitude $|v^*|$ and not with the elastic and viscous part. For two mixes, an AC 11 PmB 25/55-65 and an AC 11 70/100 the three material parameters were evaluated and analyzed.

It can be stated that also the Poisson's Ratio has an elastic and viscous component, although the viscous component only becomes dominant at a high temperatures (50°C). It is state of the art to use a constant value of around 0.35 for calculations, modeling and simulation for bituminous bound materials. From the results presented in this chapter it becomes obvious that this value is only true for intermediate temperatures at low frequencies or high temperatures and high frequencies. Thus, this chapter provides a deeper insight into the material behavior of bituminous bound mixes. In addition the results can account for a more realistic and exact modeling and simulation.

Further, in this section, impacts of binder and void content on the evolution of the dynamic Poisson's Ratio were analyzed. It can be shown that an increasing binder content leads to increasing Poisson's Ratios at 10°C and 30°C and stable conditions at 50°C for the modified HMAs. For the mixes with the unmodified bitumen 70/100, the material parameter behaves differently when the binder content is raised. At 10°C, there is a clear maximum at the medium binder content. At 30°C the tested mix shows an increasing Poisson's Ratio with increasing binder content at low frequencies and stable conditions at higher frequencies. An increasing air void content has a conclusive effect at high temperatures for both binder types. The higher the void content the lower is the dynamic Poisson's Ratio. It seems that the soft binder at high temperatures enables mineral aggregates to slide past each other (reducing air voids) more easily and therefore most of the deformation energy is put into filling the air voids rather than producing radial deformation.

By combining the dynamic modulus and the dynamic Poisson's Ratio another viscoelastic parameter, the dynamic shear modulus $|G^*|$ is used and the impact of temperature, binder and void content on this parameter are investigated. Compared to the dynamic modulus $|E^*|$, the analysis show that the temperature and frequency sensitivity of the dynamic shear modulus is higher. $|G^*|$ decreases more strongly with increasing temperature compared to $|E^*|$. On the other hand the dynamic shear modulus exhibits a higher increase with increasing frequency than $|E^*|$.

Furthermore, a conclusive link between the viscoelastic behavior of HMAs, in this case the dynamic shear modulus $|G^*|$, and the deformation behavior (linear and logarithmic creep rate) was established. A power function links the dynamic shear modulus to the linear and logarithmic creep rate. This relationship is presently based on AC 11 mixes with 70/100 for three binder contents and with PmB 25/55-65 for one binder content.

Further investigations on mixes with a variation of mix design parameters in the future will be sensible to validate this interrelation.

VI A VISCOELASTIC MODEL OF THE BEHAVIOR OF HMA UNDER COMPRESSIVE LOADING

VI.1 Introduction

Modeling of the material behavior of HMA has become an important field of road research in recent years, especially those models following a mechanistic approach. In this case, the model parameters are related to the physical material behavior. Within the Christian Doppler Laboratory for Performance-based Optimization of Flexible Road Pavements great effort was put into advanced rheological models based on the multiscale approach, where macroscopic material parameters are determined on the basis of properties of the constituents and the mix design. These parameters are obtained by means of upscaling procedures, bridging the scales from finer levels to the macroscale. Thus, the so developed models cover a wide range of asphalt mixtures with different mix designs and constituents. (Lackner, et al., 2006) presents the general procedure. Powerful models have been developed for different application: For example, (Aigner, 2010) worked on the multiscale approach for modeling of the stiffness behavior at higher temperatures. (Fuessl, 2010) used the same approach to describe stiffness and strength behavior at low and intermediate temperatures. Although the procedure mentioned above provides successful tools for research, the application of each of those models is still limited to a small range of temperatures and frequencies.

Based on these findings, a different approach will be taken in this study. The interest was put on modeling the viscoelastic behavior of HMA and its evolution over temperatures and frequencies occurring in the field. The model to be developed within this chapter shall cover a wide range of mix design parameters to make it versatile and powerful. Since the macroscopic HMA tests result in a quasi-3-d state of strain, it was also tried to describe the viscoelastic behavior of the material in all three dimensions. To reach these goals, the model is based on master curves and the stiffness ratio between binder and the mix. Thus, a model characterizing the viscoelastic behavior of HMA (*B-A Model*) has been developed and will be presented in the following chapter.

A similar approach is used for the *Witczak E* Predictive Model for HMA*. (Bari, et al., 2006) presents a revised version of this model, which is based on data from tests on 346 HMAs. According to (Bari, et al., 2006), the *Witczak Model* is capable of accurately estimating changing in E^* of HMA as a function of changes in mixture volumetrics, material properties, temperature and loading frequency. (Di Benedetto, et al., 2004) presents a rheological model (*2S2P1D Model*) to describe the linear viscoelastic properties of bitumen and HMA. In addition a transformation method is described in (Di Benedetto, et al., 2004) that allows the dynamic modulus of the HMA to be predicted from the dynamic modulus of the binder.

VI.1.1 The Time Temperature Superposition Principle

For the derivation of the dynamic viscoelastic material parameters only the impact of time in terms of frequency of loading has been taken into consideration so far. Since the behavior of viscoelastic materials is also influenced by the temperature, its characteristics like the dynamic modulus, compliance and phase lag are a function of time and temperature, e.g. $|E^*| = |E^*|(t,T)$. One way to describe this function is to carry out a number of cyclic dynamic tests at various temperatures and frequencies. For so-called thermo-rheological simple materials, there is a more efficient way. For these materials time and temperature can be combined to one parameter. If cyclic dynamic tests are carried out on a thermo-rheological simple material at different temperatures T_i, the results show the same function in terms of time if the time domain is scaled. A lower temperature leads to a shift of material parameters to shorter times or a higher frequencies. A so called master curve can be derived from a limited number of tests at different temperatures and frequencies by taking advantage of the time-temperature superposition principle (TTSP). This means that the viscoelastic parameters can be obtained for any arbitrary temperature and a wide range of frequencies, without having to test the material at this specific temperature or frequency. (Findley, et al., 1989)

The TTSP can be stated in a general way:

$$P(t,T) = P(\zeta,T_0) \tag{6.1}$$

$$\zeta = \frac{t}{a_T} \tag{6.2}$$

$$P(t,T) = P(\zeta,T_0) = P(a_T) \tag{6.3}$$

P .... Viscoelastic parameter
t .... Time [s]
T .... Temperature [K]
T_0 .... Reference Temperature [K]
ζ .... Reduced or scaled time [s]
a_T .... Shift factor [-]

In other words, the TTSP states that a change in the test temperature has the same impact on a material parameter of a thermo-rheological simple material as a change in the measured test time (or loading frequency for dynamic tests). Compared to a reference temperature T_0, a higher temperature leads to lower frequencies and vice versa.

Figure VI-1 contains an example for test data from a dynamic shear rheometer (DSR) test carried out on a mastic from a PmB 25/55-65 and mineral filler with a ratio 1:1. This means that the mass ratio of binder to filler (aggregates < 0.063 mm) is 1:1. The top diagram, shows the mean values of $|G^*|$ for each test frequency and temperature. In the lower diagram the test data was shifted with a factor to obtain the master curve for a reference temperature of 30°C.

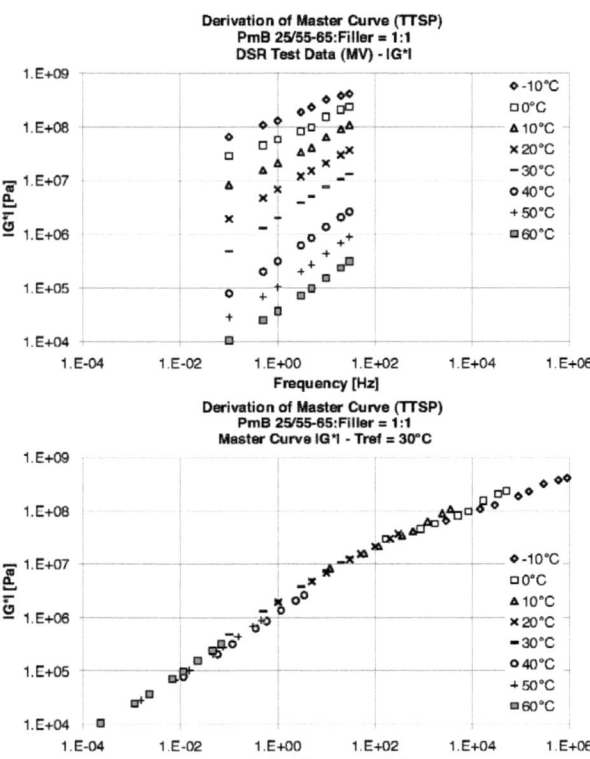

Figure VI-1. Derivation of the master curve from test data at various temperatures and frequencies (top) to complete mater curve (bottom) for a T_{ref} of 30°C.

Different approaches have been developed to derive the shift factor a_T. A well-known model is described by Williams, Landel and Ferry (Ferry, 1980):

$$\log a_T(t) \equiv \log \frac{t}{\zeta} = \frac{-k_1(T-T_0)}{k_2+(T-T_0)} \tag{6.4}$$

a_T .... Shift factor
k_1, k_2 ...... Constants
T .... Absolute temperatur [K]
T_0 .... Reference Temperature [K]

Another wide spread definition of the shift factor is according to Arrhenius Law (Steinmann, et al., 2010):

$$\ln a_T = \frac{E_a}{R} \cdot \left(\frac{1}{T} - \frac{1}{T_0} \right) \tag{6.5}$$

a_T .... Shift factor

E_a..........Activation energie [J/mol]
R............Universal gas constant $R = 8.314$ [J/K·mol]
T............Absolute temperatur [K]
T_0..........Reference Temperature [K]

The formula above is used in chemistry to describe the kinetics of chemical reactions by the activation energy and will be used in the further research presented in this thesis.

The test data that have been altered to represent a master curve can be described by means of an analytical function derived by an approximation algorithm which will be shown in the next section VI.1.2.

The advantages of thermo-rheological simple materials and the TTSP are:
- The only difference between tests results derived at different temperatures is the scale of time. By a horizontal shift of n single curves from n temperatures T_i, one master curve for a reference temperature T_{ref} or T_0 can be obtained for a large frequency range.
- The long term behavior can be determined for a reference temperature T_0 without having to carry out long term tests.
- Material parameters can be derived for temperatures and frequencies for which no tests have been carried out, if the reference temperature is set to the target temperature.

VI.1.2 Master Curve Fitting

To investigate relevant correlations between viscoelastic characteristics of binder and HMA, master curves of the dynamic moduli and the elastic and viscous components were obtained taking advantage of the TTSP. For linking time and temperature the factor a_T according to Arrhenius' Law in formula (6.5) was taken as the basic function.

The general approach taken here to describe the master curve analytically consists of
- systematically varying the shift factor a_T and
- fitting an analytical master curve function to the data for each a_T until the coefficient of correlation R^2 reaches a maximum.

The master curve of bitumen, mastic and asphalt mixes can be described by different functions (Kappl, 2007). For this analysis, the first approach to describe the master curve analytically was the following formula:

$$f(x) = y_0 + \frac{a}{1 + \left(\dfrac{x}{x_0}\right)^b} \tag{6.6}$$

a, b, y_0, x_0 ..Parameters of the logistic function

It is a logistic function with four fit parameters. The fitting procedure is carried out for one arbitrary reference temperature $T_{0,i}$ and a certain unit of stiffness U_i (e.g. Pa). For this reference temperature and unit of stiffness, the logistic function is fitted to the test data and the fit parameters $x_0(T_{0,i})$, $y_0(U_i)$, a and b are obtained at the optimal a_T.

The problem with the approximation function in (6.6) is that it did not manage to describe the complete range of data with sufficient quality. For the fitting process, different weighting functions were considered to optimize the quality of the fit. For weighting, each functional value y was multiplied with a weighting factor (none, $1/y$, $1/y^2$), so that the smaller the functional value, the larger the weighted value becomes. Deviations between approximated and test data become larger and are taken into account when the quality of the fit is obtained. This is important when the range of functional values is large. In this case, small functional values are not fitted with the same quality as large functional values, if the sum of squared errors is used for optimizing the approximation. Depending on the weighting factor used for fitting either the lower or the higher frequency domain show significant deviations between test data and approximation function – although the coefficient of correlation is at a high level (> 0.99). Figure VI-2 presents an example of three master curve fits for the same test data with different weighting. For the example, test data from a DSR test of a PmB 25/55-65 are used, but the procedure is not limited to this material. It works just as well for paving grade bitumen, mastic and HMA. The top diagram shows the fit if no weighting is considered. Obviously, the high stiffness domain is approximated well, whereas at low frequencies the quality of the fit is insufficient. Going just by the coefficient of correlation, a value of 0.993 would indicate a near to perfect fit. The diagram in the middle contains data for the same fit with a weighting of $1/y$. Still, the lower frequency domain shows significant deviations. Since both axes are shown in log-scale, the deviations might not appear to be crucial, but the fit is nearly 3 times above the test data in the low stiffness range. The lower diagram in Figure VI-2 shows the situation if a weighting of $1/y^2$ is considered. Thus, the low stiffness data get more dominant and the fit quality is high in this range. Yet, when it comes to the upper stiffness domain, the deviation – again appearing small in the log-log scale – is more than 50%.

The main problem of the data is that the values of the dynamic shear modulus (y-values) range from some 10^3 to over 10^8 Pa, at least for bitumen and mastic. This leads to unsatisfying fitting processes even if weighting is used for the regression.

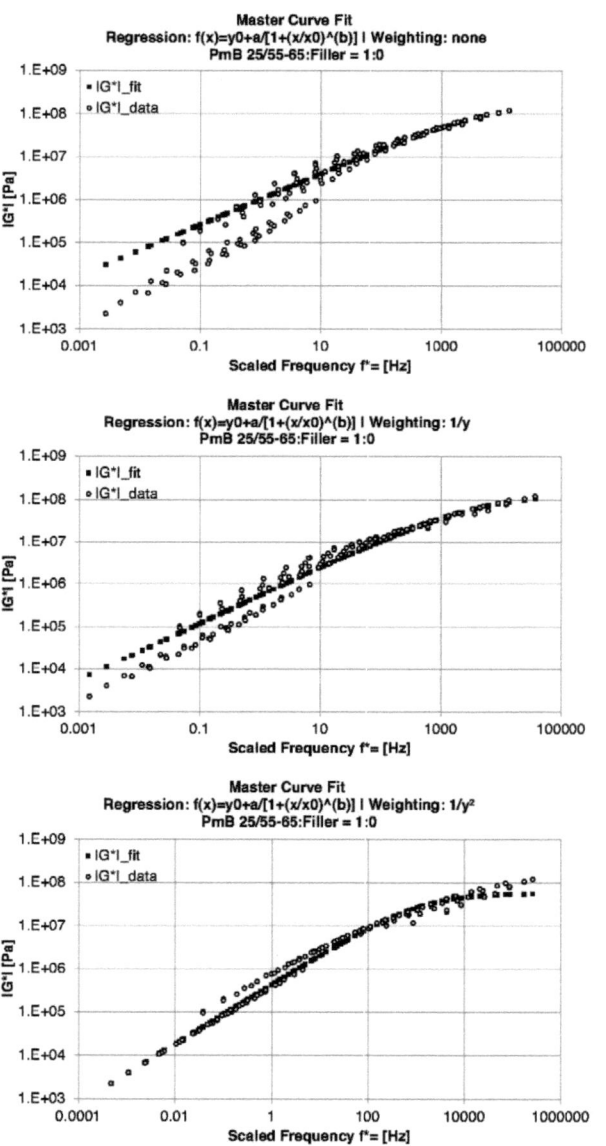

Figure VI-2. Master curve fit with approximation function according to (6.6) with no weighting (top), weighting 1/y (middle) and 1/y² (bottom).

To overcome this problem with a range of 10^5 for the y-data, different other approaches were tried. The most successful attempt is presented in the following. For the fit, the x-values (scaled frequency) were kept unchanged, but for the y-data the natural logarithm of the stiffness parameter was taken as an input for the fitting procedure using the same approximation function as shown in (6.6). Thus the actual approximation function for a stiffness parameter S is an exponential term:

$$S(x) = \exp\left(y_0 + \frac{a}{1 + \left(\frac{x}{x_0}\right)^b} \right) \quad (6.7)$$

Figure VI-3 shows the master curve fit for the procedure mentioned above. There is hardly any impact of the used weighting, but if the fit is weighted with 1/y, the results are slightly better in terms of quality of fit. The coefficient of correlation is again at a high level with a value of 0.996. But as it was shown above, R^2 does not seem to be an ideal parameter measuring the fit quality when the y-data ranges over a number of decades. High y-values are fitted with a much lower relative error with regard to its absolute values than low y-values. Thus different statistical quality benchmarks will be used to describe the fit quality of master curves. To eliminate the influence of the absolute value of a data point, the sum of the mean relative error (SMRE) is introduced as follows:

$$SMRE = \frac{\sum_{i=1}^{n} \left| \frac{y_i - \hat{y}_i}{y_i} \right|}{n} \quad (6.8)$$

y_i............Y-value of data item i at x_i
$\hat{y}_i$............Fitted y-value at x_i
n............Number of data items

In addition, more statistical values are presented for the master curve fits:
- The maximum and minimum relative error (RE).
- The 5% and 95% quantile of the RE and its 90% probability range.
- The probability of the RE lying between -0.1 and 0.1 (in some cases -0.05 and 0.05).

Since the number of data points n for the DSR tests is around 1250 and for the CCTs at least 80, the samples seem large enough for a statistical analysis.

The table in Figure VI-3 shows an example for the statistical parameters for the fit in the diagram to above. An SMRE of 13.2% can be found. The extreme values of the RE are −29.8% and +39.1%, its 5% and 95% quantile are −21.9% and +30.5% respectively. With a 44.3% chance the RE lies between −10% and +10%.

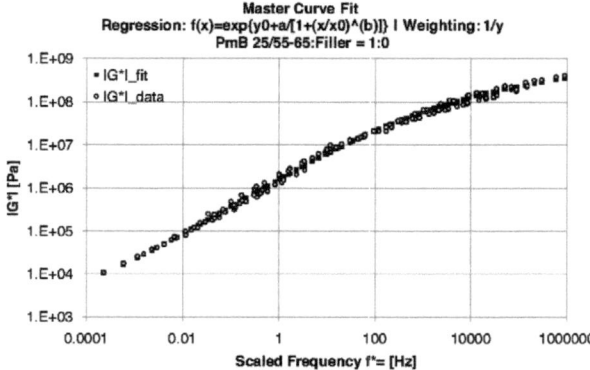

Parameter	Value
SMRE	0.132
Min. RE	-0.298
Max. RE	0.391
5% quantile of RE	-0.219
95% quantile of RE	0.305
90% probability range of RE	-0.219 to 0.305
Probability of the RE between -0.1 and 0.1	0.443

Figure VI-3. Master curve fit with approximation function according to (6.7) and weighting 1/y (top) and the statistical parameters describing the fit quality (bottom).

Table VI-1 shows the quality parameters of the approximation for the master curve fit with the standard procedure with function (6.6) shown in Figure VI-2. Even the best of the three approximations with a weighting of $1/y^2$ approximates the stiffness data clearly worse than the exponential approximation function. The 90% probability range for the RE runs from −44.9% to +22.9% for the standard function compared to −21.9% to +30.5% for the exponential approximation.

Table VI-1. Parameters describing the fit quality of the fits with the approximation function according to (6.6) for the examples given in Figure VI-2.

Parameter	weighting: 1	weighting: 1/y	weighting: 1/y²
SMRE	2.044	0.592	0.187
Min. RE	-0.506	-0.574	-0.560
Max. RE	12.590	2.276	1.157
5% quantile of RE	-0.253	-0.399	-0.449
95% quantile of RE	8.221	1.884	0.229
90% probability range of RE	-0.253 to 8.221	-0.399 to 1.884	-0.449 to 0.229
Probability of the RE between -0.1 and 0.1	0.284	0.296	0.393

To better understand the fit parameters x_0, y_0, a and b of the exponential approximation function in equation (6.7), some considerations are taken in the following:

$$f(x)|x \to 0 = \exp(y_0) \text{ for } b < 0 \text{ and } x > 0$$
$$f(x)|x \to \infty = \exp(y_0 + a) \text{ for } b < 0 \text{ and } x > 0 \quad (6.9)$$

If b is assumed to be negative and x to be positive, which is the case for all fits due to the nature of the test data, there is a lower and an upper asymptote. When x (being the scaled frequency f^*) approaches 0, the fit function approaches $exp(y_0)$ which is the lower asymptote of the fitted stiffness parameter. For the opposite case when the frequency approaches infinity, the upper asymptote $exp(y_0+a)$ is reached. Parameters x_0 and b influence the shape of the curve between those two extremes.

Parameter x_0 was found to be a function of the reference temperature and parameter y_0 a function of the unit of the test data. Parameter x_0 at the reference temperature $T_{0,1}$ is labeled $x_{0,1}$. For any other reference temperature $T_{0,i}$, $x_{0,i}$ can be obtained from the following relationship:

$$x_{0,i} = x_{0,1} \cdot e^{\frac{E_a}{R}\left(\frac{1}{T_{0,1}} - \frac{1}{T_{0,i}}\right)} \quad (6.10)$$

Parameter y_0 with the stiffness unit U_1 used for fitting is labeled $y_{0,1}$. For any other unit U_i, $y_{0,i}$ can be calculated using the following function:

$$y_{0,i} = y_{0,1} + \ln\left(\frac{U_i}{U_1}\right) \quad (6.11)$$

Thus, the master curve can be described for any arbitrary reference temperature and stiffness unit in an analytical way without having to fit the master curve each time the reference temperature or unit of stiffness changes.

VI.2 Test Program

In addition to the data produced from CCTs at HMA specimens presented in the preceding chapter V, a comprehensive test program was also run on binder (and mastic) to analyze correlations between the viscoelastic characteristics of binder and mastic on the one hand. Results of these investigations can be found in Annex B. On the other hand analogue interrelations between bitumen and HMA are presented in the following. Binder and mastic tests were carried out with a dynamic shear rheometer (DSR) depicted in Figure VI-4. From the DSR the dynamic shear modulus $|G^*|$ and the phase lag between stress τ and strain γ are obatined. The basic theory is analogue to the contents about the complex modulus E^* in section V.1.2.

The binder and mastic tests included both binders, as well as mixes of the binder with filler (particle size < 0.063 mm) to create a mastic. The tested mixes are shown in Table VI-3. Table VI-2 shows the test conditions in terms of temperatures and frequencies of testing. In addition, it gives information about the strain amplitude and the diameter of the load plate, as well as the specimen height. For low to intermediate temperatures from -10°C to 35°C the small load plate with a diameter of 8 mm and a specimen height (or gap between both load plates) of 2 mm was used. In the high temperature range, the load plate had a diameter of 25 mm and the specimen height was set to 1 mm.

Figure VI-4. DSR device used for binder and mastic tests.

Table VI-2. Test procedure for the determination of viscoelastic properties of binder and mastic in the DSR.

Temperature Range	Strain Amplitude [°]	Diameter of Load Plate [mm]	Specimen Height [mm]	Frequencies [Hz]	Number of Load Cycles
-10°C to 35°C $\Delta T = 5K$	0.573	8	2	0.1	4
				0.5	5
				1	10
				3	10
				5	10
				10	15
				20	15
				30	15
40°C to 60°C $\Delta T = 5K$		25	1	0.1	4
				0.5	5
				1	10
				3	10
				5	10
				10	15
				20	15
				30	15

Table VI-3. Binders and mastic tested with DSR.

Binder	Binder:Filler (by mass)
70/100	1:0
	1:0.5
	1:1.3
	1:1.4
	1:1.6
PmB 25/55-65	1:0
	1:0.5
	1:1
	1:1.6
	1:1.7
	1:1.9

VI.3 Modeling the Viscoelastic Behavior of Binder and HMA

For the bituminous binders and the mastic, master curves for the shear modulus $|G^*|$ and its elastic and viscous parts were obtained from DSR tests carried out at different temperatures and frequencies. The same was done for test data from CCTs from different HMAs. The master curves from different levels of scale (bitumen, mastic, HMA) were used to scale up material parameters from bitumen to mastic and from bitumen to HMA. Volumetric parameters of the mixes were taken into consideration to develop two models, one connecting the viscoelastic parameters of bitumen and mastic (*B-M Model*) and one connecting the material behavior of bitumen and HMA (*B-A Model*). The results for the *B-A Model* are explained in the following sections. The *B-M-model* is presented in Annex B.

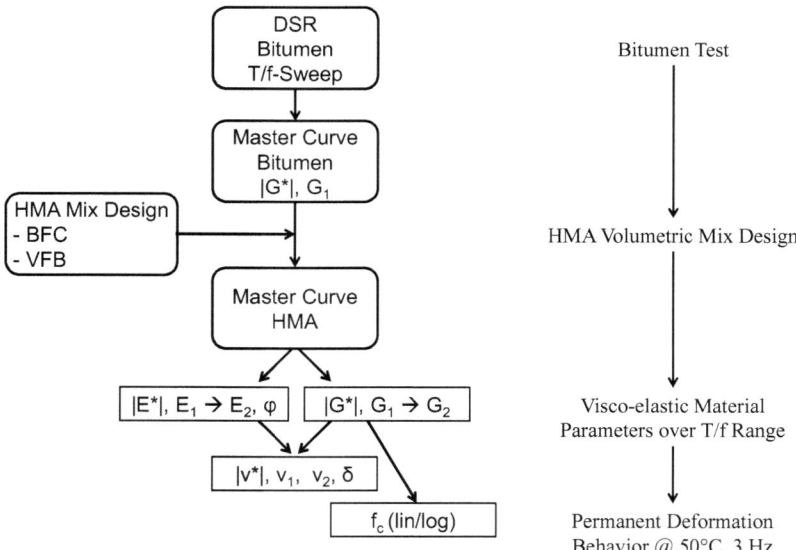

Figure VI-5. Principle steps of the *B-A Model*.

The principle steps and results of the *B-A Model* are presented in Figure VI-5. Starting point are DSR tests on bitumen with a temperature and frequency sweep to derive master curves for the viscoelastic material parameters of the binder. A second input parameter are volumetric characteristics of the HMA for which the viscoelastic behavior shall be predicted. In detail the content of binder and filler of the mix in mass percentage (*BFC*) and the voids filled with binder in volumetric percentage (*VFB*) are necessary to work with the model. A set of 9 model parameters link the master curves of the binder with the master curves of the HMA. By applying the *B-A Model*, the dynamic modulus $|E^*|$ and its elastic component E_1, as well as the dynamic shear modulus $|G^*|$ and its elastic component G_1 can be obtained directly for the mix. From these four viscoelastic material parameters, the viscous components and the phase lag φ can be derived. In addition the dynamic Poisson's Ratio and its elastic and viscous part can be obtained as well. Formulas to calculate these parameters are given in sections V.1.2 and V.1.3. Furthermore with the link between $/G^*/$ and the linear and logarithmic creep rate found in section V.7.2 the *B-A Model* is also capable of predicting the permanent deformation behavior of the mix. Thus with little testing effort on bitumen together with the volumetric mix design parameters of the HMA the complete viscoelastic behavior the mix can be described by the *B-A Model* over a large range of temperatures and frequencies, together with a prediction of the permanent deformation behavior at TCCT standard conditions (50°C, 3 Hz).

In the further course of the chapter it will be shown in detail how the *B-A Model* has been established. A three step approach was taken:
- First of all, the test data from DSR with frequency and temperature sweep is analyzed and master curves for the stiffness parameters are derived and described by analytical functions (test data → master curve) at a T_{ref} of 30°C.
- In a second step, the master curves of tests on bitumen (DSR) and HMA (CCTs) are compared and the ratio between bitumen stiffness and stiffness of each HMA type k ($S_{HMA,k}/S_{bit}$) is calculated analytically (master curve → stiffness ratio). This step can be seen as an auxiliary towards the model.
- Finally, the functions describing the stiffness ratio for each mix are investigated to derive the *B-A Model* by taking into consideration the mix design. The model works for the complete frequency range of the master curve and a wide range of mix designs (volumetric characteristics).

From the large number of CCTs carried out on different asphalt mixes (see section V.5), four mixes with different volumetric characteristics were identified for both binders, the unmodified 70/100 and the PmB 25/55-65. For these eight HMAs, CTTs had been carried out successfully at all three temperatures (10°C, 30°C and 50°C) and for the complete frequency sweep.

For the PmB 25/55-65, mixes with a varying binder content, different binder/filler ratios and different content of air voids were available. In detail, data were available for mixes with
- a binder content of 4.8% (m/m) and a filler ratio of 1.9 (= PmB_HMA_1),
- a binder content of 5.3% (m/m) and a filler ratio of 1.7 (= PmB_HMA_2),
- a binder content of 5.3% (m/m) and a filler ratio of 1.7 but a significantly higher content of air voids (= PmB_HMA_3) and
- a binder content of 5.8% (m/m) and a filler ratio of 1.6 (= PmB_HMA_4).

Two of the mixes show the same binder content but differ in the content of air voids, there are three different binder contents and three different filler ratios.

For the unmodified (= umB) 70/100 bitumen, the four mixes differed in binder content, filler ratios and content of air voids. HMAs with
- a binder content of 5.0% (m/m) and a filler ratio of 1.7 (umB_HMA_1),
- a binder content of 5.3% (m/m), a filler ratio of 1.3 and a significantly higher content of air voids (umB_HMA_2),
- a binder content of 5.3% (m/m) and a filler ratio of 1.4 (umB_HMA_3) and
- a binder content of 5.6% (m/m) and a filler ratio of 1.7 (umB_HMA_4)

were employed in this study. Thus, there are two comparable mixes with different content of air voids, three different binder contents and two different filler ratios available.

Table VI-4 to Table VI-7 give detailed information on the volumetric characteristics of the specimens for the PmB 25/55-65 mixes and Table VI-8 to Table VI-11 for the 70/100 mixes. Besides binder content, maximum density, bulk density, content of air voids, *BFC* and binder:filler ratio, the voids in the mineral aggregate (*VMA*) and the voids filled with binder (*VFB*) are given for each specimen, as well as the mean value and the standard deviation.

Table VI-4. Volumetric characteristics of PmB_HMA_1 specimens (AC 11 PmB 25/55-65 with a binder content of 4.8% (m/m)).

Specimen	Test	Binder content [%(m/m)]	Max. density [kg/m³]	Bulk density [kg/m³]	Air voids [%(v/v)]	Binder+filler [%(m/m)]	binder:filler (1:x)	VMA [%(v/v)]	VFB [%]	VFB [%(v/v)]
T374C	UCCT @ 10°C	4.8	2564.0	2515.8	1.9	14.0	1.9	13.7	86.2	11.8
T374D	UCCT @ 30°C	4.8	2564.0	2504.3	2.3	14.0	1.9	14.1	83.7	11.8
T374B	TCCT @ 50°C	4.8	2564.0	2505.5	2.3	14.0	1.9	14.1	83.7	11.8
MV	---	4.8	2564.0	2508.5	2.2	14.0	1.9	14.0	84.5	11.8
SD	---	0.0	0.0	6.3	0.2	0.0	0.0	0.2	1.4	0.0
MV+SD	---	4.8	2564.0	2514.9	2.4	14.0	1.9	14.2	85.9	11.8
MV-SD	---	4.8	2564.0	2502.2	1.9	14.0	1.9	13.8	83.1	11.8

Table VI-5. Volumetric characteristics of PmB_HMA_2 specimens (AC 11 PmB 25/55-65 with a binder content of 5.3% (m/m)).

Specimen	Test	Binder content [%(m/m)]	Max. density [kg/m³]	Bulk density [kg/m³]	Air voids [%(v/v)]	Binder+filler [%(m/m)]	binder:filler (1:x)	VMA [%(v/v)]	VFB [%]	VFB [%(v/v)]
T371D	UCCT @ 10°C	5.3	2542.2	2501.4	1.6	14.5	1.7	14.6	89.0	13.0
T373D	UCCT @ 30°C	5.3	2542.2	2485.1	2.2	14.5	1.7	15.1	85.4	12.9
T371C	TCCT @ 50°C	5.3	2542.2	2508.0	1.3	14.5	1.7	14.3	90.9	13.0
MV	---	5.3	2542.2	2498.2	1.7	14.5	1.7	14.7	88.5	13.0
SD	---	0.0	0.0	11.8	0.5	0.0	0.0	0.4	2.8	0.1
MV+SD	---	5.3	2542.2	2510.0	2.2	14.5	1.7	15.1	91.3	13.0
MV-SD	---	5.3	2542.2	2486.4	1.2	14.5	1.7	14.3	85.7	12.9

Table VI-6. Volumetric characteristics of PmB_HMA_3 specimens (AC 11 PmB 25/55-65 with a binder content of 5.3% (m/m) and high content of air voids).

Specimen	Test	Binder content [%(m/m)]	Max. density [kg/m³]	Bulk density [kg/m³]	Air voids [%(v/v)]	Binder+filler [%(m/m)]	binder:filler (1:x)	VMA [%(v/v)]	VFB [%]	VFB [%(v/v)]
T390B	UCCT @ 10°C	5.3	2552.5	2363.0	7.4	14.3	1.7	19.7	62.4	12.3
T382B	UCCT @ 30°C	5.3	2552.5	2417.1	5.3	14.5	1.7	17.9	70.3	12.6
T390D	TCCT @ 50°C	5.3	2552.5	2357.2	7.7	14.3	1.7	19.9	61.4	12.2
MV	---	5.3	2552.5	2379.1	6.8	14.4	1.7	19.2	64.7	12.4
SD	---	0.0	0.0	33.1	1.3	0.1	0.0	1.1	4.9	0.2
MV+SD	---	5.3	2552.5	2412.2	8.1	14.5	1.7	20.3	69.6	12.5
MV-SD	---	5.3	2552.5	2346.0	5.5	14.3	1.7	18.0	59.8	12.2

Table VI-7. Volumetric characteristics of PmB_HMA_4 specimens
(AC 11 PmB 25/55-65 with a binder content of 5.8% (m/m)).

Specimen	Test	Binder content [%(m/m)]	Max. density [kg/m³]	Bulk density [kg/m³]	Air voids [%(v/v)]	Binder+filler [%(m/m)]	binder:filler (1:x)	VMA [%(v/v)]	VFB [%]	VFB [%(v/v)]
T378C	UCCT @ 10°C	5.8	2534.0	2473.9	2.4	15.0	1.6	16.5	85.4	14.1
T379D	UCCT @ 30°C	5.8	2534.0	2451.5	3.3	15.0	1.6	17.2	80.9	13.9
T379B	TCCT @ 50°C	5.8	2534.0	2463.7	2.8	15.0	1.6	16.8	83.3	14.0
MV	---	5.8	2534.0	2463.0	2.8	15.0	1.6	16.8	83.2	14.0
SD	---	0.0	0.0	11.2	0.5	0.0	0.0	0.4	2.3	0.1
MV+SD	---	5.8	2534.0	2474.2	3.3	15.0	1.6	17.2	85.5	14.1
MV-SD	---	5.8	2534.0	2451.8	2.4	15.0	1.6	16.5	80.9	13.9

Table VI-8. Volumetric characteristics of umB_HMA_1 specimens
(AC 11 70/100 with a binder content of 5.0% (m/m)).

Specimen	Test	Binder content [%(m/m)]	Max. density [kg/m³]	Bulk density [kg/m³]	Air voids [%(v/v)]	Binder+filler [%(m/m)]	binder:filler (1:x)	VMA [%(v/v)]	VFB [%]	VFB [%(v/v)]
T364C	UCCT @ 10°C	5.0	2570.7	2526.4	1.7	13.5	1.7	14.1	87.9	12.4
T364A	UCCT @ 30°C	5.0	2570.7	2484.5	3.4	13.5	1.7	15.6	78.2	12.2
T364D	TCCT @ 50°C	5.0	2570.7	2517.0	2.1	13.5	1.7	14.4	85.5	12.3
MV	---	5.0	2570.7	2509.3	2.4	13.5	1.7	14.7	83.9	12.3
SD	---	0.0	0.0	22.0	0.9	0.0	0.0	0.8	5.1	0.1
MV+SD	---	5.0	2570.7	2531.3	3.3	13.5	1.7	15.5	88.9	12.4
MV-SD	---	5.0	2570.7	2487.3	1.5	13.5	1.7	13.9	78.8	12.2

Table VI-9. Volumetric characteristics of umB_HMA_2 specimens
(AC 11 70/100 with a binder content of 5.3% (m/m) and a high content of air voids).

Specimen	Test	Binder content [%(m/m)]	Max. density [kg/m³]	Bulk density [kg/m³]	Air voids [%(v/v)]	Binder+filler [%(m/m)]	binder:filler (1:x)	VMA [%(v/v)]	VFB [%]	VFB [%(v/v)]
T421D	UCCT @ 10°C	5.3	2572.9	2411.0	6.3	12.2	1.3	18.8	66.5	12.5
T422C	UCCT @ 30°C	5.3	2572.9	2458.2	4.5	12.2	1.3	17.3	73.9	12.8
T422A	TCCT @ 50°C	5.3	2572.9	2424.2	5.8	12.2	1.3	18.4	68.5	12.6
MV	---	5.3	2572.9	2431.1	5.5	12.2	1.3	18.2	69.7	12.6
SD	---	0.0	0.0	24.4	0.9	0.0	0.0	0.8	3.8	0.1
MV+SD	---	5.3	2572.9	2455.5	6.5	12.2	1.3	19.0	73.5	12.8
MV-SD	---	5.3	2572.9	2406.7	4.6	12.2	1.3	17.4	65.8	12.5

Table VI-10. Volumetric characteristics of umB_HMA_3 specimens
(AC 11 70/100 with a binder content of 5.3% (m/m)).

Specimen	Test	Binder content [%(m/m)]	Max. density [kg/m³]	Bulk density [kg/m³]	Air voids [%(v/v)]	Binder+filler [%(m/m)]	binder:filler (1:x)	VMA [%(v/v)]	VFB [%]	VFB [%(v/v)]
T333B	UCCT @ 10°C	5.3	2576.1	2516.8	2.3	12.9	1.4	15.4	85.0	13.1
T336B	UCCT @ 30°C	5.3	2576.1	2525.8	2.5	12.9	1.4	15.6	84.0	13.1
T334C	TCCT @ 50°C	5.3	2576.1	2521.3	2.1	12.9	1.4	15.2	86.2	13.1
MV	---	5.3	2576.1	2521.3	2.3	12.9	1.4	15.4	85.1	13.1
SD	---	0.0	0.0	4.5	0.2	0.0	0.0	0.2	1.1	0.0
MV+SD	---	5.3	2576.1	2525.8	2.5	12.9	1.4	15.6	86.2	13.1
MV-SD	---	5.3	2576.1	2516.8	2.1	12.9	1.4	15.2	84.0	13.1

Table VI-11. Volumetric characteristics of umB_HMA_4 specimens
(AC 11 70/100 with a binder content of 5.6% (m/m)).

Specimen	Test	Binder content [%(m/m)]	Max. density [kg/m³]	Bulk density [kg/m³]	Air voids [%(v/v)]	Binder+filler [%(m/m)]	binder:filler (1:x)	VMA [%(v/v)]	VFB [%]	VFB [%(v/v)]
T368C	UCCT @ 10°C	5.6	2547.5	2503.8	1.7	14.9	1.7	15.4	89.0	13.7
T368B	UCCT @ 30°C	5.6	2547.5	2504.2	1.7	14.9	1.7	15.4	89.0	13.7
T365D	TCCT @ 50°C	5.6	2570.7	2518.6	2.0	14.1	1.5	15.8	87.4	13.8
MV	---	5.6	2555.2	2508.9	1.8	14.6	1.6	15.6	88.5	13.8
SD	---	0.0	13.4	8.4	0.2	0.4	0.1	0.2	0.9	0.0
MV+SD	---	5.6	2568.6	2517.3	2.0	15.1	1.7	15.8	89.4	13.8
MV-SD	---	5.6	2541.8	2500.4	1.6	14.2	1.5	15.4	87.5	13.7

Test Data → Master Curve

For bitumen, master curves were derived from DSR tests for the dynamic shear modulus $|G^*|$ and its storage part G_1. The master curves were approximated by means of regression analysis according to equation (6.7) and the quality of the fit was described statistically. One example of this procedure can be seen in Figure VI-6 for the polymer-modified binder PmB 25/55-65. From the results shown in Figure VI-6 it can be stated that the one boundary in terms of test conditions (-10°C and 30 Hz) is near the upper asymptote of the material stiffness, whereas the other boundary (60°C and 0.1 Hz) leaves the material still in the transition phase, some distance apart from the lower asymptote.

For the HMAs, the test data from CCTs at the three different temperatures were merged for each of the analyzed mixes. The dynamic modulus $|E^*|$ and its elastic and viscous part, as well as the dynamic shear modulus $|G^*|$, G_1 and G_2 were derived from the test data. The mathematical background for the derivation of the stiffness data can be found in sections V.1.2 and V.1.3. Master curves were obtained by means of regression analysis and the fit quality was again given in statistical values. This procedure was carried out for the dynamic stiffness parameters and their elastic parts. As already stated above, two of the three stiffness values are sufficient to calculate the third parameter from the other two. Figure VI-7 shows one example for the umB_HMA_1. The two diagrams show the test data and the respective master curves for $|G^*|$ and G_1.

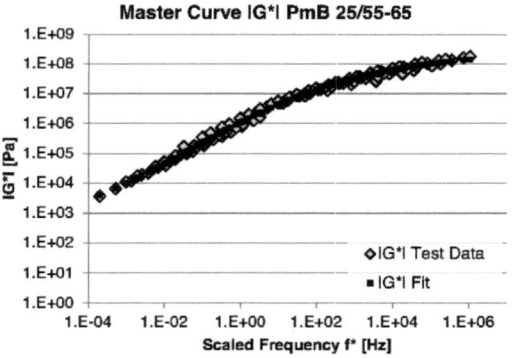

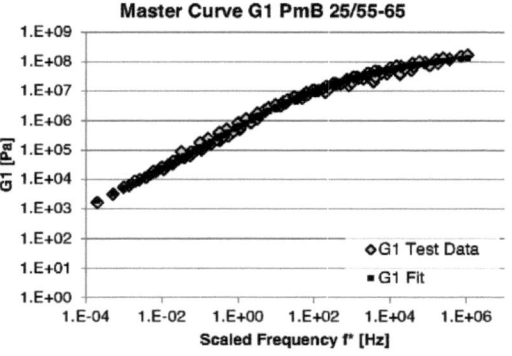

	PmB 25/55-65	
Fit Parameters	|G*|	G1
Tref,1 [°C]	30	30
Ea/R [K]	20919	20939
a [-]	1.48055E+01	1.55330E+01
b [-]	-1.88838E-01	-1.89293E-01
x0,1 [-]	6.39879E-02	9.02106E-02
y0 [-]	4.57706E+00	3.84273E+00
exp(y0) - lower asymptote	9.72279E+01	4.66525E+01
exp(y0+a) - upper asymptote	2.61668E+08	2.59878E+08
Statistical Parameters		
R^2	0.995	0.996
SMRE	0.967	0.152
Min. RE	-0.416	-0.415
Max. RE	1014.282	1.064
5% quantile of RE	-0.263	-0.233
95% quantile of RE	0.440	0.371
90% probability range of RE	-0.263 to 0.44	-0.233 to 0.371
Probability of the RE between -0.1 and 0.1	0.341	0.355

Figure VI-6. Master curves from test data and regression analysis for PmB 25/55-65 (|G*| - top and G_1 – middle) and table with regression parameters and statistical values describing the fit quality (bottom).

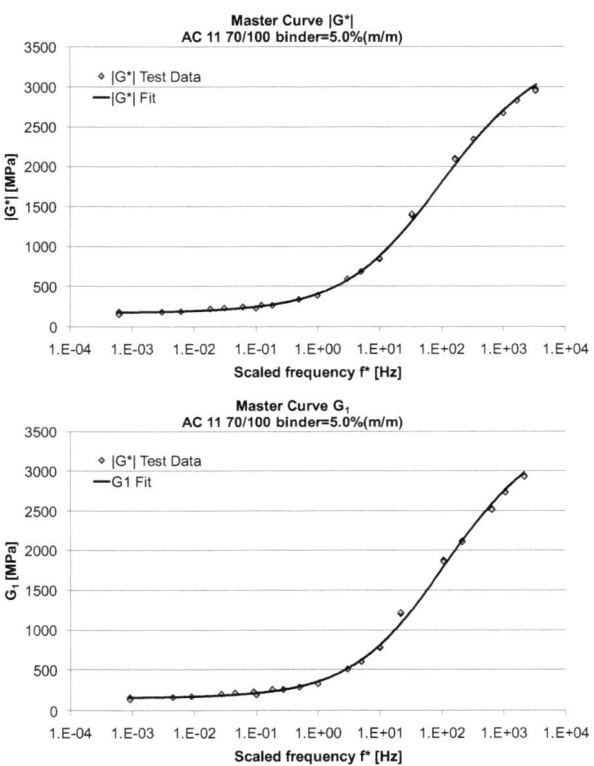

Figure VI-7. Master curves from test data and regression analysis for AC 11 70/100 with a binder content of 5.0% (m/m) (umB_HMA_1): |G*| (top) and G_1 (bottom).

Figure VI-8 shows the analytical master curves for the four PmB HMAs. The upper two diagrams represent the dynamic shear modulus and its elastic part, the lower two diagrams the dynamic modulus and its elastic part. Since all four mixes were composed of the same grading with the same aggregate type, the difference between the master curves must be related to certain design characteristics of the mix linked to the binder and filler content and the content of air voids. The dynamic shear modulus of the mixes start at a value of around $2 \cdot 10^8$ to $4 \cdot 10^8$ Pa at low frequencies (corresponding to 50°C and 0.1 Hz) and increase to around $3 \cdot 10^9$ Pa at high frequencies (equal to 10°C and 20 Hz). The dynamic modulus $|E^*|$ is at a much higher level, starting at around $6 \cdot 10^8$ to 10^9 Pa and getting as high as $6 \cdot 10^9$ Pa. The difference between the different mixes is more significant at low and intermediate frequencies and tends to decrease at higher frequencies. This indicates that impacts of the design characteristics in terms of binder/filler and void content have to be taken care especially at high and intermediate temperatures.

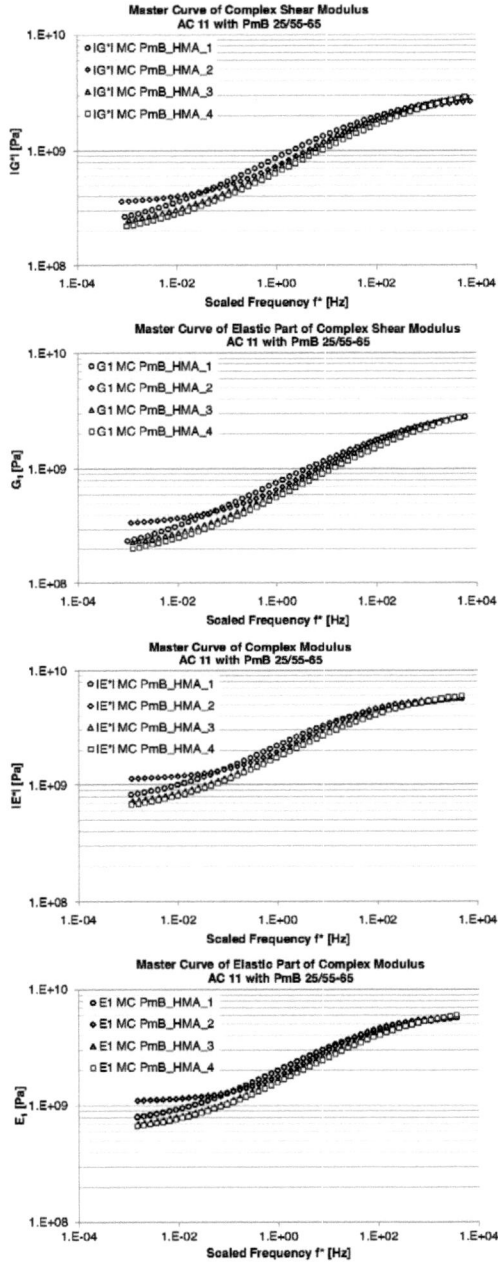

Figure VI-8. Master curves for all AC 11 PmB 25/55-65 mixes for |G*| (top), G_1 (2nd from above), |E*| (3rd from above) and E_1 (bottom).

Table VI-12 and Table VI-13 provide information on the regression parameters of the master curves for the four mixes and the four stiffness parameters. They also show the statistical values describing the fit quality. Different to the tables for the bitumen and mastic master curves, the last row of the table for HMA gives the probability of the RE lying between -0.05 and 0.05. This is due to the fact that the deviations from the test data are in most cases clearly smaller than 0.1 in terms of the relative error.

Table VI-12. Regression (according to equation (6.7)) and statistical parameters of AC 11 PmB 25/55-65 master curves for $|G^*|$ and G_1 in Pa.

Fit Parameters	PmB_HMA_1		PmB_HMA_2	
	$|G^*|$	$G_{1,max}$	$|G^*|$	$G_{1,max}$
Tref,1 [°C]	30	30	30	30
Ea/R [K]	23064	22598	23717	21821
a [-]	2.96189E+00	3.44497E+00	2.13872E+00	2.29716E+00
b [-]	-2.84781E-01	-2.35231E-01	-4.17057E-01	-4.01102E-01
x0,1 [-]	6.28758E-01	7.85848E-01	4.02227E+00	6.81953E+00
y0 [-]	1.90042E+01	1.86852E+01	1.96448E+01	1.95816E+01
exp(y0) - lower asymptote	1.79225E+08	1.30275E+08	3.40117E+08	3.19283E+08
exp(y0+a) - upper asymptote	3.46523E+09	4.08314E+09	2.88710E+09	3.17555E+09
Statistical Parameters				
n	96	96	88	88
R^2	0.996	0.998	0.991	0.993
SMRE	0.037	0.024	0.044	0.039
Min. RE	0.077	0.064	0.205	0.175
Max. RE	-0.084	-0.083	-0.072	-0.084
5% quantile of RE	-0.070	-0.055	-0.057	-0.060
95% quantile of RE	0.067	0.057	0.105	0.099
90% probability range of RE	-0.07 to 0.067	-0.055 to 0.057	-0.057 to 0.105	-0.06 to 0.099
Probability of the RE between -0.05 and 0.05	0.653	0.845	0.730	0.740

Fit Parameters	PmB_HMA_3		PmB_HMA_4	
	$|G^*|$	$G_{1,max}$	$|G^*|$	$G_{1,max}$
Tref,1 [°C]	30	30	30	30
Ea/R [K]	22135	21435	22642	21444
a [-]	2.93401E+00	3.04859E+00	3.34546E+00	3.58810E+00
b [-]	-3.10046E-01	-3.02614E-01	-2.61802E-01	-2.47562E-01
x0,1 [-]	2.23041E+00	3.41338E+00	2.92089E+00	4.43735E+00
y0 [-]	1.90855E+01	1.89975E+01	1.88454E+01	1.86989E+01
exp(y0) - lower asymptote	1.94422E+08	1.78034E+08	1.52916E+08	1.32072E+08
exp(y0+a) - upper asymptote	3.65572E+09	3.75395E+09	4.33877E+09	4.77643E+09
Statistical Parameters				
n	96	96	96	96
R^2	0.996	0.998	0.998	0.998
SMRE	0.039	0.028	0.029	0.031
Min. RE	0.092	0.074	0.062	0.055
Max. RE	-0.100	-0.105	-0.116	-0.110
5% quantile of RE	-0.080	-0.038	-0.047	-0.063
95% quantile of RE	0.087	0.060	0.049	0.047
90% probability range of RE	-0.08 to 0.087	-0.038 to 0.06	-0.047 to 0.049	-0.063 to 0.047
Probability of the RE between -0.05 and 0.05	0.688	0.879	0.916	0.901

Table VI-13. Regression (according to equation (6.7)) and statistical parameters of AC 11 PmB 25/55-65 master curves for |E*| and E_1.

	PmB_HMA_1		PmB_HMA_2	
Fit Parameters	|E*|	E1,max	|E*|	E1,max
Tref,1 [°C]	30	30	30	30
Ea/R [K]	21865	20577	21757	20530
a [-]	2.18772E+00	2.24669E+00	1.66345E+00	1.71811E+00
b [-]	-3.72524E-01	-3.69633E-01	-5.49726E-01	-5.57006E-01
x0,1 [-]	6.92917E-01	1.08990E+00	3.34878E+00	4.91392E+00
y0 [-]	2.03495E+01	2.03195E+01	2.08237E+01	2.08037E+01
exp(y0) - lower asymptote	6.88160E+08	6.67792E+08	1.10569E+09	1.08372E+09
exp(y0+a) - upper asymptote	6.13489E+09	6.31492E+09	5.83526E+09	6.04067E+09
Statistical Parameters				
n	88	88	88	88
R^2	0.997	0.998	0.987	0.990
SMRE	0.027	0.022	0.049	0.044
Min. RE	0.059	0.050	0.213	0.180
Max. RE	-0.063	-0.055	-0.068	-0.084
5% quantile of RE	-0.060	-0.048	-0.057	-0.059
95% quantile of RE	0.046	0.046	0.114	0.094
90% probability range of RE	-0.06 to 0.046	-0.048 to 0.046	-0.057 to 0.114	-0.059 to 0.094
Probability of the RE between -0.05 and 0.05	0.826	0.955	0.668	0.746

	PmB_HMA_3		PmB_HMA_4	
Fit Parameters	|E*|	E1,max	|E*|	E1,max
Tref,1 [°C]	30	30	30	30
Ea/R [K]	21527	20369	21752	20579
a [-]	2.33123E+00	2.41066E+00	2.55960E+00	2.59945E+00
b [-]	-3.77327E-01	-3.75776E-01	-3.31603E-01	-3.34611E-01
x0,1 [-]	1.43575E+00	2.26614E+00	1.87974E+00	3.28837E+00
y0 [-]	2.02515E+01	2.02177E+01	2.01305E+01	2.01404E+01
exp(y0) - lower asymptote	6.23901E+08	6.03186E+08	5.52785E+08	5.58302E+08
exp(y0+a) - upper asymptote	6.42028E+09	6.72030E+09	7.14784E+09	7.51269E+09
Statistical Parameters				
n	88	88	88	88
R^2	0.992	0.996	0.997	0.997
SMRE	0.046	0.035	0.026	0.026
Min. RE	0.099	0.085	0.067	0.040
Max. RE	-0.098	-0.074	-0.107	-0.109
5% quantile of RE	-0.085	-0.057	-0.051	-0.046
95% quantile of RE	0.093	0.066	0.051	0.037
90% probability range of RE	-0.085 to 0.093	-0.057 to 0.066	-0.051 to 0.051	-0.046 to 0.037
Probability of the RE between -0.05 and 0.05	0.591	0.732	0.849	1.000

The four diagrams in Figure VI-9 show the master curves for the HMAs with the unmodified bitumen 70/100. Compared to the stiffness of the polymer modified mixes, the unmodified HMAs react less stiff for all frequencies, which is clear since the PmB is produced from a harder bitumen (25/55 pen) than the 70/100. One of the unmodified mixes reacts significantly different from the other three HMAs. The umB_HMA_2 is the mix with a much higher air void content with a mean value of 5.5% (v/v) instead of around 2% (v/v). The mix reacts around two times stiffer at low frequencies (or high temperatures) and exhibits less stiffness at high frequencies. Obviously the higher void content allows for more aggregate interaction at high temperatures whereas the other mixes are more influenced by the soft binder (mastic) at these conditions. At the other extreme at high frequencies (or low temperatures) the mix with the higher air voids content has relatively less additional stiffening due to the mastic. The stiffening process in frequency domain is less developed for the mix with the higher void content since the impact of the frequency independent stiffness of the aggregate is more dominant for this mix.

Table VI-14 and Table VI-15 give the regression and fit quality parameters for the master curves of the umB_HMAs.

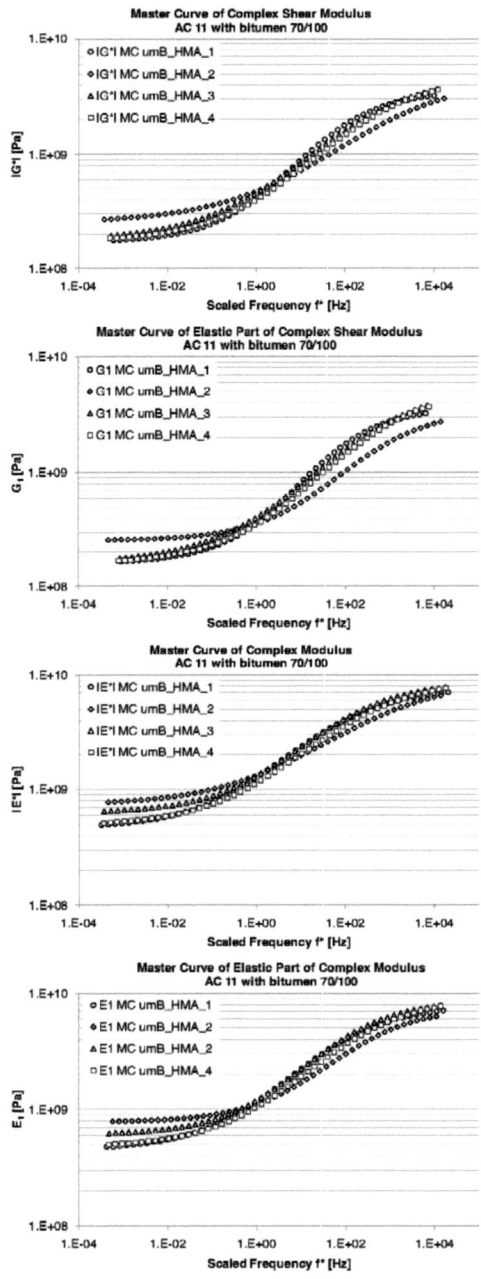

Figure VI-9. Master curves for all AC 11 70/100 mixes for |G*| (top), G_1 (2nd from above), |E*| (3rd from above) and E_1 (bottom).

Table VI-14. Regression (according to equation (6.7)) and statistical parameters of AC 11 70/100 master curves for $|G^*|$ and G_1 in Pa.

	umB_HMA_1		umB_HMA_2	
Fit Parameters	$|G^*|$	$G_{1,max}$	$|G^*|$	$G_{1,max}$
$T_{ref,1}$ [°C]	30	30	30	30
E_a/R [K]	24918	23018	27209	26320
a [-]	3.04349E+00	3.13218E+00	2.92555E+00	2.61608E+00
b [-]	-4.64599E-01	-4.82844E-01	-3.12016E-01	-4.34258E-01
$x_{0,1}$ [-]	7.01595E+00	8.45299E+00	6.48650E+01	7.32584E+01
y_0 [-]	1.89528E+01	1.88847E+01	1.93470E+01	1.93537E+01
exp(y_0) - lower asymptote	1.70261E+08	1.59043E+08	2.52532E+08	2.54219E+08
exp(y_0+a) - upper asymptote	3.57178E+09	3.64587E+09	4.70834E+09	3.47823E+09
Statistical Parameters				
n	80	80	94	94
R^2	0.997	0.995	0.985	0.972
SMRE	0.040	0.051	0.075	0.090
Min. RE	0.176	0.197	0.148	0.417
Max. RE	-0.075	-0.118	-0.212	-0.294
5% quantile of RE	-0.074	-0.093	-0.155	-0.244
95% quantile of RE	0.076	0.107	0.122	0.219
90% probability range of RE	-0.074 to 0.076	-0.093 to 0.107	-0.155 to 0.122	-0.244 to 0.219
Probability of the RE between -0.05 and 0.05	0.655	0.609	0.389	0.359

	umB_HMA_3		umB_HMA_4	
Fit Parameters	$|G^*|$	$G_{1,max}$	$|G^*|$	$G_{1,max}$
$T_{ref,1}$ [°C]	30	30	30	30
E_a/R [K]	25566	23387	25901	23688
a [-]	3.34319E+00	3.59553E+00	3.33144E+00	3.54637E+00
b [-]	-3.43427E-01	-3.45456E-01	-3.67240E-01	-3.72841E-01
$x_{0,1}$ [-]	1.48897E+01	1.84163E+01	1.90178E+01	2.41904E+01
y_0 [-]	1.89896E+01	1.88669E+01	1.89600E+01	1.88594E+01
exp(y_0) - lower asymptote	1.76635E+08	1.56232E+08	1.71489E+08	1.55070E+08
exp(y_0+a) - upper asymptote	5.00042E+09	5.69233E+09	4.79799E+09	5.37893E+09
Statistical Parameters				
n	88	88	87	87
R^2	0.990	0.987	0.987	0.982
SMRE	0.071	0.083	0.077	0.098
Min. RE	0.293	0.285	0.338	0.368
Max. RE	-0.124	-0.146	-0.145	-0.165
5% quantile of RE	-0.117	-0.139	-0.138	-0.164
95% quantile of RE	0.188	0.198	0.215	0.241
90% probability range of RE	-0.117 to 0.188	-0.139 to 0.198	-0.138 to 0.215	-0.164 to 0.241
Probability of the RE between -0.05 and 0.05	0.412	0.383	0.374	0.307

Table VI-15. Regression (according to equation (6.7)) and statistical parameters of AC 11 70/100 master curves for $|E^*|$ and E_1 in Pa.

	umB_HMA_1		umB_HMA_2	
Fit Parameters	$|E^*|$	E1,max	$|E^*|$	E1,max
Tref,1 [°C]	30	30	30	30
Ea/R [K]	28070	26824	26426	25301
a [-]	2.80784E+00	2.87072E+00	2.50934E+00	2.28212E+00
b [-]	-3.95166E-01	-4.05634E-01	-3.30374E-01	-4.44575E-01
x0,1 [-]	4.61521E+00	5.82519E+00	3.77347E+01	4.36327E+01
y0 [-]	1.99556E+01	1.99271E+01	2.04076E+01	2.04691E+01
exp(y0) - lower asymptote	4.64116E+08	4.51039E+08	7.29312E+08	7.75602E+08
exp(y0+a) - upper asymptote	7.69231E+09	7.96074E+09	8.96820E+09	7.59890E+09
Statistical Parameters				
n	84	84	91	91
R^2	0.999	0.999	0.989	0.977
SMRE	0.020	0.018	0.054	0.071
Min. RE	0.091	0.104	0.231	0.480
Max. RE	-0.095	-0.089	-0.166	-0.210
5% quantile of RE	-0.031	-0.026	-0.110	-0.157
95% quantile of RE	0.076	0.049	0.082	0.145
90% probability range of RE	-0.031 to 0.076	-0.026 to 0.049	-0.11 to 0.082	-0.157 to 0.145
Probability of the RE between -0.05 and 0.05	0.884	0.944	0.540	0.646

	umB_HMA_3		umB_HMA_4	
Fit Parameters	$|E^*|$	E1,max	$|E^*|$	E1,max
Tref,1 [°C]	30	30	30	30
Ea/R [K]	27435	26226	27514	26257
a [-]	2.63085E+00	2.66942E+00	2.94195E+00	2.99740E+00
b [-]	-4.16230E-01	-4.39787E-01	-3.67140E-01	-3.82840E-01
x0,1 [-]	1.02217E+01	1.15315E+01	1.18799E+01	1.43194E+01
y0 [-]	2.02530E+01	2.02252E+01	1.99944E+01	1.99736E+01
exp(y0) - lower asymptote	6.24816E+08	6.07685E+08	4.82457E+08	4.72545E+08
exp(y0+a) - upper asymptote	8.67596E+09	8.76984E+09	9.14386E+09	9.46670E+09
Statistical Parameters				
n	88	88	88	88
R^2	0.993	0.992	0.993	0.991
SMRE	0.053	0.055	0.053	0.057
Min. RE	0.251	0.279	0.274	0.320
Max. RE	-0.096	-0.107	-0.116	-0.125
5% quantile of RE	-0.079	-0.092	-0.104	-0.119
95% quantile of RE	0.152	0.138	0.103	0.082
90% probability range of RE	-0.079 to 0.152	-0.092 to 0.138	-0.104 to 0.103	-0.119 to 0.082
Probability of the RE between -0.05 and 0.05	0.622	0.572	0.596	0.628

Master Curve → Stiffness Ratio

Before following the line along the three step approach, a little excursus will be made that analyzes the stiffness behavior of bitumen, mastic and HMA for one example. Figure VI-10 shows the ratio of stiffness of HMA to binder and to mastic, as well as of mastic to binder. These data are shown in two diagrams for the PmB 25/55-65, the PmB_HMA_1 and the mastic with corresponding filler ratio. Since the test temperature for the bitumen and mastic samples ranged from -10°C to +60°C and for the HMA from +10°C to +50°C, the analytical regression of the master curve of HMA was extrapolated to compare the complete range from -10°C to +60°C. The domain that is proofed by actual test data is presented with symbols colored in grey. It is interesting to take a closer look at the difference between stiffness of mastic and HMA because it allows statements about the contribution of the mastics to the total stiffness of the mix and to which extent the aggregates increase the stiffness depending on the frequency of loading or the temperature. Starting at high temperatures (50°C to 60°C) or low frequencies (quasi-static loading), the stiffness of the HMA is up to 10^4 times higher than the modulus of the pure mastic, whereas the increase in the shear modulus from binder to mastic is only by a factor of 2 to 5 without being strongly influenced by the temperature/frequency. Due to the strong increase in stiffness of the mastic with increasing frequency, its share in the overall shear modulus of the mix also increases quickly. At intermediate temperatures (30°C) or frequencies, the HMA is only 10^2 to $8 \cdot 10^2$ times stiffer than the mastic, and this factor drops down to 50 at around 10°C and to 5 at around -10°C. The load bearing capacity of the mastic increases dramatically with falling temperature. Thus, it is obvious that the aggregate skeleton is of major importance at high temperatures when the stiffness of the mastic is insignificant compared to the overall stiffness. The lower the temperature or the higher the frequency the more dominant becomes the impact of the mastic. When the upper asymptote of both master curves is taken into consideration, both the mastic and the HMA end up producing similar shear moduli, at least for this mix. The upper asymptote of the dynamic shear modulus e^{a+y0} of the mastic is about 3815 MPa compared to 3465 MPa for the HMA, whereas the binder only reaches 262 MPa. From this, it can be stated that the upper stiffness threshold of the mastic is directly related to the threshold of the HMA.

Since the model to be developed in this chapter links the behavior of bitumen to HMA, the next step towards the model is to investigate the stiffness ratio between those two materials $S_{HMA,k}/S_{bit}$ and describe this relationship analytically. In detail the stiffness ratios are $|G^*|_{HMA,k}/|G^*|_{bit}$, $|E^*|_{HMA,k}/|G^*|_{bit}$, $G_{1,HMA,k}/G_{1,bit}$ and $E_{1,HMA,k}/G_{1,bit}$. Figure VI-11 shows the ratios for the PmB_HMAs. Analogue to the stiffness parameters of the HMA, the ratio of different mixes is quite similar in the high frequency domain. $|G^*|$ and G_1 of the HMA are around 40 times higher than the parameters of the bitumen, at low frequencies this value rises to $2 \cdot 10^4$ to $6 \cdot 10^4$. As the dynamic modulus and its elastic part of the HMA are higher than the shear stiffness parameters, the stiffness ratio runs from around 10^2 at low frequencies to $6 \cdot 10^4$ to $2 \cdot 10^5$ at high frequencies.

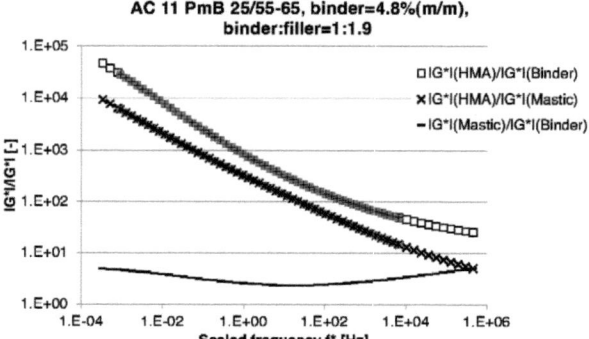

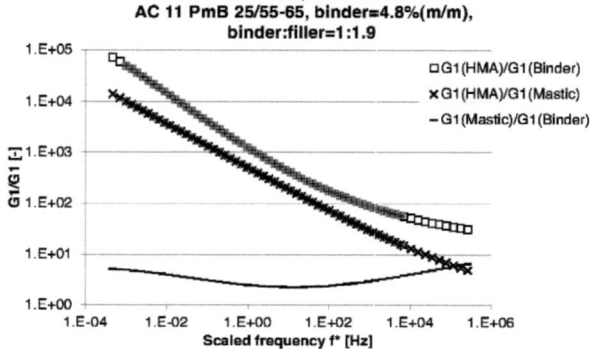

Figure VI-10. Ratios of $|G^*|$ (top) and G_1 (bottom) of binder, mastic and HMA for AC 11 PmB 25/55-65 with a binder content of 4.8% (m/m) (PmB_HMA_1).

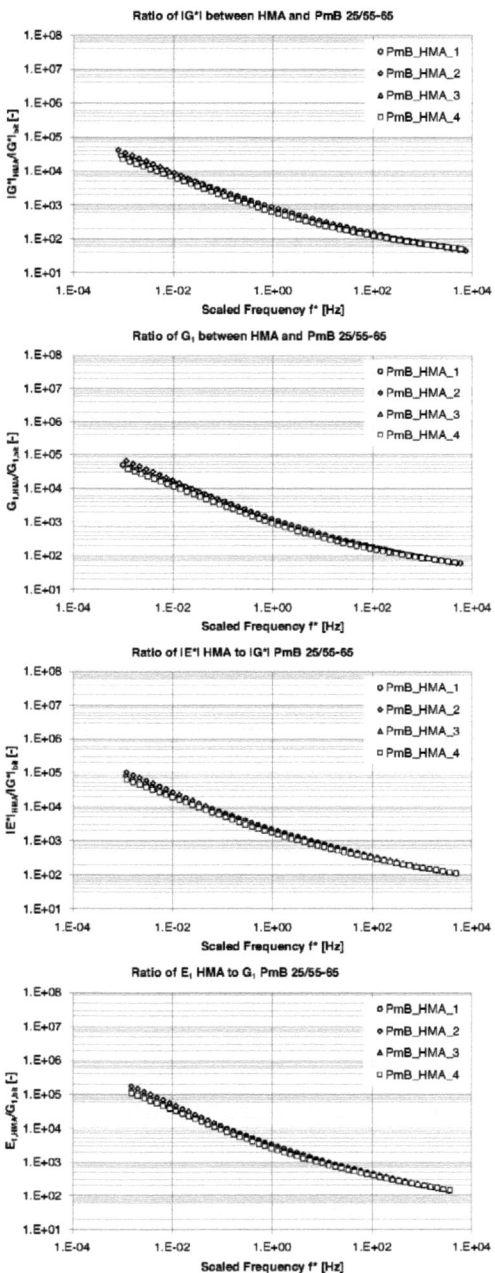

Figure VI-11. Stiffness ratio between HMA and binder PmB 25/55-65 for $|G^*|$ (top), G_1 (2nd from above), $|E^*|$ (3rd from above) and E_1 (bottom).

The derived stiffness ratios between binder and HMA can be described analytically by polynomial of 2nd order expressed in an exponential relationship. The mathematical expression is given in (6.12). For each HMA type k and each stiffness parameter, three parameters are derived. For the PmB HMAs, Table VI-16 gives the parameters for the analytical relationship between bitumen and the mix. With this analytical expression, the stiffness of an HMA can be described over the complete frequency range from the stiffness of the bitumen. In addition, all other viscoelastic parameters, like the viscous part of the dynamic stiffness, the phase lags and – more important – the dynamic Poisson's ratio can be derived. But the fit has to be carried out for each HMA type k. Thus, the benefit is limited and as stated above, this step can be seen as an auxiliary step towards the *B-A Model*.

$$\frac{S_{HMA,k}}{S_{bit}}(f^*) = \exp\left[\sum_{i=0}^{2} a_{b-a_k,i} \cdot \ln(f^*)^i\right] \qquad (6.12)$$

$S_{HMA,k}$....Stiffness modulus ($|G^*|$, G_1, $|E^*|$, E_1) of HMA type k
S_{bit}.........Stiffness modulus ($|G^*|$, G_1) of bitumen used for HMA type k
f^*...........Scaled frequency [Hz]
$a_{b-ak,i}$.....Parameters of the functional relationship describing the stiffness ratio of HMA type k to bitumen, $i = 0..2$

Table VI-16. Parameters for function (6.12) describing the stiffness ratio between HMA and binder PmB 25/55-65 analytically for $|G^*|$ (top), G_1 (2nd from above), $|E^*|$ (3rd from above) and E_1 (bottom).

$|G^*|$	PmB_HMA_1	PmB_HMA_2	PmB_HMA_3	PmB_HMA_4
$a_{b-ak,0}$	6.72767E+00	6.61103E+00	6.53578E+00	6.44253E+00
$a_{b-ak,1}$	-4.31929E-01	-4.55705E-01	-4.26581E-01	-4.23650E-01
$a_{b-ak,2}$	1.19604E-02	1.64594E-02	1.49599E-02	1.55050E-02
R^2	1.000	1.000	1.000	1.000
G1	PmB_HMA_1	PmB_HMA_2	PmB_HMA_3	PmB_HMA_4
$a_{b-ak,0}$	7.10578E+00	7.01221E+00	6.91524E+00	6.82755E+00
$a_{b-ak,1}$	-4.67951E-01	-4.91763E-01	-4.63778E-01	-4.60015E-01
$a_{b-ak,2}$	1.26893E-02	1.83577E-02	1.58979E-02	1.62255E-02
R^2	0.999	1.000	1.000	1.000
$|E^*|$	PmB_HMA_1	PmB_HMA_2	PmB_HMA_3	PmB_HMA_4
$a_{b-ak,0}$	7.66451E+00	7.60412E+00	7.50237E+00	7.43950E+00
$a_{b-ak,1}$	-4.55051E-01	-4.75706E-01	-4.44961E-01	-4.42565E-01
$a_{b-ak,2}$	1.17021E-02	1.64930E-02	1.36468E-02	1.45560E-02
R^2	1.000	0.999	1.000	1.000
E1	PmB_HMA_1	PmB_HMA_2	PmB_HMA_3	PmB_HMA_4
$a_{b-ak,0}$	8.08734E+00	8.03717E+00	7.91867E+00	7.86011E+00
$a_{b-ak,1}$	-4.91260E-01	-5.13385E-01	-4.80968E-01	-4.81542E-01
$a_{b-ak,2}$	1.26522E-02	1.80896E-02	1.49225E-02	1.60092E-02
R^2	1.000	0.999	1.000	1.000

Following the fit of the stiffness ratio, it is important to check whether the analytical function is able to actually describe the stiffness data with sufficient quality. Taking into account the stiffness values from the bitumen master curve $S_{bit,MC}$ employing (6.12) in the following way

$$S_{HMA,k,SR}\left(f^*\right) = \exp\left[\sum_{i=0}^{2} a_{b-a_k,i} \cdot \ln\left(f^*\right)^i\right] \cdot S_{bit,MC} \qquad (6.13)$$

$S_{HMA,k,SR}$ *Stiffness modulus of HMA type k derived from functional relationship describing the stiffness ratio of HMA type k to bitumen*

the stiffness of the HMA type k can be derived. Further on, the so calculated stiffness $S_{HMA,k,SR}$ is compared to the stiffness of the HMA from the master curve $S_{HMA,k,MC}$ which was derived directly from test data.

Figure VI-12 shows the graphical representation of $S_{HMA,k,MC}/S_{HMA,k,SR}$ for the PmB_HMAs. A value of 1.0 would indicate that both stiffness values are the same. The bars in the diagram show the 95% confidence interval of the deviation of the stiffness from the analytical calculation in (6.13) to the stiffness from the master curve. The smaller this interval, the better fits the calculated stiffness the actual master curve stiffness. The deviations range from -7.5% to +9.0% in the worst case. It can therefore be concluded that the analytical representation by the stiffness ratio produces excellent fit quality.

The same procedure that was carried out for the PmB_HMAs was also run for the umB_HMAs, as shown in the following figures and tables. Compared to the PmB_HMAs, the ratio between stiffness parameters of umB_HMAs and the respective binder (Figure VI-13) is similar in the high frequency range. This shows that within the domain of high bitumen stiffness, the behavior of the mix is mainly influenced by the binder. On the other hand, at low frequencies, the stiffness ratio is much higher for the unmodified HMAs. The range for $|G^*|$ at low frequencies, for example, runs from around 10^6 to $2*10^6$. This is different to the situation for the modified HMAs where these values are at some 10^4. In this high temperature domain, where the binder is very soft and hardly contributes to the load bearing, the aggregates seem to dominate the overall stiffness of the mix.

It is also interesting to compare the stiffness ratio of the dynamic modulus and its elastic part. Both ratios were similar for the HMAs with modified binder at low frequencies, meaning that the viscoelastic behavior of binder and HMA are comparable. When unmodified binders are used, the stiffness ratio of, e.g., G_I is much higher (some 10^7) than the ratio of $|G^*|$ (some 10^6) indicating that the mix reacts much more elastic under these conditions (low frequencies corresponding to around 50°C and 0.1 Hz) than the binder. It must be stated that the unmodified binder exhibits a phase lag of nearly 90° under these conditions and can be seen as a viscous liquid, whereas the mix is still a stable solid structure. From this, it can be seen that the influence of aggregates is more dominant at low frequencies or high temperatures when the mix is produced with unmodified binder.

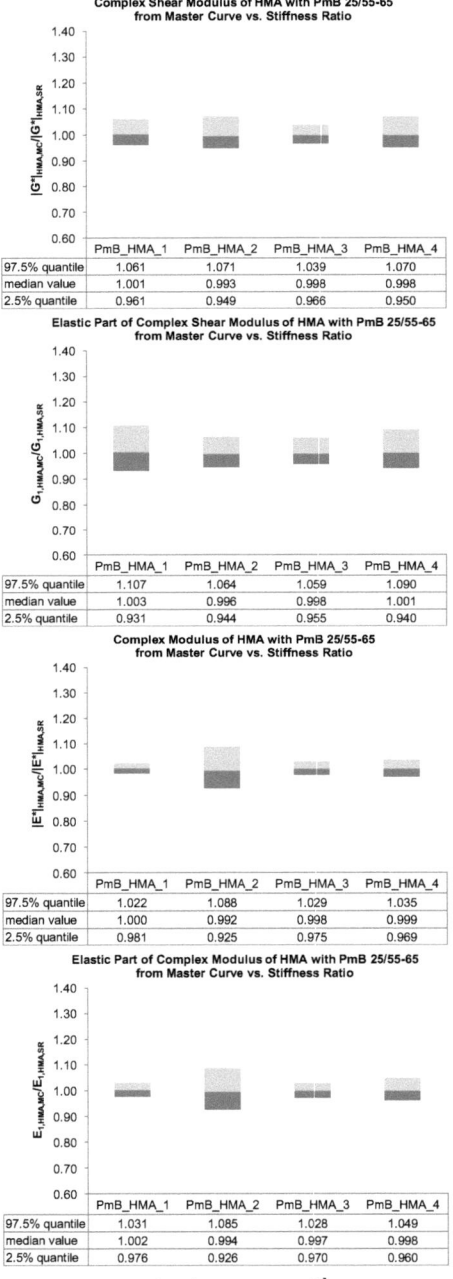

Figure VI-12. Comparison of $|G^*|$ (top), G_1 (2nd from above), $|E^*|$ (3rd from above) and E_1 (bottom) of master curve and stiffness ratio for AC 11 PmB 25/55-65 mixes.

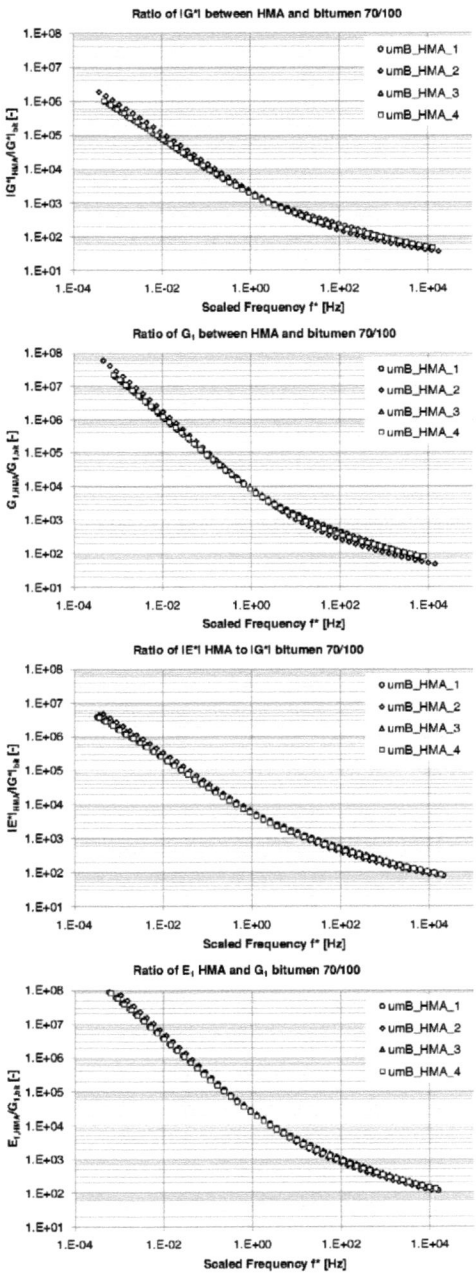

Figure VI-13. Stiffness ratio between HMA and binder 70/100 for $|G^*|$ (top), G_1 (2nd from above), $|E^*|$ (3rd from above) and E_1 (bottom).

The stiffness ratios of the unmodified HMAs were also successfully analytically by (6.12) and the parameters are listed in Table VI-17. The statement made above about larger difference between the dynamic modulus and its elastic part at low frequencies can also be observed when the parameter $a_{b-ak,0}$ is taken into consideration. From the nature of (6.12) parameter $a_{b-ak,0}$ represents the stiffness ratio at a frequency of 1 Hz when the logarithm of the scaled frequency f^* is 0. The difference of this parameter between the dynamic stiffness parameters and the elastic part for the modified HMAs is around 0.4 (see Table VI-16). In the case of umB_HMAs the difference is up to 1.5.

Table VI-17. Parameters for function (6.12) describing the stiffness ratio between HMA and binder 70/100 analytically for $|G^*|$ (top), G_1 (2nd from above), $|E^*|$ (3rd from above) and E_1 (bottom).

$\|G^*\|$	umB_HMA_1	umB_HMA_2	umB_HMA_3	umB_HMA_4
$a_{b-ak,0}$	7.68300E+00	7.73173E+00	7.72671E+00	7.60989E+00
$a_{b-ak,1}$	-6.32449E-01	-6.91449E-01	-6.46720E-01	-6.45291E-01
$a_{b-ak,2}$	2.53076E-02	2.66340E-02	2.60104E-02	2.72128E-02
R^2	0.999	0.999	0.999	0.999
G_1	umB_HMA_1	umB_HMA_2	umB_HMA_3	umB_HMA_4
$a_{b-ak,0}$	9.11895E+00	9.04935E+00	9.16567E+00	9.03958E+00
$a_{b-ak,1}$	-8.56006E-01	-9.23291E-01	-8.66644E-01	-8.67415E-01
$a_{b-ak,2}$	3.74055E-02	3.93826E-02	3.78861E-02	3.94724E-02
R^2	0.999	0.998	0.999	0.999
$\|E^*\|$	umB_HMA_1	umB_HMA_2	umB_HMA_3	umB_HMA_4
$a_{b-ak,0}$	8.75311E+00	8.74830E+00	8.79889E+00	8.64272E+00
$a_{b-ak,1}$	-6.51494E-01	-7.04582E-01	-6.68083E-01	-6.58708E-01
$a_{b-ak,2}$	2.15331E-02	2.57140E-02	2.42225E-02	2.43775E-02
R^2	1.000	0.999	0.999	0.999
E_1	umB_HMA_1	umB_HMA_2	umB_HMA_3	umB_HMA_4
$a_{b-ak,0}$	1.02391E+01	1.01719E+01	1.02904E+01	1.01318E+01
$a_{b-ak,1}$	-8.73501E-01	-9.35872E-01	-8.90824E-01	-8.81912E-01
$a_{b-ak,2}$	3.20525E-02	3.80905E-02	3.48299E-02	3.51396E-02
R^2	0.999	0.999	0.999	0.999

Also the quality of the fit of the stiffness ratio function was checked for the umB_HMAs. Analogous to Figure VI-12 for the modified HMAs, Figure VI-14 compares the stiffness parameters from the functional relation of the stiffness ratio to the stiffness of the HMA master curves. The 95% confidence interval of the deviations between stiffness from analytical representation of the stiffness ratio (equation (6.13)) is clearly larger in the case of umB_HMAs than of PmB_HMAs. For the mixes 1, 3 and 4, the deviation ranges from -17.8% to +27.7% in the worst case. Mix number 2 exhibits deviations from -23.0% to +43.1%. The reasons for could due to the fact that all CCTs with HMAs produced with 70/100 were carried out at a time when the old control unit of the test machine (see section II.1.1 for details) was still used and the problem with the multiplexing ADC was overcome by an external data logger. The quality of the recorded test data was lower when the tests were carried out with the old control unit.

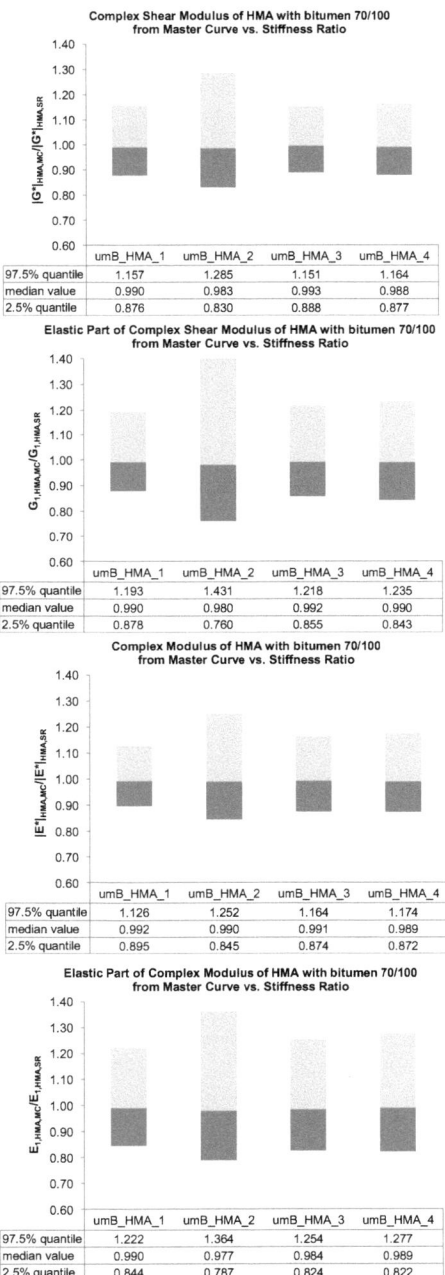

Figure VI-14. Comparison of $|G^*|$ (top), G_1 (2nd from above), $|E^*|$ (3rd from above) and E_1 (bottom) of master curve and stiffness ratio for AC 11 70/100 mixes.

Stiffness Ratio → B-A Model

The next step is to generalize the functional relationship of the stiffness ratio and achieve more benefit from the analysis carried out for this chapter. Therefore, the parameters $a_{b-ak,i}$ of the different HMA types k were investigated for correlations between volumetric characteristics of the different mixes and the parameters. In a further step it was analyzed which of the parameters depend on the binders used for the mix. By doing this, a more general *B-A Model* could be realized. It was found that

- parameter $a_{b-a,0}$ and $a_{b-a,1}$ are connected to the content of binder and filler in the mix in percentage of the mass (*BFC*) and
- parameter $a_{b-a,2}$ is linked to the volume of voids in the mineral aggregate skeleton (*VMA*) filled with binder in volumetric percentage (*VFB*). Usually, *VFB* is given in percentage of the *VMA*. In this case, *VFB* was characterized a little differently. If a *VMA* of 15 % (v/v) is given, then an air void content of 3 % (v/v) would leave a *VFB* of 12 % (v/v). This number would be used in this relationship.

All three relationships between $a_{b-a,j}$ and the volumetric mix design parameters can be described by means of a polynomial of 2nd order. Equation (6.14) presents the links between the parameters $a_{b-a,j}$ and the volumetric characteristics of the mix.

$$\hat{a}_{b-a,j}(BFC) = \sum_{i=0}^{2} b_{b-a,i,j} \cdot BFC^i \text{ for } j = 0,1$$
$$\hat{a}_{b-a,j}(VFB) = \sum_{i=0}^{2} b_{b-a,i,j} \cdot VFB^i \text{ for } j = 2$$
(6.14)

The $b_{b-a,i,j}$ are the 9 parameters of the *B-A Model*. When going further into the investigation of the model parameters, it was found that only one of three parameters $b_{b-a,i,j}$ depends on the binder used for the mix. Figure VI-15 shows three diagrams with the three parameter $a_{b-a,0}$ and $a_{b-a,1}$ vs. *BFC* and $a_{b-a,2}$ vs. *VFB* for mixes with both binders, the umB_HMAs and the PmB_HMAs. The diamonds represent the PmB_HMAs, the squares stand for the umB_HMAs. It becomes clear that polynomial relationships between the model parameters and the volumetric mix parameters are similar for mixes with both binders; the curvature (2nd order term) and the slope (1st order term) are the same, just the constant part of the function is different. The curve of a mix with one binder type can be derived from the mix with the other binder type by moving it along the y-axis. This is true for all three parameters $a_{b-a,i}$ and all viscoelastic parameters $|E^*|$, E_1, $|G^*|$ and G_1.

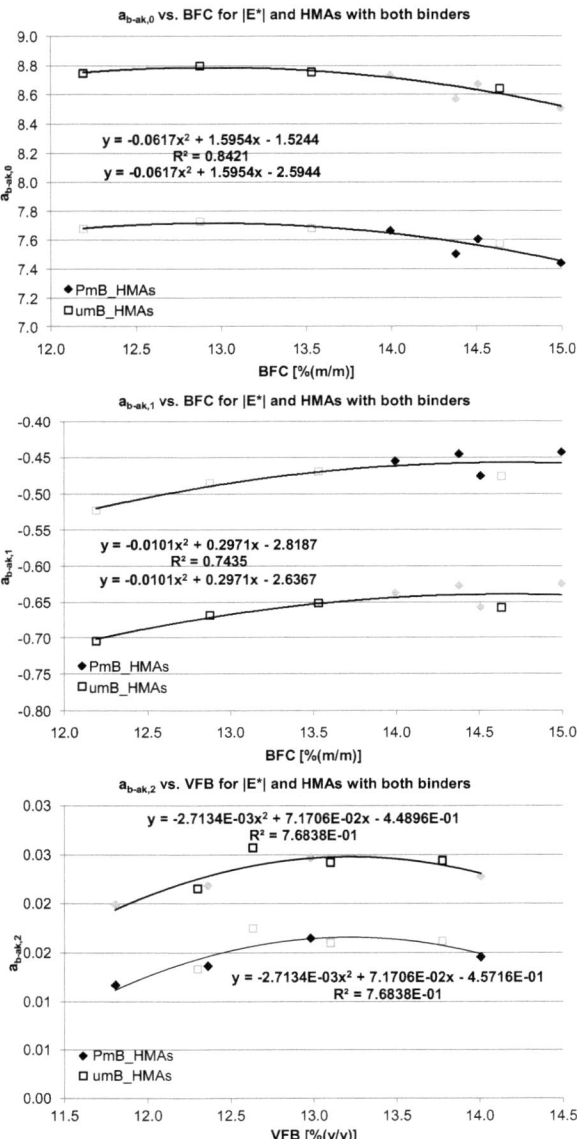

Figure VI-15. **Graphical representation of the parameter fit for the *B-A* Model** ($a_{b\text{-}a,0}$ for $|E^*|$ top, $a_{b\text{-}a,1}$ for $|E^*|$ middle and $a_{b\text{-}a,2}$ for $|E^*|$ bottom) showing that only one of three parameters depends on the binder used for the mix.

All mixes were composed of the same mineral type and similar gradation curves. The binder type, binder and filler content and the content of air voids were varied for the mixes. Therefore the *B-A Model* is not calibrated for different mineral types or gradation curves and only for a limited number of binder types.

By inserting (6.14) into (6.12) and rearranging it, the following formula (6.15) can be derived. It describes the *B-A Model* which enables the user to calculate the stiffness parameters directly for the complete frequency range and thus for any arbitrary temperature if the stiffness parameter of the bitumen (e.g. from DSR testing with temperature and frequency sweep) and the volumetric characteristics of the mix (*BFC, VFB*) are available. Further, all other viscoelastic material parameters can be calculated from the stiffness values derived above, i.e. the viscous part of the dynamic stiffness, the phase lags and even the dynamic Poisson's Ratio. It can therefore be stated that the all relevant viscoelastic patrameters of the mix can be described by the *B-A Model*. In addition, it must be limited to the type of aggregate and the gradation curve used at this point in research. Future test programs can easily verify and expand the model to further materials and gradations.

$$S_{HMA}(BFC, VFB, f^*) = \exp\left[\sum_{j=0}^{1}\left(\sum_{i=0}^{2} b_{b-a,i,j} \cdot BFC^i\right) \cdot \ln(f^*)^j + \sum_{i=0}^{2} b_{b-a,i,2} \cdot VFB^i \cdot \ln(f^*)^2\right] \cdot S_{bit,MC}$$

$$S_{HMA}(BFC, VFB, f^*) = \exp\left[\sum_{j=0}^{2}\underbrace{(b_{b-a,0,j})}_{f(binder)} \cdot \ln(f^*)^j + \sum_{j=0}^{1}\left(\sum_{i=1}^{2} b_{b-a,i,j} \cdot BFC^i\right) \cdot \ln(f^*)^j + \sum_{i=1}^{2} b_{b-a,i,2} \cdot VFB^i \cdot \ln(f^*)^2\right] \cdot S_{bit,MC}$$

(6.15)

Table VI-18 summarizes all parameters for the model for both mixes. Of the nine model parameters only three are obviously a function of the binder type. The other six parameters are constant for both mixes with regard to binder characteristics and may be connected to the gradation curve and the type of mineral used. Only two different binder types were used within this research program. Therefore, no functional relationships between the parameters $b_{b-a,i,0}$ and a certain characteristic binder material parameter could be established at this time. Still, the table provides some interesting relationships between these parameters for both binder types and also for the different stiffness parameters: The rightmost column shows the difference between parameter $b_{b-a,i,0}$ of the umB and the PmB_HMAs. For a constant i, the difference between umB and PmB mixes are similar for $|E^*|$ and $|G^*|$, and E_l and G_l respectively. Also, the ratio between the dynamic stiffness and its elastic part is around 2 for both stiffness parameters. It seems that this relative difference between mixes with different binder types is linked to the difference in stiffness between the binders. Again, as only two binder types were available for the analysis in this case, no conclusive connection could be found or rather any arbitrary parameter describing the stiffness ratio between both binders could be linked to the difference in the *B-A Model* parameters. A further test program with at least two more binders will be necessary to investigate this relationship and thus make the model completely independent of the chosen binder type. The quality of the fit of the parameters shown here is the combined R^2 for both mixes. The data points (see Figure VI-15 for example) of one mix type were kept constant and the points for the other mix type were moved parallel to the y-axis until the R^2 of the fit reached a maximum. This coefficient of determination is given in Table VI-18. Although the quality of the fit of some parameters does not seem to be high (R^2 below 0.8 in some cases), the following investigation on the quality of fit of the model will show that the model still describes the stiffness parameters with good quality.

Table VI-18. Parameters for function (6.15) describing the *B-A Model* for tested AC 11 mixes.

$b_{a-b,i,j}$	$|G^*|$					
i/j	0, umB	0, PmB	1	2	R^2,comb	$\Delta b_{a-b,i,0}$(umB,PmB)
0	-1.81051E+00	-2.78051E+00	1.49717E+00	-5.86703E-02	0.816	-0.9700
1	-3.33139E+00	-3.14639E+00	3.75948E-01	-1.30446E-02	0.721	0.1850
2	-3.12634E-01	-3.23134E-01	5.10097E-02	-1.91479E-03	0.897	-0.0105

$b_{a-b,i,j}$	G1					
i/j	0, umB	0, PmB	1	2	R^2,comb	$\Delta b_{a-b,i,0}$(umB,PmB)
0	-6.67153E+00	-8.71153E+00	2.38778E+00	-9.00470E-02	0.782	-2.0400
1	-4.05480E+00	-3.68380E+00	4.47603E-01	-1.56009E-02	0.750	0.3710
2	-4.23339E-01	-4.44539E-01	6.99659E-02	-2.64466E-03	0.828	-0.0212

$b_{a-b,i,j}$	$|E^*|$					
i/j	0, umB	0, PmB	1	2	R^2,comb	$\Delta b_{a-b,i,0}$(umB,PmB)
0	-1.52440E+00	-2.59440E+00	1.59543E+00	-6.17177E-02	0.842	-1.0700
1	-2.81869E+00	-2.63669E+00	2.97149E-01	-1.01287E-02	0.743	0.1820
2	-4.48955E-01	-4.57155E-01	7.17061E-02	-2.71337E-03	0.768	-0.0082

$b_{a-b,i,j}$	E1					
i/j	0, umB	0, PmB	1	2	R^2,comb	$\Delta b_{a-b,i,0}$(umB,PmB)
0	-3.51611E+00	-5.64611E+00	2.09285E+00	-7.94711E-02	0.781	-2.1300
1	-3.66781E+00	-3.29881E+00	3.88058E-01	-1.34224E-02	0.772	0.3690
2	-5.60335E-01	-5.77535E-01	9.04905E-02	-3.43378E-03	0.651	-0.0172

Prediction Quality of the B-A Model

Using the *B-A Model*, the stiffness parameters of the mixes were computed from the bitumen master curve. This predicted stiffness from the model $S_{HMA,mod}$ was compared to the stiffness of the respective HMA master curve $S_{HMA,MC}$ (derived from test data). Deviations between both values were calculated for the complete frequency range. Figure VI-16 shows the results in four diagrams for the PmB_HMAs. The deviations have not changed strongly from the deviations between stiffness ratio and master curve. For some mixes a shift of the median value can be observed, e.g. for PmB_HMA_3. In the worst case, which is PmB_HMA_4 and G_1, the deviations range from -13.6% to +27.8%. This indicates a good representation of the reality by the *B-A Model*.

The same analysis as for the PmB_HMAs above was also performed for the unmodified HMAs summarized in Figure VI-17. When the stiffness derived from the functional relationship of the stiffness ratio was compared to the master curve for the umB_HMAs (Figure VI-14), the deviations were at a higher level than for the PmB_HMAs even at this stage. The deviation gets hardly smaller by employing a more general *B-A Model*, but in the best case stays stable. By looking at the bars and numbers in the four diagrams, it can be seen that the deviations did not get much larger. In the worst case, the umB_HMA2 and G_1, the deviation runs from -26.7% to +32.4%. In this case the *B-A Model* describes the behavior of the mix even better than the functional relationship of the stiffness ratio. The quality of the fit of the *B-A Model* for umB_HMAs is good and still powerful for use for the prediction of viscoelastic parameters of HMAs.

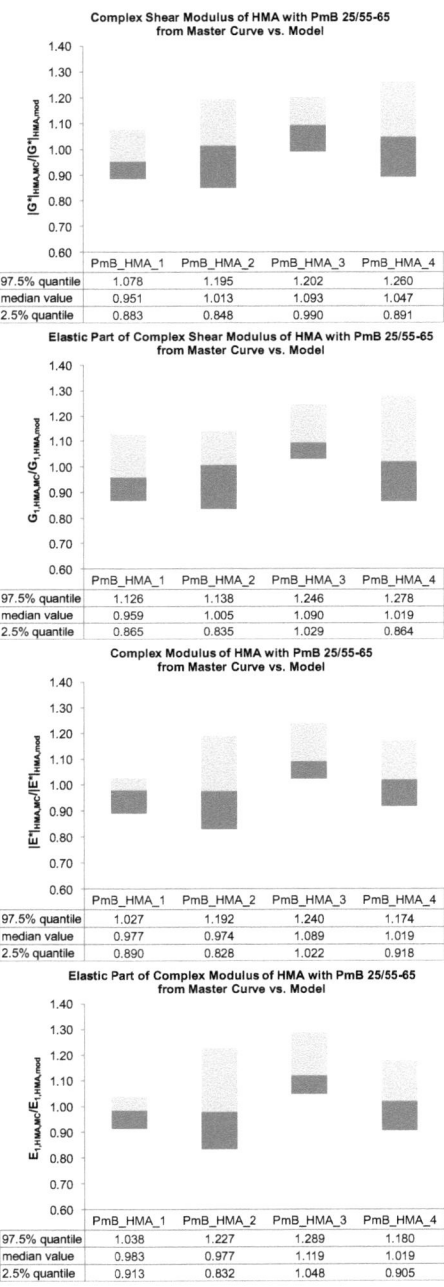

Figure VI-16. Comparison of $|G^*|$ (top), G_1 (2nd from above), $|E^*|$ (3rd from above) and E_1 (bottom) of master curve and *B-A Model* for AC 11 PmB 25/55-65 mixes.

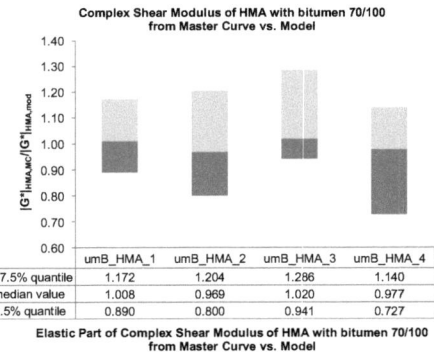
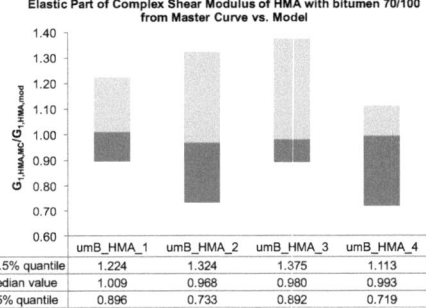
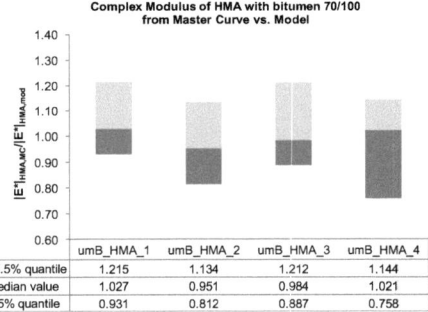
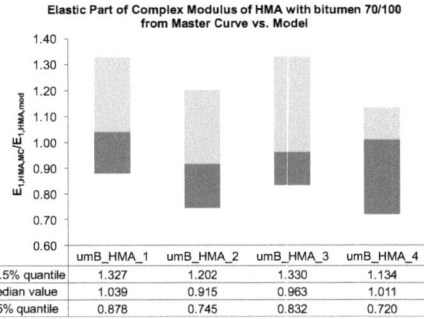

Figure VI-17. Comparison of |G*| (top), G_1 (2nd from above), |E*| (3rd from above) and E_1 (bottom) of master curve and *B-A Model* for AC 11 70/100 mixes.

Sensitivity Analysis for B-A Model Parameters

A sensitivity analysis was performed for the model parameters in (6.14) to gather information on the reaction of the model to deviated parameters. The actual model parameters ($\hat{a}_{b-a,i}$) were calculated exemplarily for one mix, the AC 11 70/100 with 5.0% (m/m) binder content. The dynamic shear modulus was derived for these model parameters ($|G^*|_{mod}$). Then the model parameters were varied between 75% and 125% of the actual model parameters and the dynamic shear modulus was again obtained for these deviated parameters ($|G^*|_{mod,dev}$). Then, the ratio of the deviated to the actual modulus was analyzed over the complete frequency range and the 95% confidence interval of this ratio was derived. Figure VI-18 shows the results graphically; the upper left diagram deals with $\hat{a}_{b-a,0}$. If this parameter deviates from the actual parameter, then no scattering of the results occurs but the results shift and an offset occurs. This effect is strong. A deviation of $\hat{a}_{b-a,0}$ of +5% leads to results that are only 69% of the actual values. Parameter $\hat{a}_{b-a,1}$ (upper right diagram) brings the opposite. The results with deviated model parameters show a large scatter from the original results. Again, the effect is strong. A deviation of the parameter of $\hat{a}_{b-a,1}$ of +5% leads to a scatter of results between 80% and 130% of the actual values. The tamest parameter is $\hat{a}_{b-a,2}$ with a clearly smaller effect of deviations on the results than the other two parameters.

Deviations from the model parameters have a strong effect on the derived material properties. This is caused by the exponential relationship between the material behavior of binder and HMA. Model parameters must therefore be derived with great care since the model is highly sensitive to small errors in its parameters.

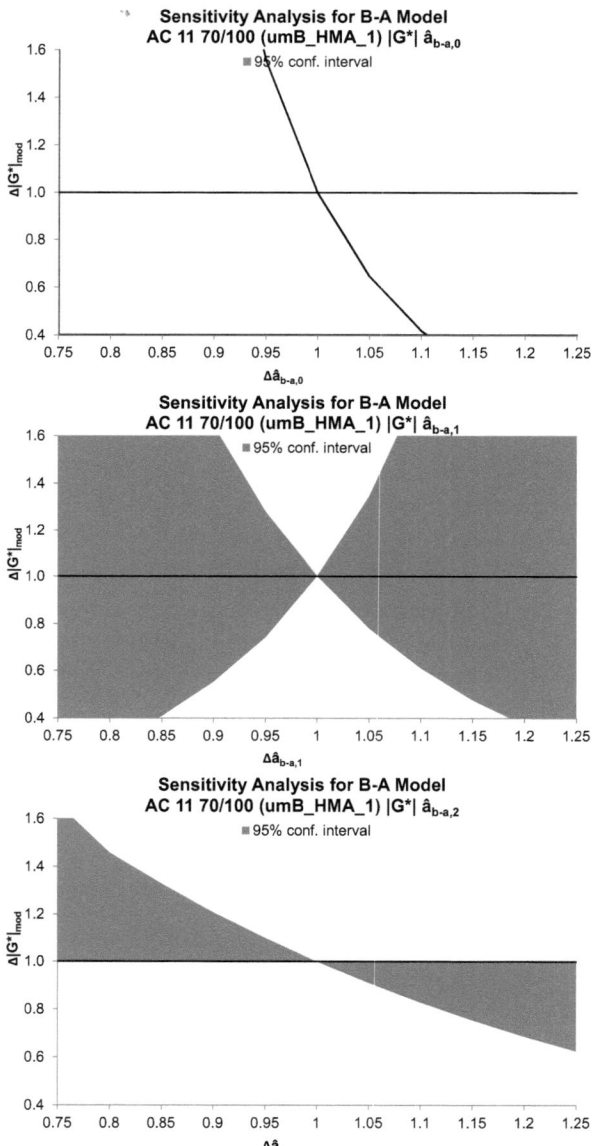

Figure VI-18. Sensitivity analysis for the *B-A Model* parameters for $|G^*|$ for the AC 11 70/100 mix with 5.0% (m/m) binder content (umB_HMA_1).

Practical Implementation

To demonstrate the practical implementation of the *B-A Model*, the viscoelastic material parameters of PmB_HMA_1 are predicted by the model in the following. The predicted values are compared to values derived from test data and from analytical expression of the master curve (by equation (6.7))

The following input data are necessary to work with the model:
- Reference temperature
- Binder type, since the model is dependent on the type of binder at the present stage
- Master curves of $|G^*|$ and G_1 of the binder at the chosen reference temperature expressed by equation (6.7)
- Volumetric characteristics of the mix (*BFC, VFB*)

For the example of PmB_HMA_1, the input data are shown in Table VI-19.

Table VI-19. Input data for prediction of the viscoelastic material parameters of PmB_HMA_1 by use of the *B-A Model*.

Reference temperature	30°C
Binder type	PmB 25/55-65
Master curves of binder	see Figure VI-6
BFC of HMA	14.0% (m/m)
VFB of HMA	11.8% (v/v)

The next steps for the prediction of the viscoelastic parameters are:
- The parameters $b_{b-a,i,j}$ for the *B-A Model* from Table VI-18 for the correct binder type are inserted into equation (6.15).
- The volumetric characteristics of the mix are inserted into equation (6.15).
- The master curve of the binder $S_{bit,MC}$ in equation (6.15) is substituted by the right side of equation (6.7) with the four parameters of the analytical representation of the master curve from the table in Figure VI-6. For prediction of the dynamic moduli $|G^*|$ and $|G^*|$ of HMA, the bitumen master curve of $|G^*|$ has to be used. For the prediction of the elastic part of the moduli G_1 and E_1, the bitumen mater curve of G_1 has to be used.
- The master curves of $|G^*|$, G_1, $|E^*|$ and E_1 of the HMA can be calculated.

For the example of PmB_HMA_1, viscoelastic parameters predicted by the model are compared to the viscoelastic parameters derived from the analytical expression of the HMA master curve and from the CCT test data. The viscoelastic parameters are presented for a scaled frequency of 0.1 Hz, 1.0 Hz and 10.0 Hz. Table VI-19 contains data for $|E^*|$ (left table) and E_1 (right table). The first column of each table shows the scaled frequency at the reference temperature of 30°C. Column 2 shows values predicted by the *B-A Model*, column 3 values derived from the analytical expression of the HMA master curve and column 4 values from CCT test data. The two rightmost columns contain deviations between *B-A Model* and master curve, and *B-A Model* and CCT test data respectively. The deviations are given in percentage. The largest difference between model and test data is -4.5% for $|E^*|$ at 10.0 Hz and +2.1% for E_1 at 1.0 Hz.

Table VI-21 presents data for $|G^*|$ (left table) and G_1 (right table). In the worst case $|G^*|$ predicted by the model is 11.6% smaller than $|G^*|$ derived from CCT at 0.1 Hz. In case of G_1 the maximum deviation is -8.8% at 0.1 Hz.

Table VI-20. $|E^*|$ **(top) and** E_1 **(bottom) derived from test data (CCT), from analytical expression of the HMA master curve (MC) and predicted from the** *B-A Model* **(mod).**

T_{ref}=30°C	|E*| [Pa]			Deviation [%]	
f* [Hz]	mod	MC	CCT	mod/MC	mod/CCT
0.1	1.3995E+09	1.4077E+09	1.4508E+09	-0.6	-3.5
1.0	2.1837E+09	2.2139E+09	2.1593E+09	-1.4	1.1
10.0	3.2636E+09	3.3981E+09	3.4183E+09	-4.0	-4.5
T_{ref}=30°C	E1 [Pa]			Deviation [%]	
f	mod	MC	CCT	mod/MC	mod/CCT
0.1	1.2807E+09	1.2886E+09	1.2811E+09	-0.6	0.0
1.0	2.0174E+09	2.0172E+09	1.9751E+09	0.0	2.1
10.0	3.1033E+09	3.1760E+09	3.1640E+09	-2.3	-1.9

Table VI-21. $|G^*|$ **(top) and** G_1 **(bottom) derived from test data (CCT), from analytical expression of the HMA master curve (MC) and predicted from the** *B-A Model* **(mod).**

T_{ref}=30°C	|G*| [Pa]			Deviation [%]	
f	mod	MC	CCT	mod/MC	mod/CCT
0.1	5.0974E+08	5.3943E+08	5.7650E+08	-5.5	-11.6
1.0	8.3232E+08	8.6896E+08	8.7206E+08	-4.2	-4.6
10.0	1.3154E+09	1.3727E+09	1.4283E+09	-4.2	-7.9
T_{ref}=30°C	G1 [Pa]			Deviation [%]	
f	mod	MC	CCT	mod/MC	mod/CCT
0.1	4.4570E+08	4.8419E+08	4.8861E+08	-7.9	-8.8
1.0	7.3532E+08	7.6582E+08	7.6815E+08	-4.0	-4.3
10.0	1.1933E+09	1.2030E+09	1.2442E+09	-0.8	-4.1

In a next step E_2 can be obtained from $|E^*|$ and E_1 by using equation (5.7). Analogous to that, G_2 can be derived from $|G^*|$ and G_1 from the last equation in formula (5.29). Table VI-22 contains data for E_2 (left table) and G_2 (right table). The prediction quality of the model is lower than for the dynamic moduli and their elastic parts. This fact is represented by the deviations between model and test data. In case of E_2 the maximum deviation is -21.9% at 10 Hz. For G_2, the value predicted by the model is 21.1% smaller than derived from test data.

Table VI-22. E_2 (top) and G_2 (bottom) derived from test data (CCT), from analytical expression of the HMA master curve (MC) and predicted from the *B-A Model* (mod).

T_{ref}=30°C	E2 [Pa]			Deviation [%]	
f	mod	MC	CCT	mod/MC	mod/CCT
0.1	5.6417E+08	5.6681E+08	6.8090E+08	-0.5	-17.1
1.0	8.3606E+08	9.1234E+08	8.7263E+08	-8.4	-4.2
10.0	1.0102E+09	1.2084E+09	1.2938E+09	-16.4	-21.9
T_{ref}=30°C	G2 [Pa]			Deviation [%]	
f	mod	MC	CCT	mod/MC	mod/CCT
0.1	2.4736E+08	2.3779E+08	3.0595E+08	4.0	-19.2
1.0	3.8994E+08	4.1062E+08	4.1284E+08	-5.0	-5.5
10.0	5.5347E+08	6.6118E+08	7.0140E+08	-16.3	-21.1

The phase lag between axial loading and axial deformation $\varphi_{ax,ax}$ is calculated from E_1 and E_2 by employing equation (5.13). Table VI-23 shows this phase lag. Since the phase lag is periodic with an angle of 360°, the basis for the calculation of deviations is 180°. A difference of 180° would represent 100% deviation. Thus, the maximum deviation between model and CCT test data is -2.3% at 0.1 Hz and 10.0 Hz.

Table VI-23. $\varphi_{ax,ax}$ derived from test data (CCT), from analytical expression of the HMA master curve (MC) and predicted from the *B-A Model* (mod).

T_{ref}=30°C	φax,ax [°]			Deviation [%]	
f	mod	MC	CCT	mod/MC	mod/CCT
0.1	23.8	23.7	28.0	0.0	-2.3
1.0	22.5	24.3	23.8	-1.0	-0.7
10.0	18.0	20.8	22.2	-1.6	-2.3

Next, the dynamic Poisson's Ratio $|v^*|$ can derived from the last equation in formula (5.29) by using $|E^*|$ and $|G^*|$. The values for $|v^*|$ are given in Table VI-24. In the worst case, the predicted value by the model is 20.9% larger than the value derived from test data.

Table VI-24. $|v^*|$ derived from test data (CCT), from analytical expression of the HMA master curve (MC) and predicted from the *B-A Model* (mod).

T_{ref}=30°C	|v*| [-]			Deviation [%]	
f	mod	MC	CCT	mod/MC	mod/CCT
0.1	0.373	0.305	0.258	22.3	18.0
1.0	0.312	0.274	0.238	13.9	15.1
10.0	0.240	0.238	0.197	1.2	20.9

Since $|v^*|$ and $\varphi_{ax,ax}$ have already been calculated in the steps above, the deformation phase lag $\delta_{ax,rad}$ can be obtained by using the second equation in formula (5.29) with G_1 and $|E^*|$. Table VI-25 contains data for the deformation phase lag. The deviations between model and CCT are below 1.0% for all frequencies, being largest for 0.1 Hz with -0.9%.

Table VI-25. $\delta_{ax,rad}$ derived from test data (CCT), from analytical expression of the HMA master curve (MC) and predicted from the *B-A Model* (mod).

T_{ref}=30°C	δax,rad [°]			Deviation [%]	
f	mod	MC	CCT	mod/MC	mod/CCT
0.1	5.3	2.4	4.1	1.6	-0.9
1.0	5.4	3.9	4.4	0.9	-0.3
10.0	6.9	8.0	7.2	-0.6	0.4

In a last step, v_1 and v_2 are calculated from equation (5.28). Table VI-26 presents the values for v_1 (left table) and v_2 (right table). For v_1 the maximum difference between model and CCT is 20.7% at 10.0 Hz. Due to the small values of v_2, the deviations between model and CCT are larger than for all other viscoelastic parameters: -29.9% at 0.1 Hz and 34.4% at 10.0 Hz.

Table VI-26. v_1 (top) and v_2 (bottom) derived from test data (CCT), from analytical expression of the HMA master curve (MC) and predicted from the *B-A Model* (mod).

T_{ref}=30°C	v1 [-]			Deviation [%]	
f	mod	MC	CCT	mod/MC	mod/CCT
0.1	0.371	0.305	0.258	21.9	18.2
1.0	0.310	0.273	0.237	13.6	15.1
10.0	0.239	0.235	0.195	1.4	20.7
T_{ref}=30°C	v2 [-]			Deviation [%]	
f	mod	MC	CCT	mod/MC	mod/CCT
0.1	0.034	0.013	0.018	166.0	-29.9
1.0	0.029	0.018	0.018	59.8	0.6
10.0	0.029	0.033	0.025	-12.9	34.1

Conclusions and Outlook

This chapter presents a three step approach towards an analytical model that links the material behavior of HMA with the respective bitumen (*B-A Model*). The model establishes a relationship between volumetric characteristics of the mix, the combined binder and filler content in % (m/m) (BFC) and the volume of voids of the aggregate skeleton filled with bitumen in % (v/v) (VFB). It consists of nine parameters that can easily be obtained from the master curves by fitting process. Three of the parameters depend on the binder type of the mix. Since the CCTs were carried out with sensors obtaining the reaction of the HMA in axial and radial direction, the *B-A Model* describes all relevant macroscopic, viscoelastic material parameters of the mix over the complete range of frequencies and temperatures in the compressive domain. Although the CCTs were carried out in the non-linear viscoelastic domain according to literature (see section V.1) in the higher temperature and low frequency range, the model works for the complete range. This is related to the fact that if the sensor data from the maximum of the oscillations is used for data evaluation, the effect of non-linear viscoelasticity was found to be of minor importance. The model parameters may not have a direct physical relation but the *B-A Model* has the vast advantage to manage the prediction of the HMA behavior in axial and radial direction over the complete range of frequencies/temperatures.

Although it seems logical that the model would work for the dynamic modulus and its elastic part as well as for the dynamic shear modulus and its elastic part, it must be kept in mind, that the data for these parameters were derived from three independent measuring systems (load cell, LVDTs (axial), strain gauges (circumferential – radial)). The success of the model also supports the test method and the employed sensors to deliver correct readings.

Furthermore, a conclusive link between the viscoelastic behavior of HMAs, in this case of the dynamic shear modulus $/G^*/$, and the deformation behavior (linear and logarithmic creep rate) has already been established earlier in section V.7.2. A power function links the dynamic shear rheometer to the linear and logarithmic creep rate. In connection with the *B-A Model*, it is a powerful tool to predict rutting resistance of mixes. Thus, the model cannot only predict the viscoelastic behavior of HMA from binder parameters as well as volumetric characteristics of the mix. With the link presented in section V.7.2, the viscoelastic parameters can also be used to obtain the permanent deformation behavior.

At the present stage, the *B-A Model* is verified for one gradation type (AC 11) with two different binders and one aggregate. In terms of the volumetric characteristics of the HMAs, the model is reliable for *BFC* between 12.0% (m/m) and 15.0% (m/m) and for *VFB* between 11.5% (v/v) and 14.5% (v/v).

Since the data basis for the model is presently limited, further research and testing must validate the model to a further extent. For the future, a research program should investigate the following matters:

- HMAs with the same gradation curve and mineral type should be tested with two more unmodified and modified binders to find functional relationships between the three dependent model parameters and bitumen characteristics.
- Different gradation curves (e.g. an SMA) and different mineral types should be included in the research to identify the impact of those parameters.
- For the two mixes presented in this study, further specimens should be tested with another variation of the volumetric parameters.
- Correlations between the 9 model parameters should be analyzed when test data from HMAs with more variation of the mix design are available to study further interrelations between mix design parameters and the model parameters and thus make the model even more universal.

VII Introducing an Enhanced TCCT with Cyclic Dynamic Confining Pressure

VII.1 Approach

As already explained in the introductory chapter to this thesis, a cyclic axial loading by traffic leads to cyclic radial confining pressure within the pavement structure. This has been shown in (Kappl, 2004) by finite element simulation of a pavement under a passing wheel computed in the elastic domain. Thus, the present state of the art in triaxial cyclic compression testing according to (EN 12697-25, 2005) with constant confining pressure does not simulate the state of stress within a structure in a realistic way. The question is how the permanent deformation behavior of HMA changes when the constant confining pressure is substituted by a cyclic confining pressure taking into account the viscoelastic characteristics of a mix. This test setup with cyclic confining pressure would represent the state of stress in a structure in a far more realistic way. Of course, the state of stress in a pavement made of time and temperature dependent materials varies from point to point. A laboratory test can never simulate the full complex stress situation in a loaded pavement. Thus, the impact of cyclic confining pressure is compared in the following on the basis of the standard test procedure. The magnitude of the cyclic confining pressure as well as mix design parameters of the HMA are varied to gather comprehensive information.

To incorporate the viscoelastic characteristics, it is important to know the phase lag between axial loading and radial reaction. The principal approach is depicted in Figure VII-1. Standard TCCTs are carried out for a certain material, the test data is evaluated and the radial phase lag $\varphi_{ax,rad}$ is analyzed. In the following enhanced TCCTs with cyclic confining pressure are run on the same material, incorporating the radial phase lag derived from the first tests. Finally, results of both test procedures, the standard and enhanced TCCT, are compared and discussed in terms of resistance to permanent deformation.

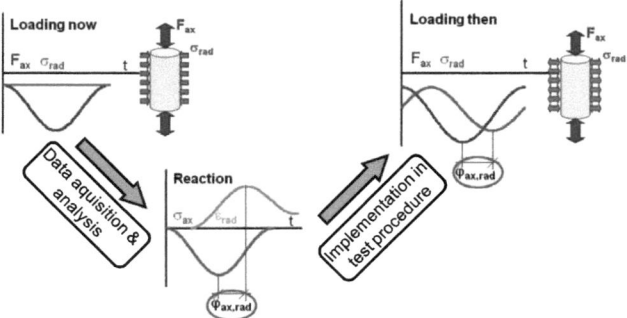

Figure VII-1. Approach to achieve an enhanced TCCT with cyclic dynamic confining pressure.

VII.2 Test Program

The same AC 11 mixes that were tested under standard TCCT conditions (see section IV.2.1) were employed in this investigation. From the AC 11 70/100, specimens with three different binder contents, 4.8% (m/m), 5.3% (m/m) and 5.8% (m/m) were produced at a target void content of 3.0% (v/v). For the AC 11 PmB 25/55-65 specimens with one binder content, 5.3% (m/m) and a target void content of 3.0% (v/v) were prepared.

To introduce an enhanced TCCT, it is necessary to determine the radial phase lag between axial loading and radial deformation and to define an amplitude for the cyclic confining pressure. It was decided to carry out enhanced TCCTs according to the conditions of the standard TCCT in (EN 12697-25, 2005) with three different radial amplitudes. Table VII-1 shows the layout of the test program. The table presents the lower and upper value of the axial stress ($\sigma_{ax,l}$ and $\sigma_{ax,u}$) and the radial stress ($\sigma_{rad,l}$ and $\sigma_{rad,u}$). The lowest radial stress amplitude is 50 kPa, the other two are set to be 75 kPa and 100 kPa

Table VII-1. Test program for the advanced characterization of the resistance to permanent deformation.

	Test conditions	$\sigma_{ax,l}$ [kPa]	$\sigma_{ax,u}$ [kPa]	$\sigma_{rad,l}$ [kPa]	$\sigma_{rad,u}$ [kPa]
Standard	50°C, 3 Hz, 25,000 load cycles	150	750	150	150
Enhanced		150	750	150	250
		150	750	150	300
		150	750	150	350

One part of the specimens of each mix was tested at standard TCCT conditions. The results have been analyzed in section IV.2. In addition, the radial phase lags $\varphi_{ax,rad}$ were also obtained from the standard TCCTs. They were analyzed as a necessary input parameter of the enhanced TCCT with cyclic confining pressure. Table VII-2 shows the mean values as well as the 2.5% and 97.5% quantiles for the four mixes. Data from three single standard TCCTs carried out for each mix were merged together and statistically analyzed. The mean values are input values for the enhanced TCCTs with cyclic confining pressure at different radial amplitudes.

Table VII-2. Radial phase lags $\varphi_{ax,rad}$ derived from the standard TCCTs.

Radial phase lag $\varphi_{ax,rad}$ from standard TCCT	AC 11 70/100 4.8	AC 11 70/100 5.3	AC 11 70/100 5.8	AC 11 PmB 25/55-65 5.3
2.5 % quantile	26.3	12.1	22.7	16.4
MV	28.2	21.2	24.7	19.1
97.5 % quantile	30.0	27.6	26.2	20.9

VII.3 Results from Enhanced TCCTs

Specimens were subjected to TCCTs with cyclic confining pressure taking into consideration the viscoelastic properties of the material by incorporating the radial phase lag. Figure VII-2 gives an example of the recorded test data. It shows three oscillations of the axial stress on the top, the confining pressure in the middle and the resulting radial deformation on the bottom. The two vertical lines indicate a minimum of the axial and the radial loading, demonstrating that the confining pressure actually lags behind the axial loading.

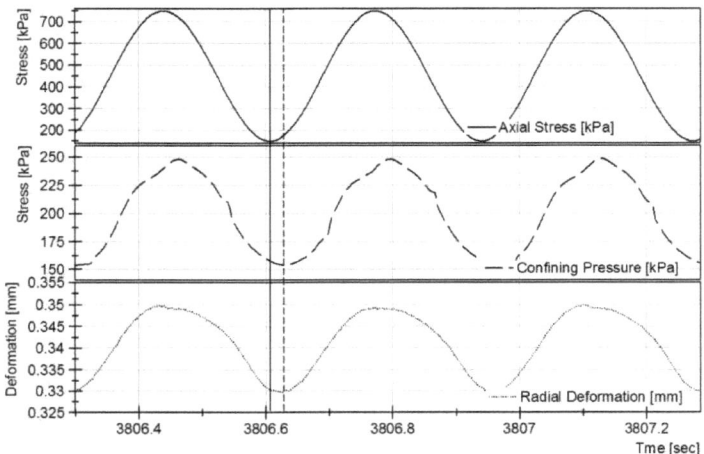

Figure VII-2. Example of recorded test data from enhanced TCCT with dynamic confining pressure.

From each mix, three specimens were tested at each test condition. All tests were run at 50°C and 3 Hz for 25,000 load cycles and an axial stress ranging from 150 kPa to 750 kPa. Different from the standard TCCT according to (EN 12697-25, 2005) the radial stress amplitude is varied in three ranges. The test procedure starts from a hydrostatic state of stress on the low level where both the radial and axial stress are at the same level. This pre-loading phase is held constant for 120 s. Then, the axial sinusoidal loading starts and with a well-defined time lag (derived from the radial phase lag) the confining pressure starts to oscillate sinusoidally as well. The stress applied to the specimen for each point in time t can be given as

$$\sigma_{dev}(t) = \sigma_{ax,m} + \sigma_{ax,a} \cdot \sin(t) - \left[\sigma_{rad,m} + \sigma_{rad,a} \cdot \sin\left(t - \varphi_{ax,rad}\right)\right] \quad (7.1)$$

σ_{dev}........ Stress deviator
$\sigma_{ax,m}$ ...... Mean axial stress
$\sigma_{ax,a}$ .... Axial stress amplitude
$\sigma_{rad,m}$ ..... Mean radial stress
$\sigma_{rad,a}$ ...... Radial stress amplitude

The expression above is also valid for standard TCCTs where the mean radial stress is set to 150 kPa and the radial stress amplitude is set to 0. One objective of the following investigation is to compare the stress introduced into the specimen within one load cycle and thus be able to compare different test conditions. For this reason equation (7.1) can be integrated over one oscillation period $T_p=2\pi/f$, or since all tests were run at the same frequency over 2π. This number is equal to an impetus and is independent of the radial phase lag $\varphi_{ax,rad}$. For the lowest radial stress amplitude from 150 kPa to 250 kPa a value of 500π results from the integration, for the medium amplitude of 450π and for the highest amplitude of 400π. The value for the standard TCCT is 600π.

AC 11 70/100

AC 11 70/100 with three different binder contents were tested in the enhanced TCCT at three different stress levels (i.e. three different amplitudes of confining pressure). The results are discussed for each binder content separately in the following chapter. The discussion is started by the lowest content of 4.8% (m/m) and followed by the other mixes.

The phase lag between axial loading and radial confining pressure was set according to the results from standard TCCTs listed in Table VII-2. To check how well the given phase lag was controlled by the test machine throughout the test, the actual phase lag between axial and radial loading was analyzed with regard to the time shift between both signals for each test and load cycle. Figure VII-3 shows these results for the three different test conditions. At the lowest radial stress amplitude all specimens were tested successfully. The worst deviation between given and actual phase lag is 19° or 10.6% (the basis is 180° since this represents the largest deviation possible). 97.5% of the test data show a lower deviation. At the second amplitude (150 kPa to 300 kPa) only two out of three tests ran successfully.

The maximum deviation is similar to the tests with the lowest amplitude. For the largest radial amplitude (150 kPa to 350 kPa) two specimens failed during the test. The error of the actual to the given radial phase lag is below 29.1° or 16.2% in 97.5 out of 100 cases. The deviation is higher than it was expected. Therefore, the test machine was optimized once more by adapting the PID (Proportional-Integral-Derivative) control of the pneumatic device responsible for the confining pressure. After the optimization the second test series with specimens made from AC 11 70/100 and a binder content of 5.3% (m/m) was carried out.

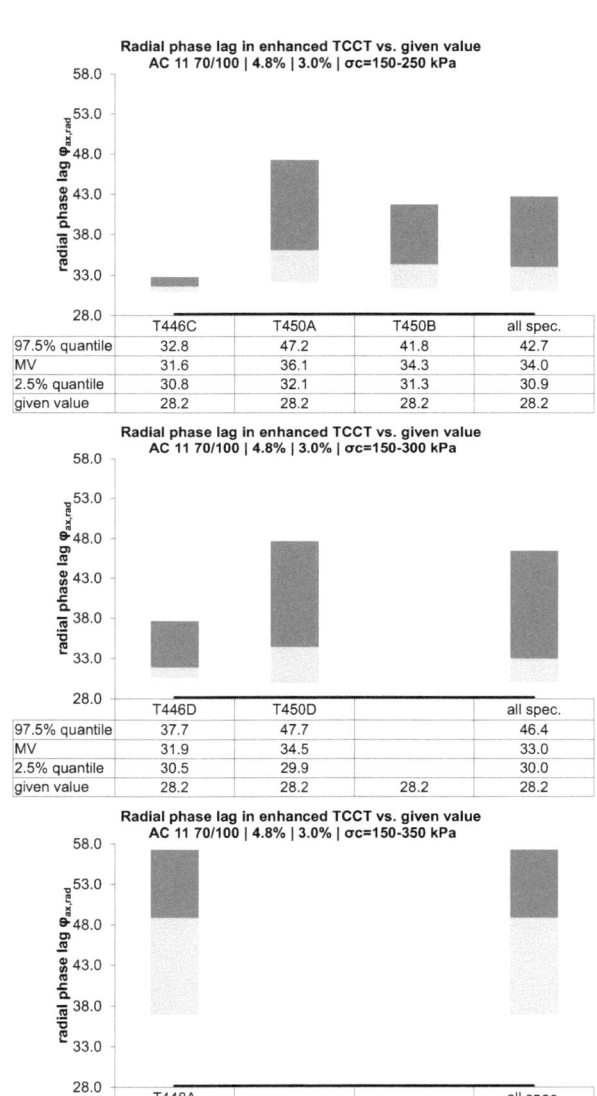

Figure VII-3. Phase lag between axial loading and radial confining pressure induced by the test machine vs. given value from the standard TCCTs for AC 11 70/100 (binder content: 4.8% (m/m)) with a confining pressure of 150 to 250 kPa (top), 150 to 300 kPa (middle) and 150 to 350 kPa (bottom).

For the further analysis, it is of great interest, how the different radial amplitudes affect the results of TCCTs in terms of resistance to permanent deformation and if there are any differences between standard and enhanced TCCTs. For this reason, a number of diagrams compare the stress deviator to various parameters which describe the deformation behavior. The stress deviator multiplied by π is congruent to the impetus put into the specimen at each load cycle and thus a proper parameter to compare different test conditions. Each diagram contains data from each single enhanced TCCT, highlighted in grey together with a linear regression. In addition, a 95% confidence interval was place around the linear regression. The confidence interval was derived by computing the relative error RE between each data point and the linear regression. In the following, the 2.5% and 97.5% quantiles were obtained for this relative error. These quantile values were then used to create the two confidence interval lines from the linear regression as follows:

$$f(x) = (a \cdot x + b) \cdot (1 + RE_{2.5\%})$$
$$f(x) = (a \cdot x + b) \cdot (1 + RE_{97.5\%})$$
(7.2)

a .... slope of the linear regression
b .... Y-intercept of the linear regression
$RE_{2.5\%}$ ... 2.5% quantile of the relative error
$RE_{97.5\%}$ .. 97.5% quantile of the relative error

Data from standard TCCTs are also shown in the diagrams marked in black to compare enhanced TCCT to standard TCCT results.

Figure VII-4 contains the total axial strain $\varepsilon_{ax,tot}$ at load cycle 10,000 in the diagram on top. There is a strong link between the stress deviator and the permanent deformation in the enhanced TCCTs. A higher radial confinement (i.e. higher radial amplitude or lower stress deviator) results in less permanent axial deformation. The results of the standard TCCT are compared to the predicted values of the enhanced TCCTs at the same stress level. From this comparison it is obvious that the standard TCCTs show significantly different results as the three data points are outside the 95% confidence interval. The MV of $\varepsilon_{ax,tot}$ in the standard TCCT is -2.60% compared to -2.17% predicted from the enhanced TCCTs. Specimens tested with constant confining pressure suffer from 20% more total axial strain after 10,000 load cycles. This benefit of the enhanced TCCT is due to the fact that the viscoelastic properties of the material are taken into account by the radial phase lag. When the volumetric and deviatoric parts of the axial strain are taken into consideration, it can be analyzed where the effect from taking into consideration the viscoelastic material reaction has its roots. The volumetric axial strain $\varepsilon_{ax,vol}$ is shown in the middle diagram. There is hardly any difference between the MV from standard TCCTs (-1.58%) and the prediction from enhanced TCCTs (-1.57%). It can therefore be stated that the change in the specimens' volume is not influence by the test type. When the focus is laid upon the deviatoric axial strain (lower diagram), the link between stress deviator and strain is not that distinct anymore. But there is a difference between standard and enhanced TCCTs. Specimens tested according to standard conditions suffer -1.02% (MV) $\varepsilon_{ax,dev}$ compared to -0.61% $\varepsilon_{ax,dev}$ in the enhanced TCCT (-40%). The reason for better performance of the material in the enhanced TCCT is due to the fact that far less deviatoric strain is activated. Due to the cyclic confining pressure which takes into account the viscoelastic characteristics and is strongest when the radial deformation is at a maximum, a smaller share of the mineral aggregates are repositioned than in the standard TCCT.

Figure VII-5 shows the interrelation between the stress deviator and parameters describing the creep curve from TCCTs. In the top diagram the linear creep rate f_c from the linear regression to the creep curve is shown. In contrary to the results of the axial strain above, the linear trend of the creep rate decreases with increasing stress deviator. When the single test results are taken into account, it seems that there is not enough data to give conclusive interpretation of results. The situation of parameter b from the logarithmic regression to the creep curve is shown in the lower diagram. The parameter is falling slightly with increasing stress deviator. Obviously, there is a significant difference between enhanced and standard TCCTs. The standard TCCT results in a MV of -0.215 compared to -0.104 from enhanced TCCTs at the same stress level. Specimens exhibit a 52% lower logarithmic creep rate when they are tested with the enhanced test setup.

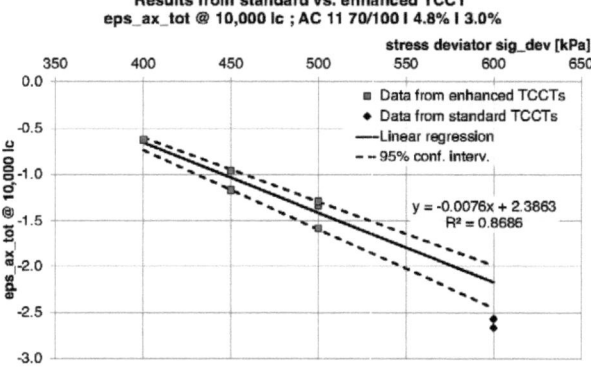
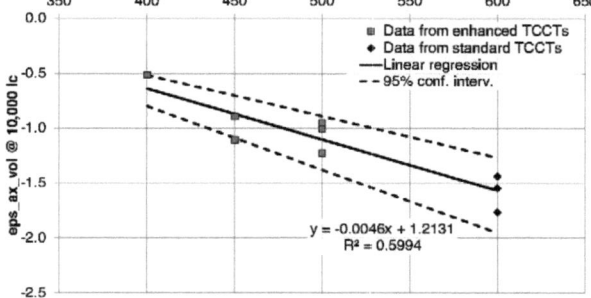
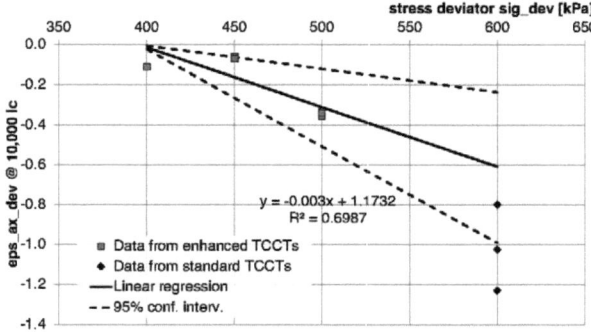

Figure VII-4. Total axial strain (top), volumetric axial strain (middle) and deviatoric axial strain (bottom) at load cycle 10,000 for AC 11 70/100 with a binder content of 4.8% (m/m) at different stress deviators from standard and enhanced TCCTs.

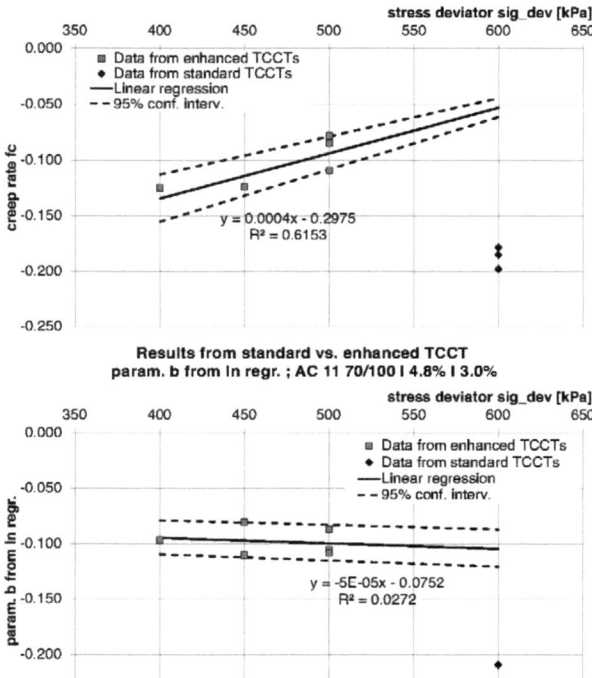

Figure VII-5. Creep rate f_c (top) and parameter b (bottom) from logarithmic regression of creep curve for AC 11 70/100 with a binder content of 4.8% (m/m) at different stress deviators from standard and enhanced TCCTs.

Before enhanced TCCTs were launched for the AC 11 70/100 mix with 5.3% (m/m) binder the test machine was optimized once more to reduce the error between given and actual radial phase lags. The two successful tests with the highest radial amplitude show the least deviation between given and actual radial phase lag (lower diagram in Figure VII-6). 97.5% of all load cycles were run with an error of 2.7° or 1.5%. For the other two test conditions the deviations are clearly higher. In the worst case (σ_c = 150 - 300 kPa), 97.5 out of 100 load cycles show a difference between given and actual value of 8.1° or 4.5%.

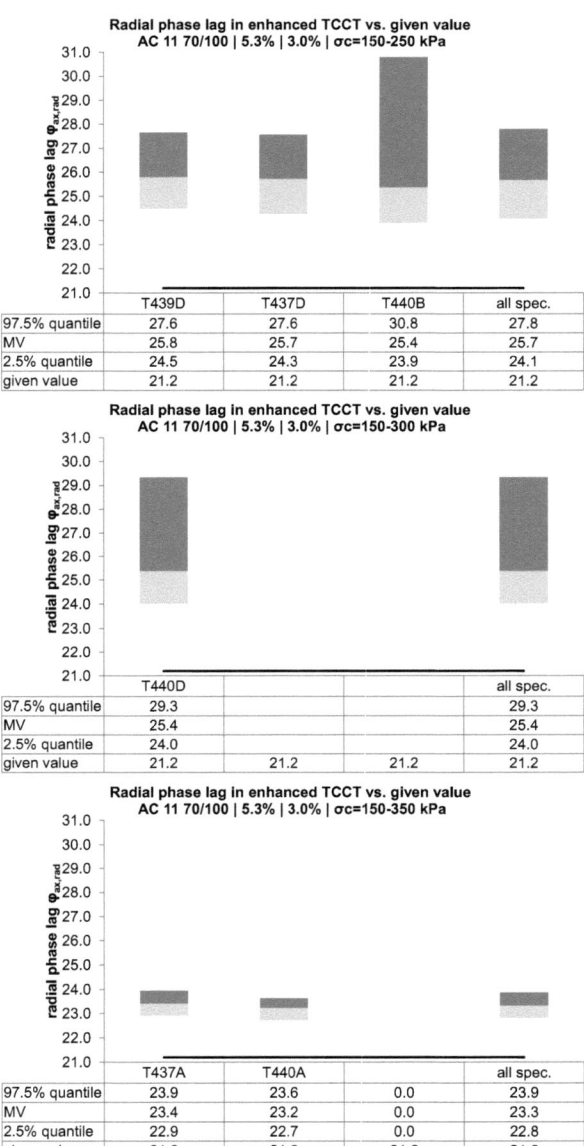

Figure VII-6. Phase lag between axial loading and radial confining pressure induced by the test machine vs. given value from the standard TCCTs for AC 11 70/100 (binder content: 5.3% (m/m)) with a confining pressure of 150 to 250 kPa (top), 150 to 300 kPa (middle) and 150 to 350 kPa (bottom).

It is also worth noting that for these test conditions only one out of three test runs were successful. Thus, the significance of this setup is limited. In case of TCCTs with cyclic confining pressure, the latex membrane protecting the specimen from water rips more often than in case of standard TCCTs. This is because the control unit of the cyclic confining pressure has not been perfectly optimized yet and thus malfunctions occurred.

Figure VII-7 shows a compilation of all test results from enhanced TCCTs in terms of axial strain at load cycle 10,000 vs. the stress deviator as well as a linear regression to these results. The top diagram shows the total axial strain. There is a decreasing trend with increasing stress deviator showing that a higher stress level leads to more deformation. Although the 95% confidence interval is quite large, the standard TCCT results in significantly more total axial strain (-3.53% vs. -2.81% predicted from enhanced TCCTs). Specimens tested in the standard TCCT setup suffer 25% more axial strain than those tested under enhanced conditions.

The diagrams in Figure VII-7 also present the situation for the volumetric and deviatoric part of the axial strain. The scatter of results is quite large. It seems that the axial volumetric strain hardly depends on the stress level. This may be due to the high degree of compaction (3.0% (v/v) voids). The standard TCCTs (MV of -2.22%) results in 32% more volumetric deformation than the prediction from enhanced TCCTs (-1.68%). From the lower diagram it is clear that the impact of the stress deviator is more significant. The standard TCCT results in a MV of -1.65% compared to a prediction from enhanced TCCT at the same stress level of -1.31%. The material exhibits 21% less deviatoric strain when tested with the enhanced test setup where the viscoelastic material properties are taken care of. It can therefore be stated that this mix reacts in a positive way (i.e. shows a better resistance to permanent deformation) when the viscoelastic material reaction is taken into account.

The same analysis is also provided for the creep parameters from different regressions to the creep curve in Figure VII-8. The top diagram presents the creep rate f_c from the standard linear regression for the creep curve, the lower diagram contains data for parameter b of the logarithmic regression.

The situation here is more significant since the scatter of results is less severe. Both parameters decrease with increasing stress deviator, showing – analogue to the total axial strain – that the material exhibits more permanent deformation when the stress level is increased. Again it is obvious that the material contains a potential of better resistance to permanent deformation when the viscoelastic material reaction is considered in the TCCT. For the creep rate, standard TCCTs result in a MV of -0.301. The prediction from enhanced TCCTs at the same stress level lies at -0.245 (-19%). For the logarithmic approximation of the creep curve, the difference between standard (-0.355) and prediction from enhanced TCCT (-0.287) is also -19%.

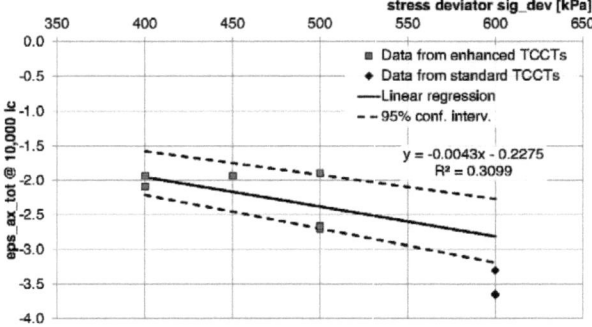
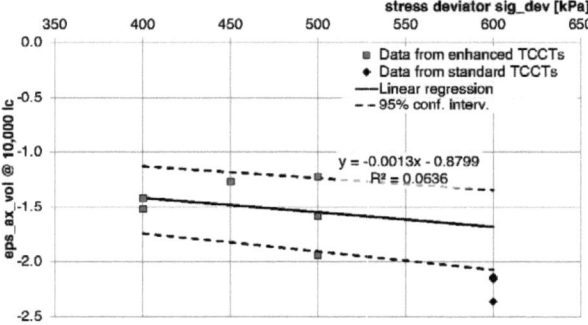
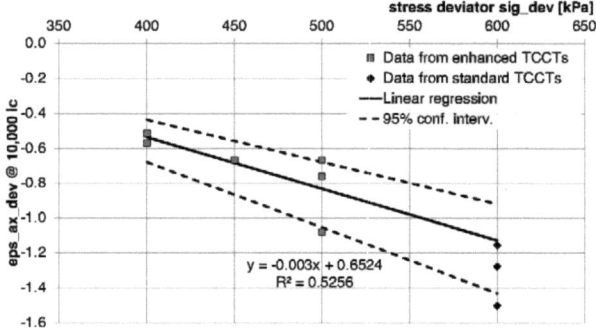

Figure VII-7. Total axial strain (top), volumetric axial strain (middle) and deviatoric axial strain (bottom) at load cycle 10,000 for AC 11 70/100 with a binder content of 5.3% (m/m) at different stress deviators from standard and enhanced TCCTs.

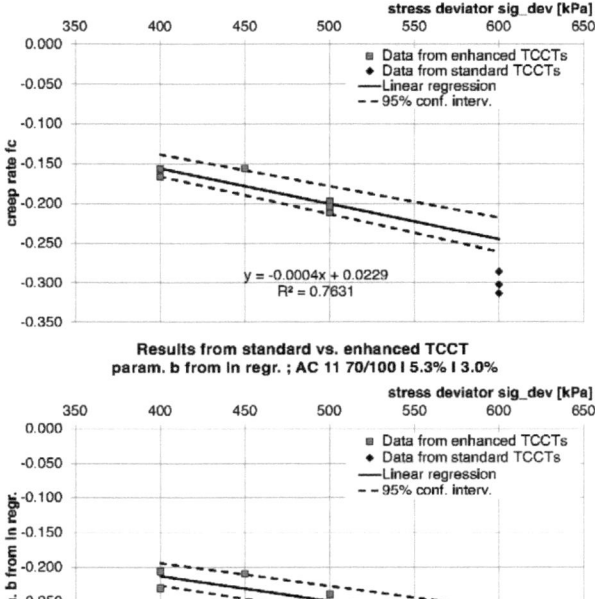

Figure VII-8. Creep rate f_c (top) and parameter b (bottom) from logarithmic regression of creep curve for AC 11 70/100 with a binder content of 5.3% (m/m) at different stress deviators from standard and enhanced TCCTs.

Last but not least, results from the AC 11 mix with 5.8% (m/m) of unmodified binder 70/100 are presented below. To check how well the test machine achieved the given value of the radial phase lag in the enhanced TCCT, the three diagrams in Figure VII-9 show a comparison of given and actual values for each single test at the three radial amplitudes. The worst case occurs for an amplitude of 150 kPa to 300 kPa with a MV of 29.1° actual phase lag produced by the test machine compared to a given value of 24.7°. This error of 4.4° or 2.4% is still acceptable. For the other two test conditions with an amplitude of 150 kPa to 250 kPa (deviation of 3.3° or 1.8%) and 150 kPa to 350 kPa (deviation of 2.0° or 1.1%) the error is even lower.

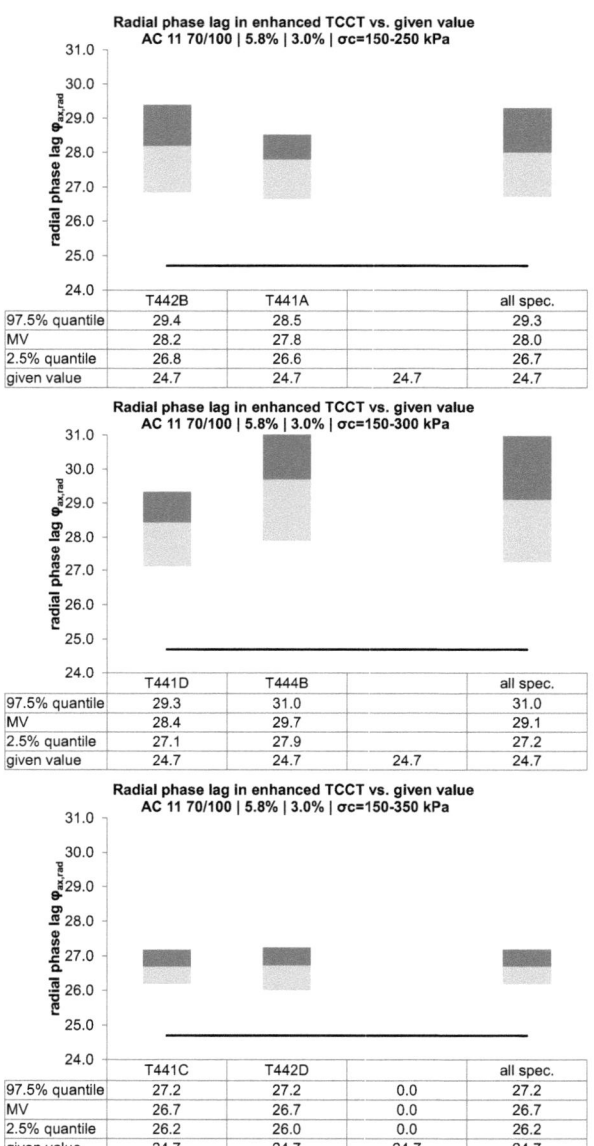

Figure VII-9. Phase lag between axial loading and radial confining pressure induced by the test machine vs. given value from the standard TCCTs for AC 11 70/100 (binder content: 5.8% (m/m)) with a confining pressure of 150 to 250 kPa (top), 150 to 300 kPa (middle) and 150 to 350 kPa (bottom).

Figure VII-10 shows the interrelation between test results from enhanced TCCTs in terms of permanent axial strain and contains results from standard TCCTs for the same material. The top diagram shows the total axial strain $\varepsilon_{ax,tot}$ vs. the stress deviator. Again, a linear regression as well as a 95% confidence interval was derived from the results of the enhanced TCCTs. Obviously, there is no significant difference between standard and enhanced TCCT when it comes to total axial strain. The MV derived from standard TCCTs is -3.84% compared to a predicted value of -3.93% from enhanced TCCTs at the same stress level.

Still, the components of the total axial strain, the volumetric and deviatoric part, shown in the diagrams of Figure VII-10 indicate that the material behavior is different in the two test setups. Specimens tested in the enhanced TCCT produce a higher ratio of volumetric strain. The standard TCCTs results in a MV of -2.13%, whereas the predicted value at the same stress level for enhanced TCCTs is -2.73% (+28%). In terms of the deviatoric strain component, the situation is reversed. Specimens in the standard TCCT suffer -1.71%. The prediction from enhanced TCCTs results in -1.20% deviatoric strain (-29%).

The creep parameters f_c (creep rate) from linear regression to the creep curve and parameter b as the incline of the logarithmic regression in the log-lin scale are compared to the stress deviator in Figure VII-11. Both diagrams show the same tendency: There is hardly any influence of the stress deviator on the results of enhanced TCCTs, but there is a significant difference between standard and enhanced tests.

The MV of the creep rate for standard TCCTs is -0.276 compared to a predicted -0.174 from enhanced tests (-40%). For parameter b from the logarithmic regression, the difference is -33%. This means that although the total axial strain after 10,000 load cycles is similar for both tests, the long term behavior expressed by the creep parameters is better when the material is tested in the enhanced TCCT. The enhanced TCCT accounts for the viscoelastic material reaction of the mix and reveals that the standard TCCT results in a conservative estimate of the rutting resistance of a mix.

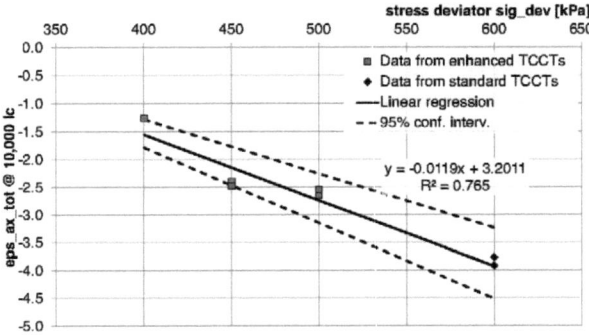
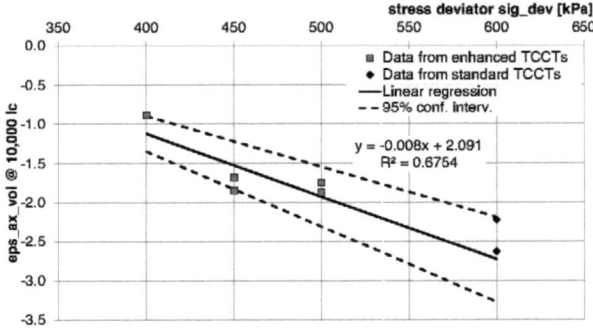
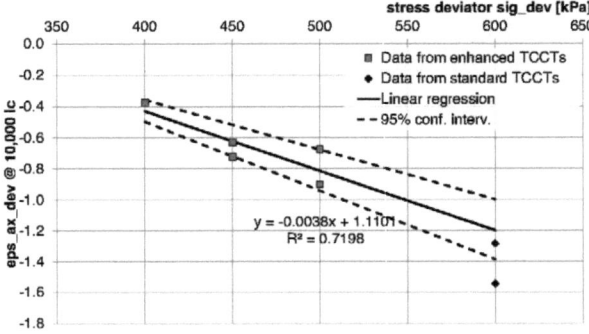

Figure VII-10. Total axial strain (top), volumetric axial strain (middle) and deviatoric axial strain (bottom) at load cycle 10,000 for AC 11 70/100 with a binder content of 5.8% (m/m) at different stress deviators from standard and enhanced TCCTs.

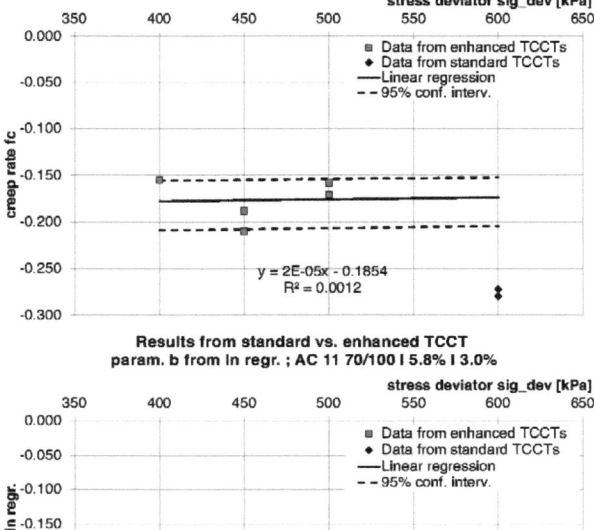

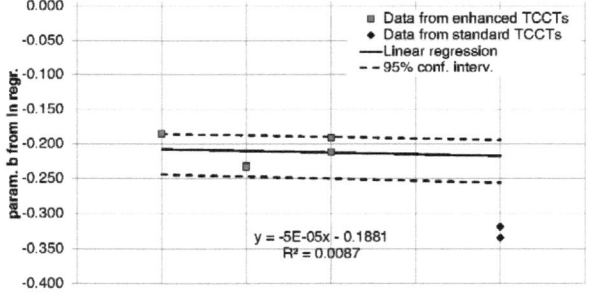

Figure VII-11. Creep rate f_c (top) and parameter b (bottom) from logarithmic regression of creep curve for AC 11 70/100 with a binder content of 5.8% (m/m) at different stress deviators from standard and enhanced TCCTs.

AC 11 PmB 25/55-65

An AC 11 mix with polymer-modified binder was tested in the enhanced TCCT as well. As stated in Table VII-2 the standard TCCT with this mix resulted in a MV of the radial phase lag of 19.1°. To make sure that the given phase lag was controlled correctly throughout the test, the actual phase lag between axial and radial stress was analyzed with regard to the time shift between both signals for each test and load cycle. Figure VII-12 presents these results for the three different radial amplitudes. The largest deviation between given and actual radial phase lag occurred for the lowest radial amplitude of 150 kPa to 250 kPa with a mean value of 22.1° instead of 19.1°. This difference of 2.0° is negligible, it represents a relative error of 1.1%. It can therefore be stated that the test control produced the cyclic confining pressure with a sufficient quality taking into consideration the viscoelastic properties of the material.

Figure VII-13 shows the total axial strain $\varepsilon_{ax,tot}$ at load cycle 10,000 on the top. There is a clear linear link between the stress deviator and the permanent deformation in the enhanced TCCTs. A stronger radial confinement (i.e. higher amplitudes of the confining pressure) results in less permanent axial deformation. When the results of the standard TCCT are compared to the results of the enhanced test, it becomes obvious that two of three data points are out of the 95% confidence interval. There seems to be a significant difference between both test types. The material suffers more axial strain when the radial confining pressure is held constant. The MV of $\varepsilon_{ax,tot}$ from standard TCCTs is -2.39% compared to a predicted -2.15% from the enhanced TCCTs at the same stress level. This shows that by considering the viscoelastic properties of the mix in the test procedure, the material exhibits 10% less axial strain. When the volumetric and deviatoric part of the axial strain $\varepsilon_{ax,vol}$ and $\varepsilon_{ax,dev}$ are taken into account, it can be investigated where the origin of this benefit lies. The middle diagram in Figure VII-13 shows the volumetric axial strain. In this case, the MV of the test data from standard TCCTs is -1.28% compared to a predicted -1.48% from the regression of the enhanced TCCTs. An enhanced TCCT would produce 16% more volumetric strain at comparable stress levels. Looking at the deviatoric axial strain in the lower diagram, the enhanced TCCTs exhibit significantly less deformation: -1.11% (standard TCCT) vs. -0.67% (prediction from enhanced TCCT) or 40% less deviatoric strain. This shows that the reason for the better performance of the material in the enhanced TCCT is due to the fact that far less deviatoric strain is activated.

Figure VII-14 shows the interrelation between the stress deviator and parameters describing the creep curve from TCCTs. In the top diagram the creep rate f_c from the linear regression of the creep curve is shown. The 95% confidence interval covers a large area around the regression. Still, there is a decreasing tendency with increasing stress deviator. The MV of f_c for the standard TCCTs is -0.158 compared to a predicted value of -0.132 with the linear regression for the data of enhanced TCCTs at the same stress level. This means that an enhanced TCCT at the same stress level as the standard TCCT would produce a 16% lower creep rate.

The situation of parameter b of the logarithmic regression to the creep curve is depicted in the lower diagram of Figure VII-14. The correlation between the parameter and the stress deviator is better than for the creep curve. The standard TCCT exhibits a value of -0.188. An enhanced TCCT at the same stress level would result in a value of -0.154 (-18%).

Both parameters show that there is a benefit in the high temperature performance of the mix if the viscoelastic properties are taken into consideration.

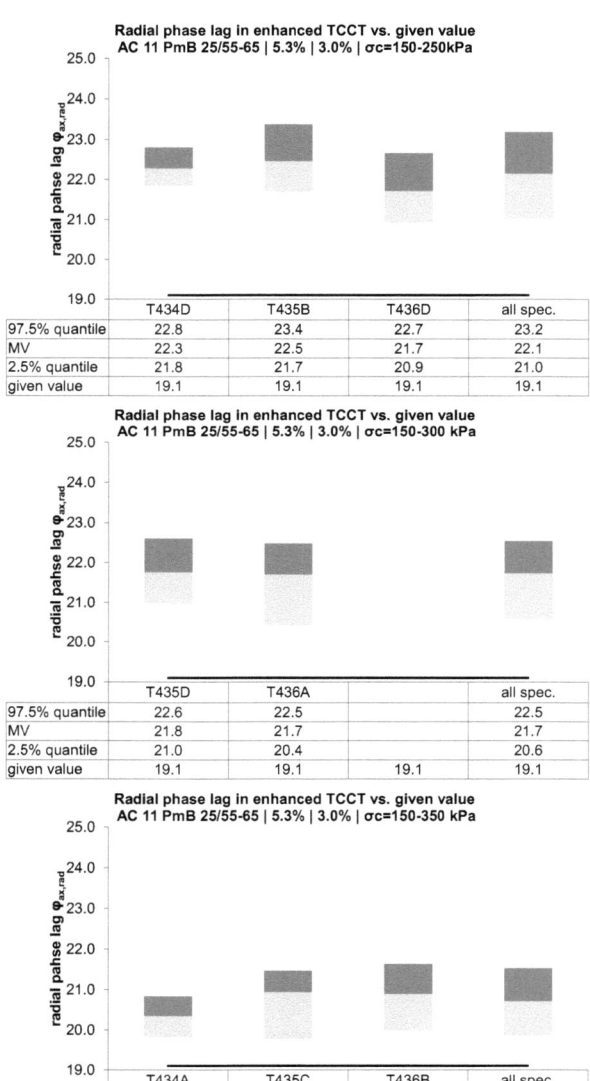

Figure VII-12. Phase lag between axial loading and radial confining pressure induced by the test machine vs. given value from the standard TCCTs for AC 11 PmB 25/55-65 with a confining pressure of 150 to 250 kPa (top), 150 to 300 kPa (middle) and 150 to 350 kPa (bottom).

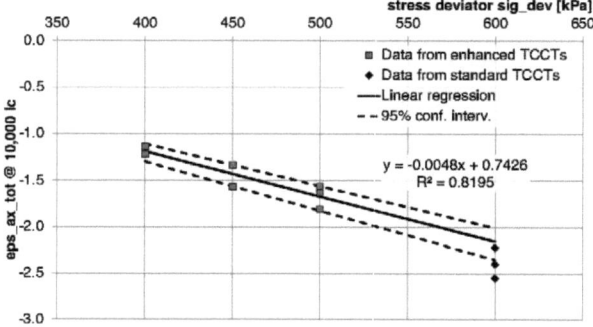
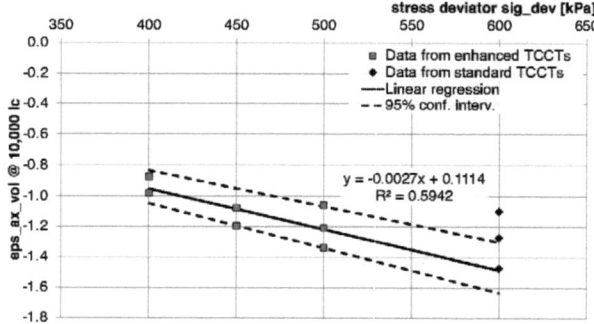
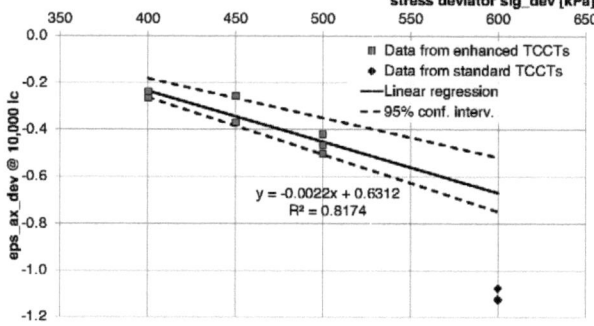

Figure VII-13. Total axial strain (top), volumetric axial strain (middle) and deviatoric axial strain (bottom) at load cycle 10,000 for AC 11 PmB 25/55-65 with a binder content of 5.3% (m/m) at different stress deviators from standard and enhanced TCCTs.

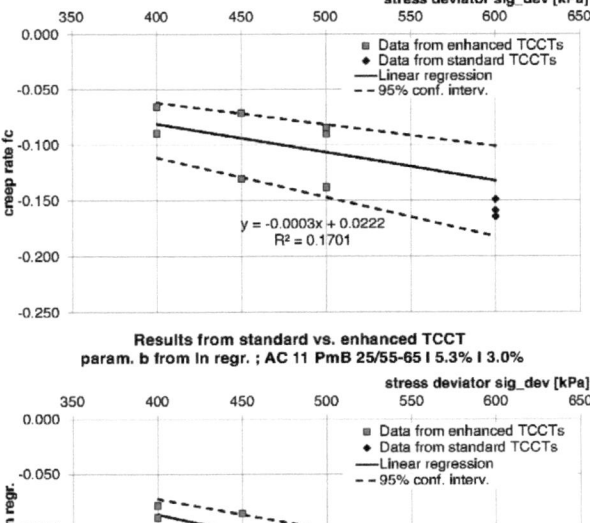

Figure VII-14. Creep rate f_c (top) and parameter b (bottom) from logarithmic regression of creep curve for AC 11 PmB 25/55-65 with a binder content of 5.3% (m/m) at different stress deviators from standard and enhanced TCCTs.

VII.4 Conclusions and Outlook

As another core objective of this thesis, **cyclic confining pressure was introduced into the TCCT** to simulate the state of stress that occurs in the field in a more realistic way. The results of standard and enhanced TCCTs are compared and interpreted. The main findings are summarized below:

- In terms of total axial strain $\varepsilon_{ax,tot}$ the four investigated mixes performed in different ways: While the mix with unmodified binders suffered from 16% to 20% less strain with binder contents of 4.8% (m/m) and 5.3% (m/m), the mix with the highest binder content did not show any significant change in total axial deformation in the enhanced TCCT. The PmB mix showed 10% less strain when specimens were tested with cyclic confining pressure.
- The volumetric part of the axial strain $\varepsilon_{ax,vol}$ showed different behaviors as well: Specimens tested in the enhanced TCCT exhibited less axial strain for the un-

- modified mix with 5.3% (m/m) of binder, no notable difference for the mix with 4.8% (m/m) binder and an increase of 28% for the 5.8% (m/m) mix. It also increased for the PmB mix by 16%.
- Clear results can be stated for the deviatoric part $\varepsilon_{ax,dev}$. It was reduced substantially for all mixes when they were tested with the enhanced test setup. The decrease varies from 14% for the AC 11 70/100 with 5.3% (m/m) binder to 30% and 40% for the other three mixes. It can thus be concluded that the cyclic confining pressure accounting for the viscoelastic characteristics leads to less deviatoric strain in all cases.
- As for the creep parameters that characterize the long-term behavior of a mix, it decreased as well for all mixes. The change varies from -16% up to -51% when the standard TCCT is compared to the enhanced TCCT.

These results show that a benefit can be activated when the material is tested by taking into account its viscoelastic properties. It seems that this benefit is mainly caused by a reduced deviatoric strain component. It also appears that mixes perform significantly better in terms of long-term behavior. The creep rates decrease by 1/6 up to 1/2 when the standard TCCT results are compared to results from enhanced TCCTs.

Of course even the standard TCCT is seen as a complex test procedure. Two independent (hydraulic) circuits are necessary which must be synchronized to achieve correct test runs. With cyclic confining pressure, the procedure gets even more complex and a test series is more time consuming and currently more prone to failure. Still, especially when it comes to the characterization of layers where a certain amount of confinement is activated (binder layers) this test setup can achieve a more realistic simulation of the long term deformation behavior and thus help to optimize mix and pavement design. It is not recommended at this stage to introduce the enhanced TCCT as standard for the characterization of the permanent deformation behavior but it is recommended to implement this option into a next version of the European standard to provide the opportunity to test materials for certain projects with special climatic or loading conditions with cyclic confining pressure in a standardized way.

For this thesis, only a limited number of mixes was tested in the enhanced TCCT. To create more findings and put the presented conclusions on an broader basis, it will be necessary to carry out a more extensive test program with a variation of the void content, gradation type, binder type and also type of aggregate.

VIII RESUME AND PERSPECTIVE

This doctoral thesis is aimed towards an advanced characterization of the material behavior of hot mix asphalt (HMA) under cyclic dynamic compressive loading. The triaxial cyclic compression test (TCCT) according to (EN 12697-25, 2005) was thoroughly reviewed. From this four main objectives were developed to enhance the output of this test type:
- Improving the assessment of the permanent deformation behavior as the core results of standard TCCTs.
- Describing the viscoelastic material parameters of HMA under cyclic dynamic loading over a wide range of temperatures and frequencies by taking into account the axial and radial deformation component.
- Developing a model to predict the viscoelastic behavior of HMA from material parameters of bitumen and volumetric characteristics of the mix.
- Introducing an enhanced TCCT with cyclic dynamic confining pressure to take into account the viscoelastic material reaction of HMA to traffic loading in a more realistic way.

Assessment of the Permanent Deformation Behavior

The present way of data evaluation and interpretation of results from standard TCCTs according to European Standards at 50°C, 3 Hz and 25,000 load cycles approximates the quasi-linear part of the creep curve (permanent axial strain ε_n vs. number of load cycles n) by a linear function. The slope of the linear (= creep rate f_c) is the benchmark of an HMA's resistance to permanent deformation (rutting) at elevated temperatures. The standard does not provide information on how to define the quasi-linear part of the creep curve and the results depend on the chosen starting point and range of the quasi-linear creep. To overcome this drawback an alternative method is developed. By investigating a large number of results from standard TCCTs it was found that the viscoelastic material parameters do not change anymore after a certain number of load cycles at the beginning of the test. A clear definition is given how this point of constant viscoelastic material reaction can be derived from test data. From this point of the test on the creep curve is linear in the log/lin-scale and thus can be approximated by a logarithmic function with high quality. TCCT data from 12 different mixes with varying grading curves, binder and aggregate types, binder content and content of air voids were evaluated with the new logarithmic approach. Not only does the logarithmic approach show equal and in some cases even better approximation quality than the linear approach. In addition an excellent correlation was found between the linear creep rate f_c and the logarithmic creep rate b, which is the slope of the logarithmic function in the log/lin-scale. Thus, the newly introduced logarithmic approach to approximate the creep curve has the major advantage that it is defined unambiguously and is not influenced by the determination of a quasi-linear range of the creep curve. In addition the correlation between the existing linear and the logarithmic creep rate is excellent. It is therefore recommended to implement the logarithmic approach into the next version of the standard (EN 12697-25, 2005) to ensure reliable and reproducible results from TCCTs.

Furthermore a significant benefit can be achieved when not only the axial deformation but also the radial deformation of the specimen is recorded in standard TCCTs: It is shown how the total axial strain $\varepsilon_{ax,tot}$ can be separated into its volumetric part $\varepsilon_{ax,vol}$ and deviatoric part $\varepsilon_{ax,dev}$. These two components characterize two different rutting mechanisms found in the field: While the volumetric strain component describes rutting by pure compression when the content of air voids is reduced without any change in shape, the deviatoric strain component stands for rutting by shear deformation without any change in volume. Again an investigation of these strain components was carried out for 12 different HMAs from which the following findings can be derived:

- The type of aggregate impacts the deviatoric strain component.
- An increase in the void content of a mix leads to an increase of the volumetric part of the deformation and thus a higher total axial strain.
- The binder type affects the total axial strain leaving the ratio of volumetric to deviatoric strain unaffected.
- When the temperature is set to a lower level, the decrease in total axial strain is due to reduced volumetric strain whereas the deviatoric part does not change.

The investigation on the volumetric and deviatoric part of the total axial strain reveals a significant potential of this approach to characterize the permanent deformation behavior of HMA. It is valuable information for road construction in the field to know how a certain mix reacts to loading. How resistant a mix is to both types of permanent deformation is important information to optimize mixes for different applications. While surface layers are more susceptible to shear deformation especially in those areas where high shear stresses are introduced into the pavement (intersections, airfields), binder layers are more confined within the structure. Thus volumetric deformations are more relevant for these parts of a pavement. With only little additional effort to record the accumulated radial strain in the TCCT (a standard LVDT-based device (e.g. Figure II-8) is perfectly suitable) a large benefit can be created in terms of results. It is therefore recommended to introduce this approach into a next revision of the European standard (EN 12697-25, 2005) to gather more data and experience on this important matter.

Viscoelastic Material Behavior of HMA under Cyclic Dynamic Compressive Loading

Since cyclic compression tests (CCT) result not only in relevant strain in axial direction but also in the plane perpendicular to the vertical axis in radial direction, it is a perfect setup to study the quasi-3-d material behavior of HMA. In a comprehensive test program HMA specimens with strain gauges (SG) attached around their circumference were tested at various temperatures and frequencies to find answers to a number of questions.

One question was, whether the test setup (i.e. whether the specimen is firmly connected to the load plates or not) impacts the material reaction and thus whether standard CCTs are a valid test method for the derivation of viscoelastic material properties. An advanced function ($F+L+1H$) was employed for regression analysis of the sinusoidal test data. This function contains not only a simple sine or fundamental oscillation term (F) and a linear term (L) to account for the accumulated strain due to the purely compressive loading but also the first harmonic of the fundamental oscillation ($1H$). The most important findings are:

- The advanced $F+L+1H$ approximation is a proper tool to quickly check the shape of sinusoidal test data. If the sine should be distorted, the amplitude ratio AR and the shift factor γ are able to describe the shape and magnitude of the distortion. Problems with the test machine control as well as shape and magnitude of the deformation oscillation can be found quickly and described easily.
- In terms of mechanical material parameters it was found that there is no difference in phase lags between both setups.
- With respect to the dynamic modulus, it makes a difference whether the specimen is glued to the load plate prior to testing or not. Unglued specimens result in a lower stiffness (80% to 90% of the glued setup depending on the test frequency).

Consequently, viscoelastic material parameters are analyzed, such as phase lags in axial and radial direction and the dynamic modulus. Results are shown for two AC 11 mixes with paving grade bitumen 70/100 and polymer-modified bitumen PmB 25/55-65. Tests were carried out with temperature and frequency sweep. In principle it was found that

- temperature and frequency have a strong and conclusive impact on the viscoelastic behavior of HMA und compression. A loss in the dynamic modulus with increasing temperature and decreasing test frequency can be stated. The axial and radial phase lags show a maximum of viscosity not at the highest test temperature of 50°C but at 30°C. Since the stiffness of the pure binder decreases dramatically between 30°C and 50°C, the impact of the bitumen on the overall mix behavior also drops and the elastic behavior of the aggregate skeleton gains more influence. Thus, the HMA viscosity decreases when the temperature is raised from 30°C to 50°C.
- In accordance to the optimal binder content after Marshall, the mixes show a maximum in stiffness at this content at 50°C. In terms of the axial viscosity expressed by the axial phase lag there is a minimum in the phase angle for all three temperatures at the optimal Marshall content for the polymer-modified HMAs. This cannot be found for the unmodified mixes where the phase lags are either stable (10°C, 50°C) or show even a maximum at the optimum binder content (30°C).
- The content of air voids affects the dynamic modulus of the mix in different ways. For the polymer-modified mixes the material seems to be rather insensi-

tive to changes in the void content at 10°C. At 30°C the dynamic modulus drops from medium to high air void content and at 50°C from low to medium content of air voids. It seems that there is a certain threshold value in terms of the void content which brings a significant loss in stiffness. This threshold value tends to shift towards lower air void contents the higher the temperature is set. Analogue to the stiffness, the viscosity of the modified HMAs also increases considerably at the same level of air voids at 30°C and 50°C. At 10°C there is also an increase in viscosity between low and medium air void content despite a loss in dynamic modulus at these conditions.

A significant difference between axial and radial phase lags in all CCTs carried out for the study was found. After a comprehensive investigation and analysis of possible explanation for the existence of a phase lag between the axial and radial deformation $\delta_{ax,rad}$ it was conclusively found that this parameter is material-inherent. It can be stated that $\delta_{ax,rad}$

- is related to material anisotropy to a minor extent,
- is not related to any uniformity of the radial deformation and
- is not related to the measuring system consisting of the SG and the adhesive.

As one result of these findings the dynamic Poisson's Ratio $|v^*|$ as well as the dynamic shear modulus $|G^*|$ are analyzed and discussed in section V.7. (Di Benedetto, et al., 2007) worked on $|v^*|$, but did not consider its loss and storage part. From the results it can be stated that the dynamic Poisson's Ratio has an elastic and viscous component, although the viscous component only becomes dominant at elevated temperatures (50°C). It is state of the art to use a constant value for v of 0.30 to 0.35 for calculations, modeling and simulation of bituminous bound materials. From the results presented in this chapter it becomes obvious that this value is only true for intermediate temperatures at low frequencies or high temperatures and high frequencies. It can be shown that an increasing binder content leads to increasing Poisson's Ratios at 10°C and 30°C and stable conditions at 50°C for modified HMAs. For the mixes with the unmodified bitumen, there is a clear maximum at 10°C and optimal Marshall binder content, at 30°C an increasing Poisson's Ratio with increasing binder content at low frequencies and stable conditions at higher frequencies occurs. An increasing air void content has a conclusive effect at high temperatures for mixes with both binder types. The higher the void content the lower is the dynamic Poisson's Ratio. It seems that soft binders at high temperatures enable mineral aggregates to slide past each other (reducing air voids) more easily and therefore most of the deformation energy is put into filling the air voids rather than producing radial deformation. By combining the dynamic modulus and the dynamic Poisson's Ratio another viscoelastic parameter, the dynamic shear modulus $|G^*|$ is introduced. The impact of temperature, binder and void content on this parameter are investigated. Compared to the dynamic modulus $|E^*|$, the analysis shows that the temperature and frequency sensitivity of the dynamic shear modulus is higher. $|G^*|$ decreases more strongly with increasing temperature compared to $|E^*|$. On the other hand the dynamic shear modulus exhibits a higher increase with increasing frequency than $|E^*|$.

Modeling the Viscoelastic Behavior of HMA under Compressive Loading

As a next step a three stage approach towards an analytical model was taken that links the material behavior of HMA with the respective bitumen (*B(inder)-A(sphalt mix) Model*). The model can predict master curves for a mix from master curves of the respective binder and the volumetric characteristics of the mix: the combined binder and filler content in % (m/m) (BFC) and the volume of voids of the aggregate skeleton filled with bitumen in % (v/v) (VFB). The prediction model consists of nine parameters, three of the parameters are dependent on the binder type of the mix. The *B-A Model* describes all relevant macroscopic, viscoelastic material parameters of the mix, from dynamic modulus, phase lags and dynamic shear modulus to dynamic Poisson's Ratio. The model parameters may not have a direct physical relation but the *B-A Model* has the vast advantage to manage the prediction of the HMA behavior in axial and radial direction over a large range of frequencies/temperatures.

Furthermore a conclusive link between the viscoelastic behavior of HMAs, in this case of the dynamic shear modulus $|G^*|$, and the deformation behavior (linear and logarithmic creep rate) has been established. A power function links the dynamic shear modulus to the creep rate. In connection with the *B-A Model* it is a powerful tool to predict rutting resistance of mixes.

The data basis for the model is presently limited and further research and testing must validate the model to a further extent. Right now the quasi-3-d viscoelastic behavior of an HMA with a specific gradation curve (AC) and one type of mineral was tested in the compressive loading domain and can therefore be modeled. The binder type, binder content and content of air voids, as well as the filler content are variables that are taken into account by the model. For the future, a research program should investigate the following matters:

- HMAs with the same grading curve and mineral type should be tested with two more unmodified and modified binders to find functional relationships between the three dependent model parameters and bitumen characteristics.
- Different grading curves (e.g. an SMA) and different mineral types should be included in the research to identify the impact of those parameters.
- Correlations between the 9 model parameters should be analyzed when test data from HMAs with more variation of the mix design are available to study further interrelations between mix design parameters and the model parameters and thus make the model even more universal.

An Enhanced TCCT with Cyclic Dynamic Confining Pressure

As another core objective of this thesis, cyclic confining pressure was introduced into the TCCT to simulate the state of stress that occurs in the field in a more realistic way.

Standard TCCTs were carried out for a different mixes, the test data was evaluated and the radial phase lag $\varphi_{ax,rad}$ was analyzed. In the following enhanced TCCTs with cyclic confining pressure were run on the same material incorporating the radial phase lag derived from the first tests. Finally results of both test procedures the standard and enhanced TCCT are compared and discussed in terms of permanent deformation behavior. Enhanced TCCTs were carried out at AC 11 mixes with paving grad bitumen 70/100 and three different binder contents and with PmB 25/55-65. The enhanced TCCTs were run at three different radial amplitudes to study impacts of these conditions. The main findings are summarized below and are always in comparison to standard TCCTs without cyclic confining pressure:

- In terms of total axial strain $\varepsilon_{ax,tot}$ the mixes perform in different ways: While the mix with unmodified binders suffers from up to 20% less strain with binder contents below or at optimal Marshall content, specimens with higher binder contents do not show any significant change in total axial deformation in the enhanced TCCT. The PmB mix shows 10% less strain when specimens are tested with cyclic confining pressure.
- The volumetric part of the axial strain $\varepsilon_{ax,vol}$ shows different behaviors as well: Specimens tested in the enhanced TCCT exhibit less axial strain for the unmodified mix at and below optimal Marshall binder content and an increase of 28% for higher binder contents. It also increases for the PmB mix by 16%.
- Clear results can be stated for the deviatoric part $\varepsilon_{ax,dev}$. It is reduced substantially for all mixes when they are tested with the enhanced test setup. The decrease varies from 14% to 40%. It can thus be concluded that the cyclic confining pressure that takes the viscoelastic characteristics into account leads to less deviatoric strain in all cases.
- As for the creep parameters that characterize the long-term behavior of a mix, it decreases as well for all mixes. The change varies from -16% up to -51% when the standard TCCT is compared to the enhanced TCCT.

These results show that a benefit with regard to permanent deformation can be activated when the material is tested by taking into account its viscoelastic properties and thus in a more realistic way. It seems that this benefit is mainly caused by a reduced deviatoric strain component and mixes perform significantly better in terms of long-term behavior. The creep rates decrease by 1/6 up to 1/2 when the standard TCCT results are compared to results from enhanced TCCTs.

Especially when it comes to the characterization of layers where a certain amount of confinement is activated (e.g. binder layers) this test setup can achieve a more realistic simulation of the long term deformation behavior and thus help to optimize mix and pavement design. It is not recommended at this stage to introduce the enhanced TCCT as the standard for the characterization of the permanent deformation behavior but it is recommended to implement this option into a next version of the European standard to provide the opportunity to test materials for certain projects with special climatic or loading conditions with cyclic confining pressure in a standardized way and gather more experience in this field of research.

IX Abbreviations and Symbols

Abbreviations

2PBB	2 Point Bending Beam Test
4PBB	4 Point Bending Beam Test
AC	Asphalt Concrete
ADC	Analogue Digital Converter
BRRC	Belgian Road Research Center
CCT	Cyclic Compression Test
DT	Direct Tension Test
DTC	Direct Tension/Compression Test
DSR	Dynamic Shear Rheometer
HGV	Heavy Goods Vehicle
HMA	Hot Mix Asphalt
HRA	Hot Rolled Asphalt
ITT	Indirect Tensile Test
LVDT	Linear Variable Differential Transformer
MV	Mean Value
PmB	Polymer-modified Binder
RE	Relative Error
SD	Standard Deviation
SG	Strain Gauge
SMRE	Sum of Mean Relative Error
TCCT	Triaxial Cyclic Compression Test
TSRST	Thermal Stress Restrained Specimen Test
TTSP	Time Temperature Superposition Principle
UCCT	Uniaxial Cyclic Compression Test
UTST	Uniaxial Tensile Stress Test
UmB	Unmodified Binder
VE	Viscoelasticity, viscoelastic
VFB	Voids Filled With Binder
VMA	Voids In The Mineral Aggregate

WTT	Wheel Tracking Test

Symbols

a	Regression parameter of logarithmic function for the approximation of the creep curve (intersection with the y-axis at x=1)
A	Regression parameter of power function for the approximation of the creep curve (intersection with the y-axis at x=1)
A_1	Regression parameter of linear function for the approximation of the creep curve (intersection with the y-axis at x=0)
AR	Ratio of amplitude of 1st harmonic to fundamental oscillation
a_T	Shift factor
b	Regression parameter of logarithmic function for the approximation of the creep curve (slope in log-lin-scale) = logarithmic creep rate
B	Regression parameter of power function for the approximation of the creep curve (slope in log-log-scale)
B_1	Regression parameter of linear function for the approximation of the creep curve (slope in lin-lin scale)
C	Circumference
γ	Shift factor between 1st harmonic and fundamental oscillation Shape factor
d or D	Diameter
δ	Phase lag between two deformation components (e.g. axial and radial)
$\|E^*\|$	Dynamic modulus
E^*	Complex modulus
E_a	Activation energy
ε	Strain
f	Frequency
f^*	Reduced or scaled frequency
f_c	Linear creep rate
F	Force Approximation function with fundamental oscillation
F+L	Approximation function with fundamental oscillation and linear term
F+L+1H	Approximation function with fundamental oscillation, linear term and 1st harmonic
$\|G^*\|$	Dynamic shear modulus

G^*	Complex shear modulus
h	Height
i	Complex number
J^*	Complex compliance
$|J^*|$	Dynamic compliance
L	Length
m	Mass
µ	Mass factor
n	Number of load cycles
$|\nu^*|$	Dynamic Poisson's Ratio
ν^*	Complex Poisson's Ratio
P	General for "parameter"
r	Radius
R	Universal gas constant
R^2	Coefficient of determination
S	Stiffness parameter (e.g. $|E^*|$)
σ	Stress
t	Time
T	Temperature
T_p	Periodic time
φ	Material phase lag between loading and deformation
Φ	Phase lag between loading and deformation with impacts of test machine (inertia effects)
V	Volume
w	Dissipated energy
ω	Angular frequency
z	Deformation

Indices

0	Initial
	Reference
1	Elastic part

2	Viscous part
a or ax	Axial Amplitude
ax,ax	Between axial (e.g. loading) and axial (e.g. deformation)
ax,rad	Between axial (e.g. loading) and radial (e.g. deformation)
bit	Regarding bitumen
c	Confining
def	Deformed
dev	Deviatoric
e	Elastic
eq	Equal
HMA	Regarding hot mix asphalt
l	Lower
m	Mean
max	At the maximum of the sine
MC	Master curve
min	At the minimum of the sine
MV+	At the mean value between minimum and maximum of the sine coming from the minimum
MV-	At the mean value between minimum and maximum of the sine coming from the maximum
norm	Normalized
p	Plastic
rad	Radial
SR	Stiffness ratio
tot	Total
tr	Transverse
u	Upper
ve	Viscoelastic
vol	Volumetric
vp	Viscoplastic

Prefixes

Δ   Change in…, Difference

X BIBLIOGRAPHY

AA542 Application of Strain Gauges to Cylindrical HMA-Specimens for Measurement of Radial Strain (in German) // Work Instruction. - Vienna : Institut of Transportation, Vienna University of Technology, 2010.

Aigner E. G. Multiscale Modeling of the Thermorheological Behavior of Building Materials // Doctoral Thesis. - Vienna, Austria : Vienna University of Technology, 2010.

Airey G. D., Collop A. C. and Zoorob S. The Influence of Laboratory Compaction Methods on the Performance of Asphalt Mixtures // Final Report. - [s.l.] : Engineering and Physical Science Research Council, 2005.

Airey G. D., Rahimzadeh B. and Collop A. C. Linear and Non-linear Rheological Properties of Asphalt Mixtures // Performance of Bitumunious and Hydraulic Materials in Pavements. - 2002.

Airey G., Rahimzadeh B. and Collop A. Viscoelastic Linearity Limits for Bituminous Materials // Proceedings of the 6th International RILEM Symposium on Performance Testing and Evaluation of Bituminous Materials. - Zurich, Switzerland : [s.n.], 2003.

Anderson D.A. [et al.] Binder Charaterization, Volume 3: Physical Properties // SHRP-A-369. - Washington D.C. : Strategic Highways Research Program, 1994.

Bari J. and Witczak M. W. Development of a new revised version of the Witczak E* predictive model for hot mix asphalt mixtures [Article] // Journal of the Association of Asphalt Pavement Technologists. - 2006. - 75.

Best R. Digitale Messwertverarbeitung (in German) // Book. - Munich, Germany : Oldenbourg Verlag GmbH, 1991.

Blab R. and Kappl K. Enhanced algorithms for the derivation of material parameter from triaxial cyclic compression tests on asphalt specimen // Proceedings of the 7th International Rilem Symposium on Acvanced Testind and Characterization of Bituminous Materials. - Island of Rhodes, Greece : [s.n.], 2009.

Blab R. and Litzka J. Progonose von Spurrinnenausbildungen in Asphaltbefestigungen (in German) // Schrfitenreihe Straßenforschung, Heft 485. - Vienna, Austria : Bundesministerium für wirtschaftliche Angelegenheiten, 1999.

Blab R. Methods for Optimization of Flexible Road Pavements. - Vienna : Vienna University of Technology, 2007.

Di Benedetto H. [et al.] Linear viscoelastic behavior of bituminous materials : from binders to mixes [Article] // Special Issue of Road Materials and Pavement Desgin. - 2004. - 5.

Di Benedetto H. [et al.] Stiffness testing for bituminous mixtures [Article] // Materials and Structures. - 2001. - Vol. 34.

Di Benedetto H., Delaporte B. and Sauzéat C. Three-Dimensional Linear Behavior of Bituminous Materials: Experiments and Modeling // International Journal of Geomechanics. - 2007. - Vol. 7.

DOLI Elektronic Gmbh Technical Details to WDC 580 [Online] // Technical Details to WDC 580. - 09 12, 2010. - 09 12, 2010. - http://www.doli.de/download/doc/EDC220_580.ZIP.

EN 12697-24 Bituminous mixtures - Test methods for hot mix asphalt - Part 24: Resistance to fatigue. - 2007.

EN 12697-25 Bituminous mixtures - Test methods for hot mix asphalt - Part 25: Cyclic compression test. - 2005.

EN 12697-26 Bituminous mixtures - Test methods for hot mix asphalt - Part 26: Stiffness. - 2004.

EN 12697-33 Bituminous mixtures - Test methods for hot mix asphalt - Part 33: Specimen prepared by roller compactor. - 2007.

EN 12697-35 Bituminous mixtures - Test methods for hot mix asphalt - Part 35: Laboratory mixing. - 2007.

EN 12697-46 Bituminous mixtures - Test methods for hot mix asphalt - Part 46: Low temperature cracking and properties. - 2009.

EN 13043 Aggregates for Bituminous Mixtures and Surface Treatments for Roads, Airfields and other Trafficked Areas. - 2002.

EN 13108-1 Bituminous mixtures - Material Specifications - Part 1: Asphalt Concrete. - 2006.

Ferry J.D. Viscoelastic Properties of Polymers. - [s.l.] : John Wiley & Sons, 1980.

Findley W. N., Kasif O. and Lai J. Creep and Relaxation of Nonlinear Viscoelastic Materials // Book. - Mineola : Dover Publications Inc., 1989.

Francken L. Permanent deformation law of bituminous road mixes in repeated triaxial compression // Proceedings of the 4th international conference on structural design of asphalt pavements. - Ann Arbor : [s.n.], 1977.

Fuessl J. Multiscale Fracture Modeling of Bituminous Mixtures // Doctoral Thesis. - Vienna, Austria : Vienna University of Technology, 2010.

Gabet T. [et al.] French wheel tracking round robin test on a polymer [Article] // Materials and Structures. - 2011. - 44. - pp. 1031-146.

Hoffmann K. An Introduction to Measurements using Strain Gages. - [s.l.] : HBM - Hottinger Baldwin Messtechnik, 1987.

Hofko B. and Blab R. Assessment of Permanent Deformation Behavior of Asphalt Concrete by Improved Triaxial Cyclic Compression Testing. - Nagoy, Japan : Proceedings of the 11th International Conference of Asphalt Pavements, 2010.

Hofko B. and Blab R. Assessment of Permanent Deformation Behavior of Asphalt Concrete by Improved Triaxial Cyclic Compression Testing. - Chicago, IL : Proceedings of the 1st Integrated Transport & Development Congress, 2011a.

Hofko B. Rheological Models to describe the Material Behavior of Hot Mix Asphalt (in German) // Master Thesis. - Vienna, Austria : Vienna University of Technology, 2006.

Hofko B., Blab R. and Ringleb A. Influence of Compaction Direction on Performance Characteristics of Roller-compacted HMA Specimens // (submitted). - [s.l.] : International Journal of Pavement Engineering, 2011b.

Höflinger G. Untersuchungen zur Probekörperherstellung von Walzasphalten mit dem Walzsegmentverdichter (in German) // Master Thesis. - Vienna : Vienna University of Technology, 2006.

Huet R. Etude par une méthode d'impendance du comportement viscoélastique des matériaux hydrocarbonés (in French) // Doctoral Thesis. - Paris, France : Faculté des Sciences de l'université de Paris, 1963.

Huschek S. Zum Verformungsverhalten von Asphaltbeton unter Druck // Doctoral Thesis. - Zurich, Switzerland : Swiss Federal Institute of Technology Zurich (ETH), 1983.

Jaeger W. Mechanisches Verhalten von Asphaltprobekörpern (in Germna) // Veröffentlichungen des Instituts für Straßenbau und Eisenbahnwesen der Universität Karlsruhe. - Karlsruhe, Germany : [s.n.], 1980.

Kappl K. Development of New Test Methods by Modeling Traffic Load with FEM // Presentation at the Evaluation of the Christian Doppler Laboratory. - Vienna : [s.n.], 2004.

Kappl K. F. Assessment and Modelling of Permanent Deformation Behaviour of Bituminous Mixtures with Triaxial Cyclic Compression Tests (in German) // Doctoral Thesis. - Vienna, Austria : Vienna University of Technology, 2007.

Keil S. Beanspruchungsermittlung mit Dehnungsmessstreifen (in German). - Zwingenberg, Germany : Cuneus-Verlag, 1995.

Krass K. Kriechverhalten an zylindrischen Asphaltprobekörpern (in German) // Veröffentlichungen des Instituts für Straßenbau und Eisenbahnwesen der Universität Karlsruhe. - Karsruhe, Germany : [s.n.], 1971.

Krass K. Modellvorstellungen zum Kriechverhalten (in German) // Colloquium 77, Plastische Verformungen. - Zürich : [s.n.], 1977.

Krebs H. G. and Jäger W. Mechanisches Verhalten von Asphaltprobekörern (in German) // Issue 334 of "Schriftenreihe Forschung Straßenbau und Straßenverkehrstechnik". - Bonn-Bad Godesberg, Germany : Minister of Transport, 1982.

Lackner R. [et al.] Characterization and Multiscale Modeling of Asphalt - Recent Developments in Upscaling of Viscous and Strength Properties // Proceedings of the 3rd European Conference on Computational Mechanics. - 2006.

Macrosensors Macro Sensors LVDT Tutorial // Website. - http://www.macrosensors.com/lvdt_tutorial.html : 2011/13/01, 2011.

Mang H. and Hofstetter G. Festigkeitslehre (in German). - Vienna, Austria : Springer Verlag, 2000.

Olard F. [et al.] Linear viscoelastic Properties of bituminous Binders and Mixtures at low and intermediate Temperatures // Road Materials and Pavement Design. - 2003. - Vol. 4.

ONR 23580 Type testing of bituminous mixtures - Rules for the implementation of OENORM EN 13108-20:2009. - 2009.

Partl M. N. Zum isothermen Kriechen eines bituminösen Mörtels unter mehrstufiger Belastung // Doctoral Thesis . - Swiss Federal Institut of Technology Zurich (ETH Zürich) : [s.n.], 1983.

Pellinen T. and Crockford B. Comparison of analysis techniques to obtain modulus and phase angle from sinusoidal test data // Proceedings of the 6th RILEM Symposium PTEBM'03. - Zurich : [s.n.], 2003.

Sachs L. and Hedderich J. Angewandte Statistik (in German) // Book. - Berlin : Springer, 2009.

Sayegh G. Variation des modules de quelques bitumes purs et bétons bitumineux // Doctoral Thesis. - Paris, France : Faculté des Sciences de l'université de Paris, 1965.

Steinmann S. and Friedrich C. Grundpraktikum Makromolekulare Chemie [Online]. - 12 10, 2010. - http://www.fmf.uni-freiburg.de/service/servicegruppen/sg_rheol/downloads/skripte/rheo.

Thamfeld H. Permanent deformation characterization of asphalt mixes by means of wheel track testing // Doctoral Thesis. - Vienna, Austria : Vienna University of Technology, 1990.

Trautwein U. Poroelastische Verformung und petrophysikalische Eigenschaften von Rotliegenden Sandsteinen (in German) // Doctoral Thesis. - Berlin, Germany : Berlin University of Technology, 2005.

Van der Poel C. A General System Describing the Viscoelastic Properties of Bitumens and its Relation to Routine Test Data // Journal of Applied Chemistry. - 1954.

Van Dijk W. and Visser W. Energy Approach to Fatigue Design // Proceedings of the Association of Asphalt Paving Technologists. - 1975.

von der Decken S. Triaxialversuch mit schwellendem Axial- und Radialdruck zur Untersuchung des Verformungswiderstandes von Asphalten (in German) // Schriftenreihe Institut für Straßenwesen Technische Universität Braunschweig. - Braunschweig, Germany : [s.n.], 1997.

Weiland N. Verformungsverhalten von Asphaltprobekörpern unter dynamischer Belastung (in German) // Veröffentlichungen des Instituts für Straßenbau und Eisenbahnwesen der Universität Karlsruhe. - Karlsruhe, Germany : [s.n.], 1986.

Weise C. and Wellner F. Determination of the Fatigue Behavior of Asphalt Mixes with the Triaxile Tensile Test // Proceedings of the 4th Eurasphalt and Eurobitume Congress. - Copenhagen, Denmark : [s.n.], 2008.

Widyatmoko I., Ellis C. and Read J.M. Energy Dissipation and the Deformation Resistance of Bituminous Mixtures // Materials and Structures. - 1999a.

Widyatmoko I., Ellis C. and Read J.M. The Application of the Dissipated Energy Method for Assessing the Performance of Polymer-modified Bituminous Mixtures // Materials and Structures. - 1999b. - Vol. 32.

Wistuba M. P. and Hauser E. Fundamental Requirements for Hot Mix Asphalt - Product Specification Sheets for selected Mixes // Presentation for the Advisory Board of the Christian Doppler Laboratory. - Vienna, Austria : [s.n.], 2007.

Witczak M.W. [et al.] Simple Performance Test for Superpave Mix Design // NCHRP Report 465. - 2002.

Zeidler E. Oxford Users' Guide to Mathematics // Book. - New York : Oxford University Press, 2004.

XI INDEX OF FIGURES AND TABLES

List of Figures

Figure I-1. Loading conditions in the TCCT according to (EN 12697-25, 2005); a) sinusoidal shaped axial loading and b) block-impulses as axial loading, both with constant confining pressure. 18

Figure I-2. Standard result of a TCCT according to (EN 12697-25, 2005); accumulated axial strain ε_n vs. number of load cycles n. 18

Figure I-3. Stress/strain situation under a passing tire in a pavement structure (top) and in the standard TCCT according to (EN 12697-25, 2005) (bottom). 24

Figure I-4. Road map to guide the reader through the thesis. 25

Figure I-5. Approach to achieve an enhanced TCCT with cyclic dynamic confining pressure 27

Figure II-1. Components and equipment of the 2-circuit triaxial test machine (Type LFV63/50). 30

Figure II-2. Phase lag induced by control unit due to multiplexing ADC. 31

Figure II-3. 5%, 50% and 95% quantiles of $\varphi_{ax,ax}$ for the aluminum specimen at 23°C. 32

Figure II-4. Main elements of the triaxial cell. 33

Figure II-5. Principle of the pneumatic device to apply cyclic confining pressure. 34

Figure II-6. Two LVDTs recording the axial deformation on both sides of the load plunger. 35

Figure II-7. "Chain Extensometer" attached to a cylindrical specimen. (Trautwein, 2005) 36

Figure II-8. Measuring frame on LVDT basis. 36

Figure II-9. Signals from force, axial deformation, and radial deformation (LVDT-based) sensor in a cyclic dynamic test. 37

Figure II-10. End of an SG glued to an HMA specimen. 38

Figure II-11. Cleaning of the adhesive area (left) and attaching the SG to the glued area (right). 40

Figure II-12. Overlapping ends of two SGs (left) and attaching two SGs temporarily to the specimen (right). 40

Figure II-13. Overlapping ends of two SGs before (left) after being glued together (right). 40

Figure II-14. HMA specimen glued to the load plates with an SG attached on both ends in axial direction. 41

Figure II-15. Static test to compare LVDT vs. SG signal data at 30°C; applied tensile force (top) and reaction of the specimen in terms of axial strain (bottom). 42

Figure II-16. Cyclic dynamic test to compare LVDT vs. SG signal data at 30°C; applied tensile force (top) and reaction of the specimen in terms of axial strain (bottom) at 0.1 Hz. 43

Figure II-17. 5%, 50% and 95% quantiles of the strain amplitude ratio SG/LVDT at 30°C. 45

Figure II-18. Classification of signals: a) Continuous signal in time and amplitude domain = analogue signal; b) signal continuous in time domain and discrete in amplitude domain; c) signal discrete in time domain and continuous in amplitude domain; d) discrete signal in time and amplitude domain = digital signal. According to (Best, 1991) 47

Figure II-19. Example of a digitalized signal with clear effects of noise (black) and analytical regression function by sinusoidal approximation (red). 47

Figure II-20. Principle of the evaluation software for the determination of material parameters and test results of cyclic dynamic material tests. 50

Figure II-21. Graphic example of the p-quantile for a normal distribution. 53

Figure II-22. Example of CCT test data with axial stress (σ_{ax}) and strain (ε_{ax}) and the analytical approximation $F+L$ – in time domain (top) and as a stress-strain relationship (bottom). 54

Figure II-23. Example of CCT test data with axial stress (σ_{ax}) and strain (ε_{ax}) and the analytical approximation $F+L+1H$ – in time domain (top) and as a stress-strain relationship (bottom). 55

Figure II-24. Definition of function values for the calculation of the four different phase lags. 56

Figure II-25. Variation of shift factor between 1st harmonic and fundamental from $a_6 = -180°$ (upper left) to $+180°$ (bottom). 58

Figure III-1. Produced grading curves for the AC 11 surf. 63
Figure III-2. Segment roller compactor used for slab compaction. 64
Figure III-3. Principle of specimen direction within HMA slab. 64
Figure IV-1. Example of a creep curve (cumulative axial deformation ε_n ($= \varepsilon_{ax}(n)$) vs. number of load cycles n) showing the different phases. 68
Figure IV-2. Differential quotient of axial strain vs. load cycles for the complete range of load cycles (top) and from $5,000^{th}$ to $25,000^{th}$ load cycle (bottom). 70
Figure IV-3. Linear (top) and exponential (bottom) approximation to the TCCT test data. 71
Figure IV-4. Evolution of $|E^*|$ (top), $\varphi_{ax,ax}$ (middle) and $\varphi_{ax,rad}$ (bottom) of AC 11 70/100 with the number of load cycles at standard TCCT conditions (50°C, 3 Hz). 74
Figure IV-5. Creep curve for AC 11 70/100; the point from where the behavior of the material is linear viscoelastic is marked (load cycle 628). 76
Figure IV-6. Creep curve for AC 11 70/100 and different approximation functions in the log/lin scale (top) and the lin/lin scale (bottom). 76
Figure IV-7. Fit quality of different approximation functions (top) and regression parameters (bottom). 77
Figure IV-8. Comparison of results from standard TCCT according to (EN 12697-25, 2005) with the linear and logarithmic regression. 79
Figure IV-9. Regression parameters from standard TCCT for AC 11 mixes. 80
Figure IV-10. Linear vs. logarithmic regression to the creep curve including the mixes tested in this chapter. 80
Figure IV-11. Different strain components from standard TCCT for an AC 11 70/100 mix vs. load cycles (top) and ratio of deviatoric and volumetric to total axial strain (bottom) at 50°C. 83
Figure IV-12. Strain ratio of deviatoric and volumetric to total axial strain from standard TCCT for an AC 11 PmB 45/80-65 diabase (a) and steel slack (b) at 50°C and steel slack (c) at 40°C. ... 84
Figure IV-13. Strain ratio of deviatoric and volumetric to total axial strain from standard TCCT for an SMA 11 PmB 45/80-65 diabase (a) and steel slack (b). 85
Figure IV-14. Strain ratio of deviatoric and volumetric to total axial strain from standard TCCT for an SMA 11 70/100 diabase (a) and steel slack (b). 86
Figure IV-15. Strain ratio of deviatoric and volumetric to total axial strain from standard TCCT for an SMA 11 PmB 45/80-65 diabase with high air void content. 87
Figure IV-16. Strain ratio of deviatoric and volumetric to total axial strain from standard TCCT for an SMA 11 70/100 diabase (a) and steel slack (b) with high air void content. 88
Figure IV-17. Strain ratio of deviatoric and volumetric to total axial strain from standard TCCT for an AC 22 PmB 45/80-65 limestone (a) and AC 22 50/70 limestone (b) at 40°C. 89
Figure IV-18. Strain ratio of deviatoric and volumetric to total axial strain from standard TCCT for AC 11 70/100 mixes with 4.8% (a), 5.3% (b) and 5.8% (c) binder content. 90
Figure IV-19. Strain ratio of deviatoric and volumetric to total axial strain from standard TCCT for AC 11 PmB 25/55-65 mix with 5.3% binder content. 91
Figure V-1. Illustration of the different deformation components. (Thamfeld, 1990) 97
Figure V-2. Stress $\sigma(t)$ and strain $\varepsilon(t)$ under sinusoidal loading for an elastic (upper left), viscoelastic (upper right) and viscous (below) material. 98
Figure V-3. Definition of function values for the calculation of the four different phase lags. 101
Figure V-4. Deformed specimen after CCT with actual shape (left) and adapted shape with uniform radial deformation (right). 104
Figure V-5. Example of test results from consolidation and main test phase at 50°C, MV of $\varphi_{ax,ax,max}$. 107
Figure V-6. Scheme of the test control in case of the specimen not being glued to the load plates. 109
Figure V-7. Two test setups to compare results of mix-controlled (a) and force-controlled (b) CTTs. 110

Figure V-8. 5%, 50% and 95% quantiles of the coefficient of determination R^2 for the force sensor for glued vs. unglued test setup; F+L approximation (top) and F+L+1H approximation (bottom). 112

Figure V-9. 5%, 50% and 95% quantiles of AR (top) and γ (bottom) for the force sensor for glued vs. unglued test setup at 30°C. 113

Figure V-10. 5%, 50% and 95% quantiles of the coefficient of determination R^2 for the deformation sensor for glued vs. unglued test setup at 30°C; F+L approximation (top) and F+L+1H approximation (bottom). 114

Figure V-11. 5%, 50% and 95% quantiles of AR (top) and γ (bottom) for the deformation sensor for glued vs. unglued test setup at 30°C. 116

Figure V-12. 5%, 50% and 95% quantiles of $\varphi_{ax,ax,max}$ (top) and $|E^*|$ (bottom) for glued vs. unglued test setup at 30°C; F+L approximation. 117

Figure V-13. 5%, 50% and 95% quantiles of $\varphi_{ax,ax,min}$ (top), $\varphi_{ax,ax,max}$ (2nd from above), $\varphi_{ax,ax,MV+}$ (3rd from above) and $\varphi_{ax,ax,MV-}$ (bottom) for glued vs. unglued test setup at 30°C; F+L+1H approximation. 119

Figure V-14. 5%, 50% and 95% quantiles of the coefficient of determination R^2 for the radial (top) and axial (bottom) deformation sensor for three deviatoric stress levels at 10°C; F+L+1H approximation. 122

Figure V-15. 5%, 50% and 95% quantiles of the radial (top) and axial (bottom) strain amplitude for three deviatoric stress levels at 10°C; F+L+1H approximation. 123

Figure V-16. 5%, 50% and 95% quantiles of $\varphi_{ax,rad,min}$ (top), $\varphi_{ax,rad,max}$ (2nd from above), $\varphi_{ax,rad,MV+}$ (3rd from above) and $\varphi_{ax,rad,MV-}$ (bottom) for three deviatoric stress levels at 10°C; F+L+1H approximation. 125

Figure V-17. 5%, 50% and 95% quantiles of $\varphi_{ax,ax,min}$ (top), $\varphi_{ax,ax,max}$ (2nd from above), $\varphi_{ax,ax,MV+}$ (3rd from above) and $\varphi_{ax,ax, MV-}$ (bottom) for three deviatoric stress levels at 10°C; F+L+1H approximation. 126

Figure V-18. 5%, 50% and 95% quantiles of $|E^*|$ for three deviatoric stress levels at 10°C; F+L+1H approximation. 127

Figure V-19. 5%, 50% and 95% quantiles of the coefficient of determination R^2 for the radial (top) and axial (bottom) deformation sensor for three deviatoric stress levels at 30°C; F+L+1H approximation. 128

Figure V-20. 5%, 50% and 95% quantiles of the radial (top) and axial (bottom) strain amplitude for three deviatoric stress levels at 30°C; F+L+1H approximation. 129

Figure V-21. 5%, 50% and 95% quantiles of $\varphi_{ax,rad,min}$ (top), $\varphi_{ax,rad,max}$ (2nd from above), $\varphi_{ax,rad,MV+}$ (3rd from above) and $\varphi_{ax,rad, MV-}$ (bottom) for three deviatoric stress levels at 30°C; F+L+1H approximation. 131

Figure V-22. 5%, 50% and 95% quantiles of $\varphi_{ax,ax,min}$ (top), $\varphi_{ax,ax,max}$ (2nd from above), $\varphi_{ax,ax,MV+}$ (3rd from above) and $\varphi_{ax,ax, MV-}$ (bottom) for three deviatoric stress levels at 30°C; F+L+1H approximation. 132

Figure V-23. 5%, 50% and 95% quantiles of $|E^*|$ for three deviatoric stress levels at 30°C; F+L+1H approximation. 133

Figure V-24. 5%, 50% and 95% quantiles of the coefficient of determination R^2 for the radial (top) and axial (bottom) deformation sensor for three deviatoric stress levels at 50°C; F+L+1H approximation. 134

Figure V-25. 5%, 50% and 95% quantiles of the radial (top) and axial (bottom) strain amplitude for three deviatoric stress levels at 50°C; F+L+1H approximation. 135

Figure V-26. 5%, 50% and 95% quantiles of $\varphi_{ax,rad,min}$ (top), $\varphi_{ax,rad,max}$ (2nd from above), $\varphi_{ax,rad,MV+}$ (3rd from above) and $\varphi_{ax,rad,MV-}$ (bottom) for three deviatoric stress levels at 50°C; F+L+1H approximation. 136

Figure V-27. 5%, 50% and 95% quantiles of $\varphi_{ax,ax,min}$ (top), $\varphi_{ax,ax,max}$ (2nd from above), $\varphi_{ax,ax,MV+}$ (3rd from above) and $\varphi_{ax,ax,MV-}$ (bottom) for three deviatoric stress levels at 50°C; F+L+1H approximation. 138

Figure V-28. 5%, 50% and 95% quantiles of $|E^*|$ for three deviatoric stress levels at 50°C; F+L+1H approximation. 139

Figure V-29. Mean values of $|E^*|$ for AC 11 70/100 vs. binder content for 10°C (top), 30°C (middle) and 50°C (bottom) and 0.1 Hz, 1 Hz and 10 Hz. 141

Figure V-30. Mean values of $|E^*|$ for AC 11 PmB 25/55-65 vs. binder content for 10°C (top), 30°C (middle) and 50°C (bottom) and 0.1 Hz, 1 Hz and 10 Hz. 142

Figure V-31. Mean values of $\varphi_{ax,ax,max}$ for AC 11 70/100 vs. binder content for 10°C (top), 30°C (middle) and 50°C (bottom) and 0.1 Hz, 1 Hz and 10 Hz. 144

Figure V-32. Mean values of $\varphi_{ax,ax,max}$ for AC 11 PmB 25/55-65 vs. binder content for 10°C (top), 30°C (middle) and 50°C (bottom) and 0.1 Hz, 1 Hz and 10 Hz. 145

Figure V-33. Mean values of $\varphi_{ax,rad,max}$ for AC 11 70/100 vs. binder content for 10°C (top), 30°C (middle) and 50°C (bottom) and 0.1 Hz, 1 Hz and 10 Hz. 147

Figure V-34. Mean values of $\varphi_{ax,rad,max}$ for AC 11 PmB 25/55-65 vs. binder content for 10°C (top), 30°C (middle) and 50°C (bottom) and 0.1 Hz, 1 Hz and 10 Hz. 148

Figure V-35. Mean values of $|E^*|$ for AC 11 70/100 vs. air void content for 10°C (top), 30°C (middle) and 50°C (bottom) and 0.1 Hz, 1 Hz and 10 Hz. 150

Figure V-36. Mean values of $|E^*|$ for AC 11 PmB 25/55-65 vs. air void content for 10°C (top), 30°C (middle) and 50°C (bottom) and 0.1 Hz, 1 Hz and 10 Hz. 151

Figure V-37. Mean values of $\varphi_{ax,ax,max}$ for AC 11 70/100 vs. air void content for 10°C (top), 30°C (middle) and 50°C (bottom) and 0.1 Hz, 1 Hz and 10 Hz. 153

Figure V-38. Mean values of $\varphi_{ax,ax,max}$ for AC 11 PmB 25/55-65 vs. air void content for 10°C (top), 30°C (middle) and 50°C (bottom) and 0.1 Hz, 1 Hz and 10 Hz. 154

Figure V-39. Mean values of $\varphi_{ax,rad,max}$ for AC 11 70/100 vs. air void content for 10°C (top), 30°C (middle) and 50°C (bottom) and 0.1 Hz, 1 Hz and 10 Hz. 156

Figure V-40. Mean values of $\varphi_{ax,rad,max}$ for AC 11 PmB 25/55-65 vs. air void content for 10°C (top), 30°C (middle) and 50°C (bottom) and 0.1 Hz, 1 Hz and 10 Hz. 157

Figure V-41. Coordinate system of compaction. 161

Figure V-42. 5%, 50% and 95% quantiles of R^2 for regression of the radial (top) and the axial deformation (bottom) for H-X-X vs. H-Z-Z orientation; F+L+1H approximation. 162

Figure V-43. 5%, 50% and 95% quantiles of AR of the radial (top) and the axial deformation (bottom) for H-X-X vs. H-Z-Z orientation; F+L+1H approximation. 163

Figure V-44. 5%, 50% and 95% quantiles of $\varphi_{ax,rad,max}$ for H-X-X vs. H-Z-Z orientation; F+L+1H approximation. 164

Figure V-45. 5%, 50% and 95% quantiles of $\varphi_{ax,ax,max}$ for H-X-X vs. H-Z-Z orientation; F+L+1H approximation. 164

Figure V-46. 5%, 50% and 95% quantiles of $\delta_{ax,rad,max}$ between radial and axial deformation for H-X-X vs. H-Z-Z orientation; F+L+1H approximation. 165

Figure V-47. Different SG setups to analyze the uniformity of radial deformation around the circumference; 1x SG 150 mm (top left = standard), 2x SG 150mm (top right) and 2x SG 100 mm (bottom). 166

Figure V-48. 50% quantiles of R^2 for regression of the radial (top) and the axial deformation (bottom) for three different SG setups; F+L+1H approximation. 168

Figure V-49. 50% quantiles of $\varphi_{ax,rad,max}$ (top) and $\varphi_{ax,rad,MV-}$ (bottom) for three different SG setups; F+L+1H approximation. 169

Figure V-50. 50% quantiles $\varphi_{ax,ax,max}$ (top) and $\varphi_{ax,ax,MV-}$ (bottom) for three different SG setups; F+L+1H approximation. 170

Figure V-51. 50% quantiles of $\delta_{ax,rad,max}$ between radial and axial phase lag for three different SG setups; F+L+1H approximation. ... 171

Figure V-52. MV and SD of $\varphi_{ax,ax,max}$ and $\varphi_{ax,rad,max}$ for an AC 11 70/100 and UCCT at 30°C. (Hofko, et al., 2011a) ... 172

Figure V-53. 50% quantiles of $\delta_{ax,rad,min}$ (top), $\delta_{ax,rad,max}$ (2nd from above), $\delta_{ax,rad,MV+}$ (3rd from above) and $\delta_{ax,rad,MV-}$ (bottom) for 3 different test temperatures and AC 11 PmB 25-55/65; F+L+1H approximation. ... 174

Figure V-54. 50% quantiles of $\delta_{ax,rad,min}$ (top), $\delta_{ax,rad,max}$ (2nd from above), $\delta_{ax,rad,MV+}$ (3rd from above) and $\delta_{ax,rad,MV-}$ (bottom) for 3 different test temperatures and AC 11 70/100; F+L+1H approximation. ... 176

Figure V-55. Recorded axial force and radial deformation with the LVDT-based measuring device at AC 11 70/100 specimen tested in UCCT at 30°C. ... 177

Figure V-56. 50% quantile of $\delta_{ax,rad,max}$ for LVDT-based radial deformation device vs. SG; F+L+1H approximation; AC 11 70/100 specimen tested in UCCT at 30°C. ... 178

Figure V-57. 5%, 50% and 95% quantiles of $|v^*|$ (top), v_1 (middle), v_2 (bottom) for the AC 11 PmB 25/55-65 at $\delta_{ax,rad,max}$; F+L+1H approximation. ... 180

Figure V-58. 5%, 50% and 95% quantiles of $|v^*|$ (top), v_1 (middle), v_2 (bottom) for the AC 11 70/100 at $\delta_{ax,rad,max}$; F+L+1H approximation. ... 181

Figure V-59. Mean values of $|v^*|$ for AC 11 PmB 25/55-65 vs. binder content for 10°C (top), 30°C (middle) and 50°C (bottom) and 0.1 Hz, 1 Hz and 10 Hz. ... 183

Figure V-60. Mean values of $|v^*|$ for AC 11 70/100 vs. binder content for 10°C (top), 30°C (middle) and 50°C (bottom) and 0.1 Hz, 1 Hz and 10 Hz. ... 184

Figure V-61. Mean values of $|v^*|$ for AC 11 PmB 25/55-65 vs. air void content for 10°C (top), 30°C (middle) and 50°C (bottom) and 0.1 Hz, 1 Hz and 10 Hz. ... 186

Figure V-62. Mean values of $|v^*|$ for AC 11 70/100 vs. air void content for 10°C (top), 30°C (middle) and 50°C (bottom) and 0.1 Hz, 1 Hz and 10 Hz. ... 187

Figure V-63. Mean values of $|G^*|$ for AC 11 PmB 25/55-65 (top) and AC 11 70/100 (bottom) vs. temperature for 0.1 Hz, 1 Hz and 10 Hz. ... 189

Figure V-64. Mean values of $|G^*|$ for AC 11 PmB 25/55-65 vs. binder content for 10°C (top), 30°C (middle) and 50°C (bottom) and 0.1 Hz, 1 Hz and 10 Hz. ... 190

Figure V-65. Mean values of $|G^*|$ for AC 11 70/100 vs. binder content for 10°C (top), 30°C (middle) and 50°C (bottom) and 0.1 Hz, 1 Hz and 10 Hz. ... 191

Figure V-66. Mean values of $|G^*|$ for AC 11 PmB 25/55-65 vs. air void content for 10°C (top), 30°C (middle) and 50°C (bottom) and 0.1 Hz, 1 Hz and 10 Hz. ... 193

Figure V-67. Mean values of $|G^*|$ for AC 11 70/100 vs. air void content for 10°C (top), 30°C (middle) and 50°C (bottom) and 0.1 Hz, 1 Hz and 10 Hz. ... 194

Figure V-68. Linking viscoelastic material behavior to the permanent deformation behavior for the linear creep rate (top) and the logarithmic creep rate (bottom) for standard TCCTs. ... 196

Figure V-69. Fit quality of the link between $/G^*/$ and the creep parameter. ... 196

Figure VI-1. Derivation of the master curve from test data at various temperatures and frequencies (top) to complete mater curve (bottom) for a T_{ref} of 30°C. ... 201

Figure VI-2. Master curve fit with approximation function according to (6.6) with no weighting (top), weighting 1/y (middle) and 1/y² (bottom). ... 204

Figure VI-3. Master curve fit with approximation function according to (6.7) and weighting 1/y (top) and the statistical parameters describing the fit quality (bottom). ... 206

Figure VI-4. DSR device used for binder and mastic tests. ... 208

Figure VI-5. Principle steps of the *B-A Model*. ... 209

Figure VI-6. Master curves from test data and regression analysis for PmB 25/55-65 ($|G^*|$ - top and G_1 – middle) and table with regression parameters and statistical values describing the fit quality (bottom). ... 214

Figure VI-7. Master curves from test data and regression analysis for AC 11 70/100 with a binder content of 5.0% (m/m) (umB_HMA_1): $|G^*|$ (top) and G_1 (bottom). 215

Figure VI-8. Master curves for all AC 11 PmB 25/55-65 mixes for $|G^*|$ (top), G_1 (2nd from above), $|E^*|$ (3rd from above) and E_1 (bottom). 216

Figure VI-9. Master curves for all AC 11 70/100 mixes for $|G^*|$ (top), G_1 (2nd from above), $|E^*|$ (3rd from above) and E_1 (bottom). 220

Figure VI-10. Ratios of $|G^*|$ (top) and G_1 (bottom) of binder, mastic and HMA for AC 11 PmB 25/55-65 with a binder content of 4.8% (m/m) (PmB_HMA_1). 224

Figure VI-11. Stiffness ratio between HMA and binder PmB 25/55-65 for $|G^*|$ (top), G_1 (2nd from above), $|E^*|$ (3rd from above) and E_1 (bottom). 225

Figure VI-12. Comparison of $|G^*|$ (top), G_1 (2nd from above), $|E^*|$ (3rd from above) and E_1 (bottom) of master curve and stiffness ratio for AC 11 PmB 25/55-65 mixes. 228

Figure VI-13. Stiffness ratio between HMA and binder 70/100 for $|G^*|$ (top), G_1 (2nd from above), $|E^*|$ (3rd from above) and E_1 (bottom). 229

Figure VI-14. Comparison of $|G^*|$ (top), G_1 (2nd from above), $|E^*|$ (3rd from above) and E_1 (bottom) of master curve and stiffness ratio for AC 11 70/100 mixes. 231

Figure VI-15. Graphical representation of the parameter fit for the *B-A Model* ($a_{b-a,o}$ for $|E^*|$ top, $a_{b-a,1}$ for $|E^*|$ middle and $a_{b-a,2}$ for $|E^*|$ bottom) showing that only one of three parameters depends on the binder used for the mix. 233

Figure VI-16. Comparison of $|G^*|$ (top), G_1 (2nd from above), $|E^*|$ (3rd from above) and E_1 (bottom) of master curve and *B-A Model* for AC 11 PmB 25/55-65 mixes. 237

Figure VI-17. Comparison of $|G^*|$ (top), G_1 (2nd from above), $|E^*|$ (3rd from above) and E_1 (bottom) of master curve and *B-A Model* for AC 11 70/100 mixes. 238

Figure VI-18. Sensitivity analysis for the *B-A Model* parameters for $|G^*|$ for the AC 11 70/100 mix with 5.0% (m/m) binder content (umB_HMA_1). 240

Figure VII-1. Approach to achieve an enhanced TCCT with cyclic dynamic confining pressure. 247

Figure VII-2. Example of recorded test data from enhanced TCCT with dynamic confining pressure. 249

Figure VII-3. Phase lag between axial loading and radial confining pressure induced by the test machine vs. given value from the standard TCCTs for AC 11 70/100 (binder content: 4.8% (m/m)) with a confining pressure of 150 to 250 kPa (top), 150 to 300 kPa (middle) and 150 to 350 kPa (bottom). 251

Figure VII-4. Total axial strain (top), volumetric axial strain (middle) and deviatoric axial strain (bottom) at load cycle 10,000 for AC 11 70/100 with a binder content of 4.8% (m/m) at different stress deviators from standard and enhanced TCCTs. 254

Figure VII-5. Creep rate f_c (top) and parameter b (bottom) from logarithmic regression of creep curve for AC 11 70/100 with a binder content of 4.8% (m/m) at different stress deviators from standard and enhanced TCCTs. 255

Figure VII-6. Phase lag between axial loading and radial confining pressure induced by the test machine vs. given value from the standard TCCTs for AC 11 70/100 (binder content: 5.3% (m/m)) with a confining pressure of 150 to 250 kPa (top), 150 to 300 kPa (middle) and 150 to 350 kPa (bottom). 256

Figure VII-7. Total axial strain (top), volumetric axial strain (middle) and deviatoric axial strain (bottom) at load cycle 10,000 for AC 11 70/100 with a binder content of 5.3% (m/m) at different stress deviators from standard and enhanced TCCTs. 258

Figure VII-8. Creep rate f_c (top) and parameter b (bottom) from logarithmic regression of creep curve for AC 11 70/100 with a binder content of 5.3% (m/m) at different stress deviators from standard and enhanced TCCTs. 259

Figure VII-9. Phase lag between axial loading and radial confining pressure induced by the test machine vs. given value from the standard TCCTs for AC 11 70/100 (binder content:

5.8% (m/m)) with a confining pressure of 150 to 250 kPa (top), 150 to 300 kPa (middle) and 150 to 350 kPa (bottom). 260

Figure VII-10. Total axial strain (top), volumetric axial strain (middle) and deviatoric axial strain (bottom) at load cycle 10,000 for AC 11 70/100 with a binder content of 5.8% (m/m) at different stress deviators from standard and enhanced TCCTs. 262

Figure VII-11. Creep rate f_c (top) and parameter b (bottom) from logarithmic regression of creep curve for AC 11 70/100 with a binder content of 5.8% (m/m) at different stress deviators from standard and enhanced TCCTs. 263

Figure VII-12. Phase lag between axial loading and radial confining pressure induced by the test machine vs. given value from the standard TCCTs for AC 11 PmB 25/55-65 with a confining pressure of 150 to 250 kPa (top), 150 to 300 kPa (middle) and 150 to 350 kPa (bottom). 265

Figure VII-13. Total axial strain (top), volumetric axial strain (middle) and deviatoric axial strain (bottom) at load cycle 10,000 for AC 11 PmB 25/55-65 with a binder content of 5.3% (m/m) at different stress deviators from standard and enhanced TCCTs. 266

Figure VII-14. Creep rate f_c (top) and parameter b (bottom) from logarithmic regression of creep curve for AC 11 PmB 25/55-65 with a binder content of 5.3% (m/m) at different stress deviators from standard and enhanced TCCTs. 267

Figure A-1. Sketch of a foil SG. (Keil, 1995) 299

Figure A-2. Typical shapes of measuring grids for linear SGs with different shapes of connections (top), a short measuring grid (0.6 mm) (middle) and a long measuring grid (150 mm) (bottom). (Hoffmann, 1987)..300

Figure A-3. The introduction of the object's strain into the measuring grid. (Hoffmann, 1987) 300

Figure A-4. Qualitative example of recorded strain data with measuring grid of different lengths. (Hoffmann, 1987)..301

Figure A-5. Electrical conductor under longitudinal loading. (Mang, et al., 2000).................................302

Figure A-6. Definition of the transverse sensitivity: a) SG stressed in its active direction and b) transversal to its active direction. (Hoffmann, 1987) 303

Figure A-7. Temperature dependency of the gauge factor for different measuring grid materials. (Hoffmann, 1987)..304

Figure A-8. Change in the SG signal with time after instantaneous loading of the object and after its instantaneous release. (Hoffmann, 1987) 306

Figure A-9. Example of the elastic after-effect on the strain during constant loading and after complete release. (Hoffmann, 1987) 306

Figure A-10. Schematic diagram showing the compensation of the SG creep by the elastic after-effect. (Hoffmann, 1987)..306

Figure A-11. Sketch of a standard measuring setup. According to (Mang, et al., 2000)........................307

Figure A-12. Different representations of the Wheatstone bridge circuit. (Hoffmann, 1987).................308

Figure A-13. Configurations of the Wheatstone bridge circuit: a) quarter bridge, b) half bridge, c) double quarter or diagonal bridge, d) full bridge. 309

Figure A-14. Principle of the shunt calibration. (Hoffmann, 1987) 310

Figure B-1. Master curves from test data and regression analysis for PmB 25/55-65 ($|G^*|$ - top, G_1 – middle and G_2 – bottom). 314

Figure B-2. Stiffness ratio of the mastic compared to the binder PmB 25/55-65 for $|G^*|$ (top), G_1 (middle), G_2 (bottom). 316

Figure B-3. Stiffness ratio of the mastic compared to the binder 70/100 for $|G^*|$ (top), G_1 (middle), G_2 (bottom). 318

Figure B-4. Comparison of $|G^*|$ (top) and G_1 (bottom) of master curve and stiffness ratio according to (B.1) for PmB 25/55-65 mastic. 321

Figure B-5. Comparison of $|G^*|$ (top) and G_1 (bottom) of master curve and stiffness ratio according to (B.1) for 70/100 mastic. 322

Figure B-6. Graphical representation of (B.3) for two cases: $a_{m-b,1}$ for G_1 of PmB 25/55-65 mastic (lower R²) and $a_{m-b,2}$ for $|G^*|$ of PmB 25/55-65 mastic (high R²). 324

Figure B-7. Comparison of $|G^*|$ (top) and G_1 (bottom) from mastic master curve and *B-M Model* for PmB 25/55-65 mastic. 326

Figure B-8. Comparison of $|G^*|$ (top) and G_1 (bottom) from master curve and *B-M Model* for 70/100 mastic. 327

Figure B-9. Comparison of $|G^*|$ from test data and master curve (top) and test data and *B-M Model* (bottom) for PmB 25/55-65 mastic. 328

Figure B-10. Sensitivity analysis for the *B-M Model* parameters for $|G^*|$ of a PmB 25/55-65 mastic with a filler ratio of 1.9. 329

List of Tables

Table II-1. Characteristics of the test machine LFV63/50. 30
Table II-2. Axial strain amplitudes of LVDT and SG for 0.1 Hz and 0.5 Hz at 30°C. 44
Table II-3. Example of an output file from the evaluation routine. 52
Table II−4. Input data used for analysis of the advanced approximation function *F+L+1H*. 57
Table II-5. Phase lags between advanced and standard approximation function at different functional values. 60
Table III-1. Characteristics of "Loja" according to (EN 13043, 2002) 61
Table III-2. Characteristics of bitumen. 62
Table IV-1. Mixes used for investigation of alternative assessment of deformation resistance. 72
Table V-1. Different deformation components. 96
Table V-2. Standard test conditions for the determination of viscoelastic properties. 107
Table V-3. Standard test characteristics for the determination of viscoelastic properties. 107
Table V-4. AC 11 specimen characteristics used for tests to compare results of glued and unglued setups. 110
Table V-5. Specimen characteristics used for tests to compare results at different deviatoric stress levels at 10°C. 122
Table V-6. Specimen characteristics used for tests to compare results at different deviatoric stress levels at 30°C. 127
Table V-7. Specimen characteristics used for tests to compare results at different deviatoric stress levels at 50°C. 133
Table V-8. Specimen characteristics used for tests to compare specimen orientation. 160
Table V-9. Specimen characteristics used for tests to compare results with different SG setups. 167
Table V-10. Specimen characteristics (AC 11 PmB 25/55-65) used for tests to investigate correlations between $|E^*|$ and δ. 172
Table V-11. Specimen characteristics (AC 11 70/100) used for tests to investigate correlations between $|E^*|$ and δ. 173
Table VI-1. Parameters describing the fit quality of the fits with the approximation function according to (6.6) for the examples given in Figure VI-2. 206
Table VI-2. Test procedure for the determination of viscoelastic properties of binder and mastic in the DSR. 208
Table VI-3. Binders and mastic tested with DSR. 208
Table VI-4. Volumetric characteristics of PmB_HMA_1 specimens (AC 11 PmB 25/55-65 with a binder content of 4.8% (m/m)). 211
Table VI-5. Volumetric characteristics of PmB_HMA_2 specimens (AC 11 PmB 25/55-65 with a binder content of 5.3% (m/m)). 211
Table VI-6. Volumetric characteristics of PmB_HMA_3 specimens (AC 11 PmB 25/55-65 with a binder content of 5.3% (m/m) and high content of air voids). 211

Table VI-7. Volumetric characteristics of PmB_HMA_4 specimens (AC 11 PmB 25/55-65 with a binder content of 5.8% (m/m)). 212

Table VI-8. Volumetric characteristics of umB_HMA_1 specimens (AC 11 70/100 with a binder content of 5.0% (m/m)). 212

Table VI-9. Volumetric characteristics of umB_HMA_2 specimens (AC 11 70/100 with a binder content of 5.3% (m/m) and a high content of air voids). 212

Table VI-10. Volumetric characteristics of umB_HMA_3 specimens (AC 11 70/100 with a binder content of 5.3% (m/m)). 212

Table VI-11. Volumetric characteristics of umB_HMA_4 specimens (AC 11 70/100 with a binder content of 5.6% (m/m)). 212

Table VI-12. Regression (according to equation (6.7)) and statistical parameters of AC 11 PmB 25/55-65 master curves for $|G^*|$ and G_1 in Pa. 217

Table VI-13. Regression (according to equation (6.7)) and statistical parameters of AC 11 PmB 25/55-65 master curves for $|E^*|$ and E_1. 218

Table VI-14. Regression (according to equation (6.7)) and statistical parameters of AC 11 70/100 master curves for $|G^*|$ and G_1 in Pa. 221

Table VI-15. Regression (according to equation (6.7)) and statistical parameters of AC 11 70/100 master curves for $|E^*|$ and E_1 in Pa. 222

Table VI-16. Parameters for function (6.12) describing the stiffness ratio between HMA and binder PmB 25/55-65 analytically for $|G^*|$ (top), G_1 (2nd from above), $|E^*|$ (3rd from above) and E_1 (bottom). 226

Table VI-17. Parameters for function (6.12) describing the stiffness ratio between HMA and binder 70/100 analytically for $|G^*|$ (top), G_1 (2nd from above), $|E^*|$ (3rd from above) and E_1 (bottom). 230

Table VI-18. Parameters for function (6.15) describing the *B-A Model* for tested AC 11 mixes. 235

Table VI-19. Input data for prediction of the viscoelastic material parameters of PmB_HMA_1 by use of the *B-A Model*. 241

Table VI-20. $|E^*|$ (top) and E_1 (bottom) derived from test data (CCT), from analytical expression of the HMA master curve (MC) and predicted from the *B-A Model* (mod). 242

Table VI-21. $|G^*|$ (top) and G_1 (bottom) derived from test data (CCT), from analytical expression of the HMA master curve (MC) and predicted from the *B-A Model* (mod). 242

Table VI-22. E_2 (top) and G_2 (bottom) derived from test data (CCT), from analytical expression of the HMA master curve (MC) and predicted from the *B-A Model* (mod). 243

Table VI-23. $\varphi_{ax,ax}$ derived from test data (CCT), from analytical expression of the HMA master curve (MC) and predicted from the *B-A Model* (mod). 243

Table VI-24. $|v^*|$ derived from test data (CCT), from analytical expression of the HMA master curve (MC) and predicted from the *B-A Model* (mod). 243

Table VI-25. $\delta_{ax,rad}$ derived from test data (CCT), from analytical expression of the HMA master curve (MC) and predicted from the *B-A Model* (mod). 244

Table VI-26. v_1 (top) and v_2 (bottom) derived from test data (CCT), from analytical expression of the HMA master curve (MC) and predicted from the *B-A Model* (mod). 244

Table VII-1. Test program for the advanced characterization of the resistance to permanent deformation. 248

Table VII-2. Radial phase lags $\varphi_{ax,rad}$ derived from the standard TCCTs. 248

Table A-1. Average gauge factors for different measuring grid materials. (Hoffmann, 1987) 302

Table A-2. Possible disturbance values impacting the recorded strain. (Hoffmann, 1987) 312

Table B-1. Parameters for function (B.1) describing the relation between a certain mastics M(j) with binder:mastic = 1:j and the respective binder PmB 25/55-65 for $|G^*|$ (top) and G_1 (bottom). 319

Table B-2. Parameters for function (B.1) describing the relation between a certain mastics M(j) with binder:mastic = 1:j and the respective binder 70/100 for $|G^*|$ (top) and G_1 (bottom). 320

Table B-3. Parameters for function (B.3) for PmB 25/55-65 mastic (left) and 70/100 mastic (right). 323

A Annex: The Principle of Strain Gauges

A.1 Basic Information

The actual measuring element of the most commonly used type of SG, the metal foil SG, is a thin electric conductor which represents the measuring grid. This grid is embedded in a carrier foil that isolates the measuring grid from the object and transfers the strain from the object to the grid. The coating layer protects the measuring grid from damage. Figure A-1 contains a sketch of a typical foil SG and its different layers.

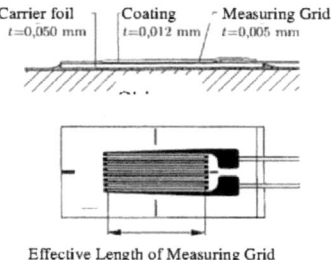

Figure A-1. Sketch of a foil SG. (Keil, 1995)

SGs with their small dimensions and weight (around 10 mg to 500 mg) hardly influence any object. Even small objects do not change their static and dynamic behavior if the right choice of SG is used. Depending on the application a vast variety of different types of SGs exist.

SG can be distinguished by the measuring grid length, as well as by the shape and location of the connections. Figure A-2 shows typical linear SGs. In terms of the orientation of the measuring grid, linear SGs are the most common type. For other applications (e.g. stress analysis), X-rosettes with two measuring grids orthogonal to each other, R-rosettes with three measuring grids in different angles, SG-chains and other configurations for special applications are available.

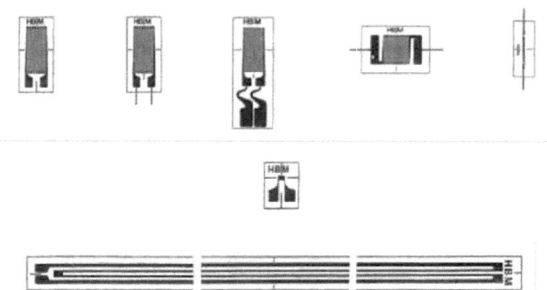

Figure A-2. Typical shapes of measuring grids for linear SGs with different shapes of connections (top), a short measuring grid (0.6 mm) (middle) and a long measuring grid (150 mm) (bottom). (Hoffmann, 1987)

An important fact is that the length of the measuring grid has no impact on the sensitivity of the device. Since the signal is correlated to the relative change in length – the strain – and not the absolute change in length, it does not matter how long the measuring grid is. It is advisable to use longer SGs rather than short ones if there is enough space on the object.

One important aspect that needs to be taken into consideration is the fact, that the strain of the object must be introduced into the measuring grid via the adhesive and the carrier of the grid. Therefore a certain length within the layers is necessary. This length is a function of the thickness of each layer and the stiffness of the layers. Figure A-3 presents a qualitative example of this effect.

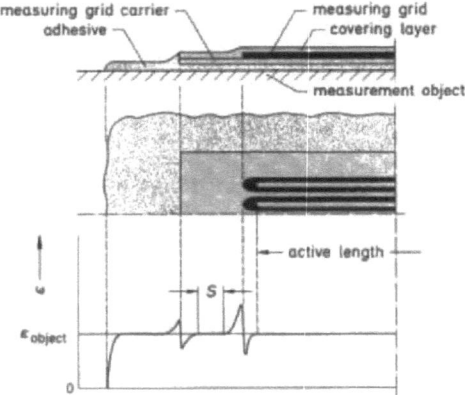

Figure A-3. The introduction of the object's strain into the measuring grid. (Hoffmann, 1987)

The length for the transfer of strain from the object to the measuring grid varies with the temperature because the stiffness of the materials (carrier foil and adhesive) is temperature-depending. At higher temperatures the transfer length will be longer than for lower temperatures. To give a number, roughly 1 to 2 mm have to be taken into account at each end.

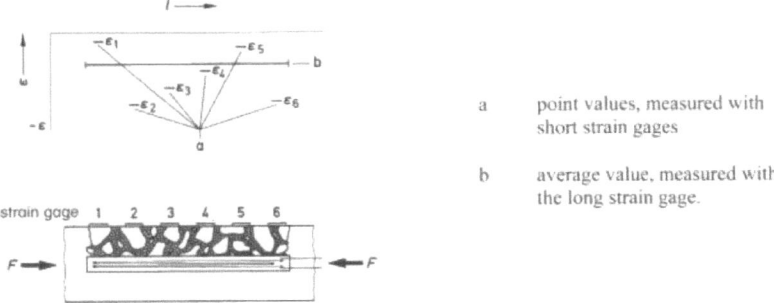

a   point values, measured with short strain gages

b   average value, measured with the long strain gage.

Figure A-4. Qualitative example of recorded strain data with measuring grid of different lengths. (Hoffmann, 1987)

Another important aspect is the minimum length of the measuring grid. Especially for inhomogeneous materials, like HMA or concrete, an SG with a too short length would only record a partial strain (e.g. between on aggregate and the mastic). Therefore it is advised to choose the length of the measuring grid to be at least 5 times the largest aggregate size of a mix (Hoffmann, 1987). If this lower limit is exceeded an average rather than a partial strain is measured.

A.2 Principle of Measurement

The physical principle of the SG is quite simple. The resistance of an electric conductor changes with a change in length of the respective conductor. The resistance R of a conductor with a circular profile is defined as

$$R = \frac{\rho \cdot l}{A} = \frac{4 \cdot \rho \cdot l}{\pi \cdot a^2} \qquad (A.1)$$

ρ............ Specific resistance [Ωm]
l............ Length of the conductor [m]
A............ Cross section of the conductor [m²]
a............ Diameter of the conductor [m]

If the conductor is subjected to mechanical loading and the length is increased by Δl, the diameter decreased by Δa (see Figure A-5) the relative change in the resistance is

$$\frac{\Delta R}{R} = \frac{\Delta \rho}{\rho} + \frac{\Delta l}{l} - 2\frac{\Delta a}{a} \qquad (A.2)$$

If it is now taken into consideration that $\Delta l/l = \varepsilon$ and $\Delta a/a = -\nu\varepsilon$ (ν being the Poisson's Ratio) and $\Delta \rho = \beta_p \varepsilon$ (β_p being a material parameter for conducting materials), (A.2) can be quoted as follows:

$$\frac{\Delta R}{R} = (1+2\nu+\beta_p)\cdot\varepsilon = k\cdot\varepsilon \qquad (A.3)$$

ε............Strain [-]
k............Gauge factor [-]

The relative change of the resistance is thus proportional to the introduced strain ε. The gauge factor k is a parameter to characterize the sensitivity of an SG. k is a value that is derived by experimental means by the manufacturer and given for each individual set of SGs. For constantan, a material commonly used for the measuring grid, the gauge factor ranges from ±1 to ±3 mm/m. (Mang, et al., 2000)

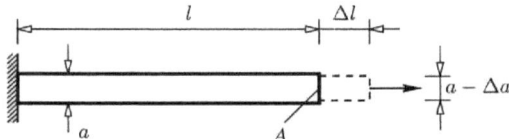

Figure A-5. Electrical conductor under longitudinal loading. (Mang, et al., 2000)

A.3 Characteristics of Strain Gauges

The most important parameters and characteristics of SGs will be presented in the following.

Stain Gauge Sensitivity

Depending on the material of the conductive measuring grid, different gauge factors occur. Table A-1 shows average k factors for common measuring grid materials.

Table A-1. Average gauge factors for different measuring grid materials. (Hoffmann, 1987)

Material	Guide values for composition	Approximate gauge factor
Constantan	57 Cu, 43 Ni	2.05
Karma	73 Ni, 20 Cr, res. F + Al	2.1
Nichrome V	80 Ni, 20 Cr	2.2
Platinum-Tungsten	92 Pt, 8 W	4.0

Transverse Sensitivity

SGs should only react to strain with a change in resistance in the active direction of the measuring grid. Still, a change in resistance can also occur when the SG is loaded transversal to its active direction. This reaction is referred to as the transverse sensitivity q derived from the ratio of the gauge factors in and transverse to the active direction:

$$q = \frac{k_q}{k_l} \quad (A.4)$$

with

$$k_l = \frac{\Delta R / R_0}{\varepsilon_l} \quad (A.5)$$

$$k_q = \frac{\Delta R / R_0}{\varepsilon_q} \quad (A.6)$$

ε_l .... Strain in the active direction of the SG (according to Figure A-6)
ε_q .... Strain transversal to the active direction of the SG (according to Figure A-6)

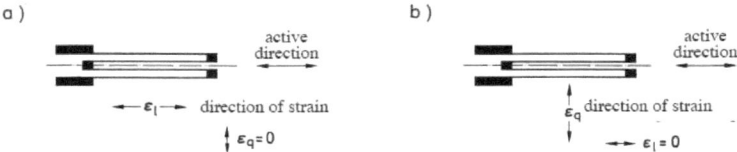

Figure A-6. Definition of the transverse sensitivity: a) SG stressed in its active direction and b) transversal to its active direction. (Hoffmann, 1987)

The transverse sensitivity given by the manufacturer is around -0.1% for standard SGs.

Temperature Response of SGs

The temperature response of an SG ε_T is a temperature-dependent change of the signal in a state free of any or at a constant level of mechanical stress of the SG. This response matters only in a situation when the temperature is changed throughout the measuring. If the temperature is set back to its initial value, the temperature response disappears. It is therefore reversible. The temperature response is a function of

- α_C the object's thermal expansion,
- α_M the thermal expansion of the material of the SG's measuring grid,
- α_R the temperature coefficient of the grid material's electrical resistance and
- ΔT the temperature change.

Since the factors named above are temperature-dependent themselves, a relation can be given only for a limited temperature range according to (Hoffmann, 1987):

$$\varepsilon_T = \left(\frac{\alpha_R}{k} + \alpha_C - \alpha_M \right) \cdot \Delta T \quad (A.7)$$

If mechanical stress is superposed by a change of temperature, a virtual strain component ε_T is measured and biases the results. There are different ways to compensate this effect, e.g. by self-compensating SGs. Self-compensating SGs can balance out the temperature response because their temperature coefficient matches the temperature coefficient of the object. Tests carried out at constant temperature show no temperature response. Within this thesis tests were run at a constant temperature.

Impact of the Temperature on the SG Sensitivity

The sensitivity of SGs is represented by the gauge factor k. This factor is a function of temperature. The information given by the manufacturer regarding k is only valid at room temperature, strictly speaking. Depending on the material of the measuring grid and the temperature, different values for k can be measured. Figure A-7 shows an example for four common materials and a wide range of temperatures. The x-axis gives the temperature in °C and the y-axis the relative deviation of the k factor compared to its standard value at room temperature (23°C). Constantan for example describes an increasing sensitivity with increasing temperature. Yet, the incline is rather small. At a temperature of 200°C, the relative change of the sensitivity is clearly below 5%. Given the fact that the highest temperature used within this research was 50°C, the temperature sensitivity can be left out of the considerations.

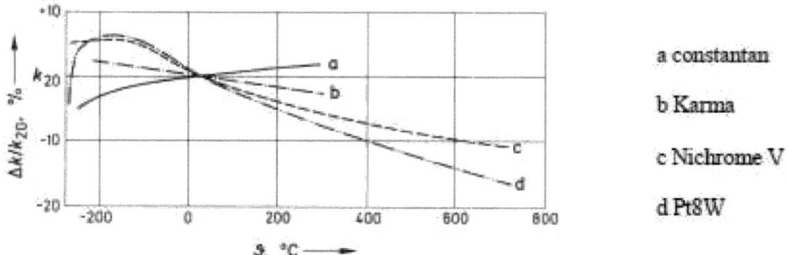

Figure A-7. Temperature dependency of the gauge factor for different measuring grid materials. (Hoffmann, 1987)

Static Elongation Limits

In general the strain of SGs is restricted to about ±3,000 μm/m which corresponds to 0.3%. For special applications (e.g. on certain synthetics) this range can be exceeded if specially configured SGs are used. These SGs can reach strains up to 20,000 μm/m or even 200,000 μm/m. Special attentions needs to be laid on the maximum extensibility and bonding strength of the adhesive at these extreme conditions. The maximum strain for the SGs used in this research is 50,000 μm/m which corresponds to 5.0%. (Hoffmann, 1987)

Dynamic Strain Measurement

SGs are perfectly capable of dynamic strain measurements since they have no significant influence on the object due to their small mass. Still, metal foil SGs react with fatigue to a dynamic strain measurement. The degree of fatigue depends on the strain amplitude and the number of load cycles. Two phenomenona are seen as the main reasons for fatigue (Hoffmann, 1987):

- An increase in electrical resistance occurs which is apparent in a dynamic zero point drift and
- with increasing disintegration of the material, microcracks occur on the grain boundaries in the metal of the measuring grid leading to a macrocrack eventually.

For commonly used SGs a stability in dynamic strain measurements for a strain amplitude of up to 2,000 µm/m can be taken as granted if the number of load cycles does not exceed 10^5.

Electric Loading

SGs usually have a small cross section of the measuring grid. Even with small electric loading, given an example of a voltage of 5 V and a grid resistance of 120 Ω, a current of 20.8 mA flows through a regular SG. At a glance, this is not very much. But if the cross section of the grid is taken into consideration a current density of 46 A/mm² occurs, which is extremely high, even in the field of power engineering (Hoffmann, 1987). To keep the heating at the SG within acceptable limits, the power transferred through the SG must be in some sort of equilibrium with the dissipated power. The following parameters have a significant influence on this ratio:
- the bridge excitation voltage level,
- the electrical resistance of the measuring grid,
- the size and geometry of the SG's measuring grid,
- the ambient temperature,
- the thermal conductance of the component material and
- the component's thermal capacity, its radiation and cooling characteristics.

The manufacturer gives a maximum permissible bridge excitation voltage level. It must be kept in mind that this number is only valid if the SG is attached to an object with good thermal conductivity properties and at room temperature. If the temperature is higher or the object's thermal conductivity is at a low level, the bridge excitation voltage must be reduced in order not to damage the SG.

Creep Effects

The reason for creep effects can be found in the rheological behavior of the strain transferring layers: the adhesive and the carrier foil of the measuring grid. When loaded the extended measuring grid is similar to a tensioned spring. Mainly in the region of the measuring grid's end loops on the contact surfaces between the measuring grid and the carrier the force of the spring produces shear stress. This shear stress is superposed by the normal stress due to the extension. The synthetics in the SG and the bonding material relax, i.e. the reactive force slackens, the measuring grid draws in and a negative error occurs. A qualitative example is shown in Figure A-8. This effect has a relatively larger effect on short measuring grids than on long ones.

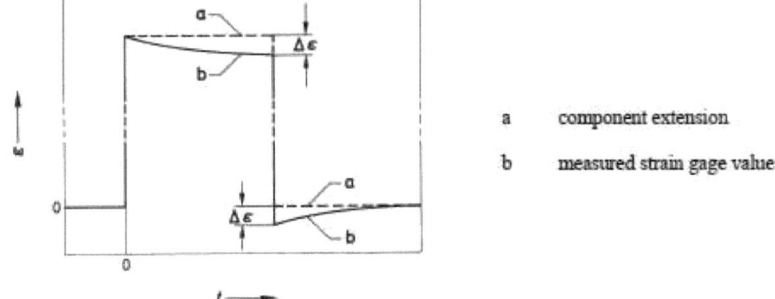

Figure A-8. Change in the SG signal with time after instantaneous loading of the object and after its instantaneous release. (Hoffmann, 1987)

In addition to the instantaneous strain of even purely elastic objects, time-dependent, asymptotic, reversible strain occurs that is commonly called the "elastic after-effect". This effect results in a positive error shown in Figure A-8.

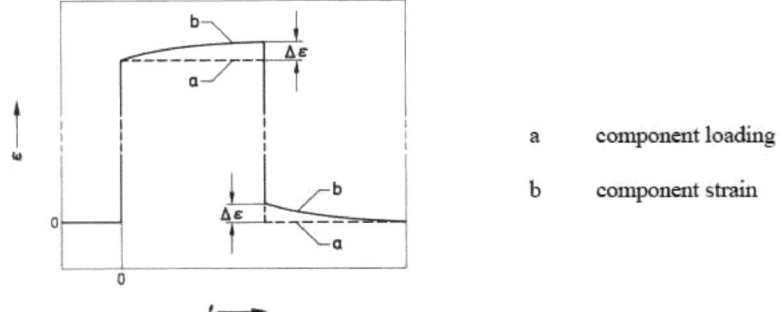

Figure A-9. Example of the elastic after-effect on the strain during constant loading and after complete release. (Hoffmann, 1987)

SG creep and elastic after-effect have opposite signs and – in the most favorable case – a fair compensation can be obtained.

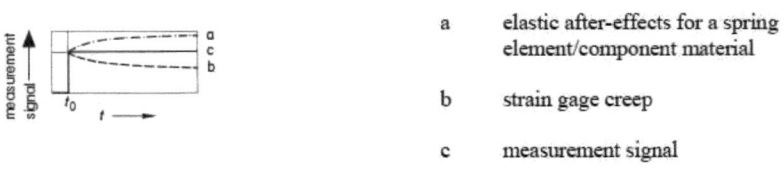

Figure A-10. Schematic diagram showing the compensation of the SG creep by the elastic after-effect. (Hoffmann, 1987)

Temperature is another factor influencing the creep of SGs. (Hoffmann, 1987) shows results of tests that were carried out on elastic objects with standard SGs attached. A constant normal force was applied and held constant for 24 h at different temperatures (23°C, 60°C, 100°C). After 24 h the load was removed and the recovering of the strain was recorded. The elastic after-effect at 60°C is about 0.2% of the measured value after 24 h and is completely reversible. Only at 100°C irreversible creep effects occur with a permanent displacement of the zero point.

A.4 Recording Strain

The strain of an object is transferred to the SG which causes a change in resistance within the SG. The change in resistance is usually small and not suited for direct recording. By bridge circuits these changes in resistance are transformed to changes in voltage or current. An additional amplifier increases the level of voltage or current. Thus the signals are easily recordable with high accuracy. The process from the strain on the object to a digital data item is presented in Figure A-11

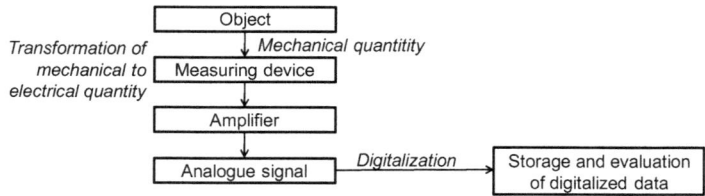

Figure A-11. Sketch of a standard measuring setup. According to (Mang, et al., 2000)

The easiest way to measure changes in electrical resistance is the Wheatstone bridge circuit. It was originally invented in the middle of the 19th century to obtain a material's resistance by comparing it with a known resistance value. When it is used with SGs, its purpose is to record changes in the resistance of the SG.

Figure A-12 explains the principle of the Wheatstone bridge circuit. Four arms or branches of the bridge are represented by four resistances R_1 to R_4. The corner points 2 and 3 are connections for the bridge excitation voltage V_S and the two other corner points 1 and 4 are connections for the bridge output voltage V_o, the measuring signal.

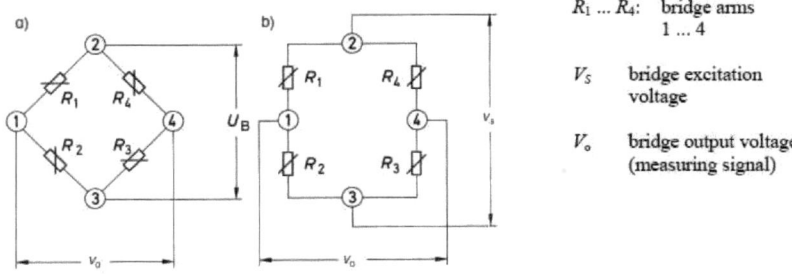

Figure A-12. Different representations of the Wheatstone bridge circuit. (Hoffmann, 1987)

When excitation voltage V_S is applied to the bridge, the voltage is divided into both bridge halves $R_1{:}R_2$ and $R_4{:}R_3$ depending on the ratio of the bridge resistances. Each half of the bridge represents a voltage divider. According to Ohm's and Kirchhoff's law the ratio between V_o and V_S is

$$\frac{V_o}{V_S} = \frac{R_1}{R_1 + R_2} - \frac{R_4}{R_3 + R_4} \tag{A.8}$$

For the two cases $R_1 = R_2 = R_3 = R_4$ and $R_1/R_2 = R_4/R_3$ a balanced bridge circuit is achieved with an output voltage of 0.

If the resistances R_1 to R_4 are changed (e.g. if the resistance of an SG changes during a measurement) the bridge circuit gets unbalanced and a certain output voltage V_o is recorded. If it is taken into consideration that the four resistances have equal nominal values $R_i = R$ and very small changes in resistance $\Delta R_i \ll R_i$ occur, it can be stated that

$$\frac{V_o}{V_S} \approx \frac{1}{4} \cdot \left(\frac{\Delta R_1}{R} - \frac{\Delta R_2}{R} + \frac{\Delta R_3}{R} - \frac{\Delta R_4}{R} \right) \tag{A.9}$$

and by considering (A.3)

$$\frac{V_o}{V_S} \approx \frac{k}{4} \cdot (\varepsilon_1 - \varepsilon_2 + \varepsilon_3 - \varepsilon_4) \tag{A.10}$$

The formula reveals that changes in resistance caused by strain have different signs depending on the location of an SG within the bridge. Adjoining resistances are being subtracted when they have the same sign and are added up when their signs are different. This fact comes in handy for certain application, especially when temperature effects should be compensated.

At least one of the four resistances in the bridge must be replaced by an SG to construct a valid measuring system. In case of one SG, R_1 represents the SG and the other three resistances are constant completion resistances. Alternatively up to all four resistances can be substituted by SGs. Depending on the setup and number of SGs in a bridge circuit four cases can be distinguished (Figure A-13):

- Quarter bridge: 1 SG, 3 completion resistances
- Half bridge: 2 SGs on adjoining branches, 2 completion resistances. Commonly used for temperature compensation
- Double quarter bridge: 2 SGs on opposite sides of the bridge, 2 completion resistances
- Full bridge: 4 SGs

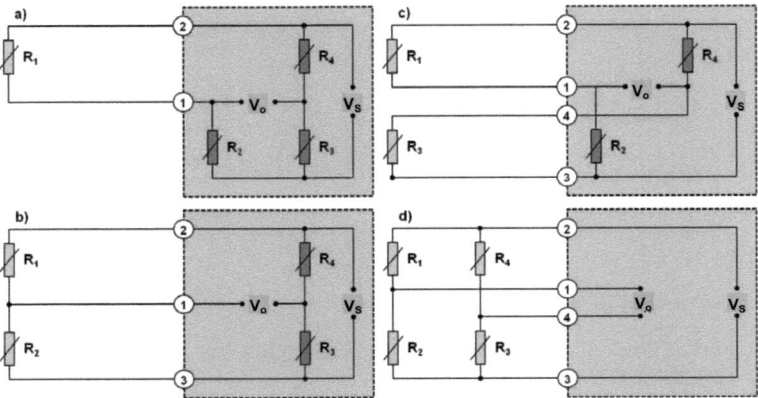

Figure A-13. Configurations of the Wheatstone bridge circuit: a) quarter bridge, b) half bridge, c) double quarter or diagonal bridge, d) full bridge.

The nominal value of all resistances should be equal in the unloaded state, so that the bridge is balanced in this state.

In this research the quarter and double-quarter bridge was used with resistances showing a nominal value of 120 Ω.

A.5 Calibration of the Wheatstone Bridge Circuit with SGs

When a Wheatstone bridge circuit consisting of one or more SGs is connected to the amplifier and all four resistances have the same nominal resistance, there will still be a minor unbalanced bridge because of small differences in the resistances. A certain output voltage will be recorded even though the object has not been subjected to loading so far. This tare or zero signal can be compensated by a zero balance function which should be standard in an amplifier. After the zero signal is removed, the measuring device has to be calibrated to get a clear relation between the recorded output voltage signal and the strain of the SG. For this task different methods can be used:

- Calibration signal from the amplifier
- Calibration device
- Shunt calibration

Some amplifiers provide a well-definied **calibration signal** that can be applied to the bridge circuit. The calibration signal can be given in μm/m or mV/V.

An external **calibration device** is attached to the measuring bridge. The device simulates a certain strain by changing the resistance of the circuit. By knowing the excitation voltage, the resistance of the circuit and the gauge factor the circuit can be calibrated.

Finally, the **shunt calibration** also works on the basis of virtual strain by adding a parallel shunt resistance R_P to the SG as shown in Figure A-14.

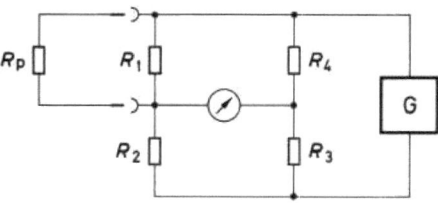

Figure A-14. Principle of the shunt calibration. (Hoffmann, 1987)

The shunt calibration is used in this research. The amplifier has an integrated shunt resistance with a known nominal value to calibrate the bridge. If a quarter bridge is assumed then $\Delta R_2/R_2$, $\Delta R_3/R_3$ and $\Delta R_4/R_4$ are 0. (A.9) results to

$$\frac{V_o}{V_S} = \frac{1}{4} \cdot \frac{\Delta R_1}{R} = \frac{1}{4} \cdot k \cdot \varepsilon \qquad (A.11)$$

For the shunt resistance the following applies:

$$R_P \| R_1 \rightarrow R_r \qquad (A.12)$$

$$R_r = \frac{R_1 \cdot R_P}{R_1 + R_P} \qquad (A.13)$$

$$\Delta R_1 = R_r - R_1 \qquad (A.14)$$

The relationships above state that the parallel connection of the shunt R_P to the SG R_1 results in a resistance R_r, and further a change in resistance ΔR_1. Coming from a balanced bridge circuit, this change in resistance leads to a measured output voltage $V_{o,p}$ and correlates with virtual strain ε_p. By taking into account the formulas above a calibration factor CF (only valid for the bridge excitation voltage V_S that is used during calibration) can be derived. This factor represents an unambiguous relation between a certain strain and the recorded voltage:

$$\varepsilon_p = \frac{1}{k} \cdot \frac{\Delta R_1}{R_1} = \frac{1}{k} \cdot \frac{-R_1}{R_1 + R_p} \qquad (A.15)$$

$$CF = \frac{\varepsilon_p}{V_{o,p}} \quad (A.16)$$

A short example: The nominal value of the resistances including the SG be 120 Ω, the gauge factor k of the SG 2.10 and the shunt resistance for calibration 59 kΩ. This setup leads to a virtual calibration strain ε_p of

$$\varepsilon_p = \frac{1}{2.10} \cdot \frac{-120}{120+59000} = -0.000966557 \quad (A.17)$$

If the output voltage for this ε_p is −0.644 V, the calibration factor is:

$$CF = \frac{-0.9665577}{-0.644} = 0.00150087 \; [1/V] \quad (A.18)$$

Any recorded output voltage $V_{o,i}$ can be transferred to the actual strain ε_i in a subsequent measurement by using the calibration factor.

A.6 Compensation of Disturbance Values

The principle of strain measurement with SG is simple, as it was shown above. There is a number of possible error sources and interference effects that need to be taken into consideration if applicable. Table A-2 presents an overview on these disturbance values.

Some of the interference effects can be removed by protecting or shielding the SG and measuring device from the source of interference, some effects have to be compensated by other means. For each of the listed effects in Table A-2 (Hoffmann, 1987) describes the reasons and possible compensations in detail. The interested reader may be referred to this source or any other basic literature on SGs. In the preparation to the research presented in this thesis and also throughout the process of the research possible interference factors were isolated and – if necessary – compensated. The most common factor is a temperature related error. Since the tests were carried out only at constant temperatures, a compensation method was not necessary. Also a compensation of too long cables which is also a common problem was not necessary since the length of the cables was kept clearly below 1 m. The cables were shielded against electro-magnet influences and the SGs were protected against water and moisture by coating them with a standard seal based on silicone gel.

Table A-2. Possible disturbance values impacting the recorded strain. (Hoffmann, 1987)

				Possible error sources and interference effects				
Mounting	Mechanical loading	Temperature	Cable effects	Exceeding the permissible limits	Pneumatic and hydraulic effects	Chemical effects	Radiating fields	Component properties
selection of the strain-gage type, fixing method, measuring point protection, quality of the mounting, soldering, insulation	shock, impact, compression, acceleration, fatigue loading	level, change, rate of change, active period	resistance, capacitance and symmetry of cable, surge impedance, insulation, screening	extensibility, temperature range, permissible load cycles	pressure, vacuum	humidity, water, chemicals, gasses, bacteria	neutron, gamma and X-ray radiation, electric and magnetic fields	inhomogeneities, anisotropic conditions, rheological properties

	Strain-gage measuring point consisting of				
Component material adhesive and welding properties	Strain-gage measuring grid material carrier material	Fixing agent organ. adhesive ceram. mounting agent, welded joint	Electrical connections joining leads, solder points, connection cable	Electrical circuit quarter bridge half bridge diagonal bridge full bridge	Protective agent covering agent screening armoring

B ANNEX: MODELING THE VISCOELASTIC BEHAVIOR OF BINDER AND MASTIC

B.1 Bitumen → Mastic

The main objective of this annex is the derivation of an analytical model that enables the user to scale up the material behavior of bitumen to the mastic level by taking into consideration the ratio r between binder and filler (binder:filler = 1:r). To reach this goal, a three step approach was taken:
- First of all, the test data from DSR with frequency and temperature sweep was analyzed and master curves for the stiffness parameters were derived and described by analytical functions (test data → master curve) at a T_{ref} of 30°C.
- In a second step, the master curves of bitumen and mastic were compared and the ratio between bitumen and mastic stiffness ($S_{mas,j}/S_{bit}$) was calculated analytically for each filler content (master curve → stiffness ratio). This step can be seen as an auxiliary towards the model.
- Finally, the functions describing $S_{mas,j}/S_{bit}$ for each filler content are used to derive the *B-M Model*, which enables the user to obtain all viscoelastic material parameters for a mastic with any arbitrary filler content for the complete frequency range of the master curve. The model parameters have to be obtained for each binder type. (stiffness ratio → *B-M Model*).

Test Data → Master Curve

DSR tests were carried out according to the test program on pure binders and mastic consisting of different ratios r (binder:filler = 1:r) from 0.5 to 1.9. Master curves were derived for the dynamic shear modulus $|G^*|$ and its storage and loss parts G_1 and G_2 for bitumen and mastic. The master curves were approximated by means of regression analysis and the fit quality was described statistically. One example of this procedure can be seen in Figure B-1 for the polymer-modified binder PmB 25/55-65. From the results shown in Figure B-1 it can be stated that one boundary in terms of test conditions (-10°C and 30 Hz) is near the upper asymptote of the material stiffness, whereas the other boundary (60°C and 0.1 Hz) leaves the material still in the transition phase a certain distance apart from the lower asymptote.

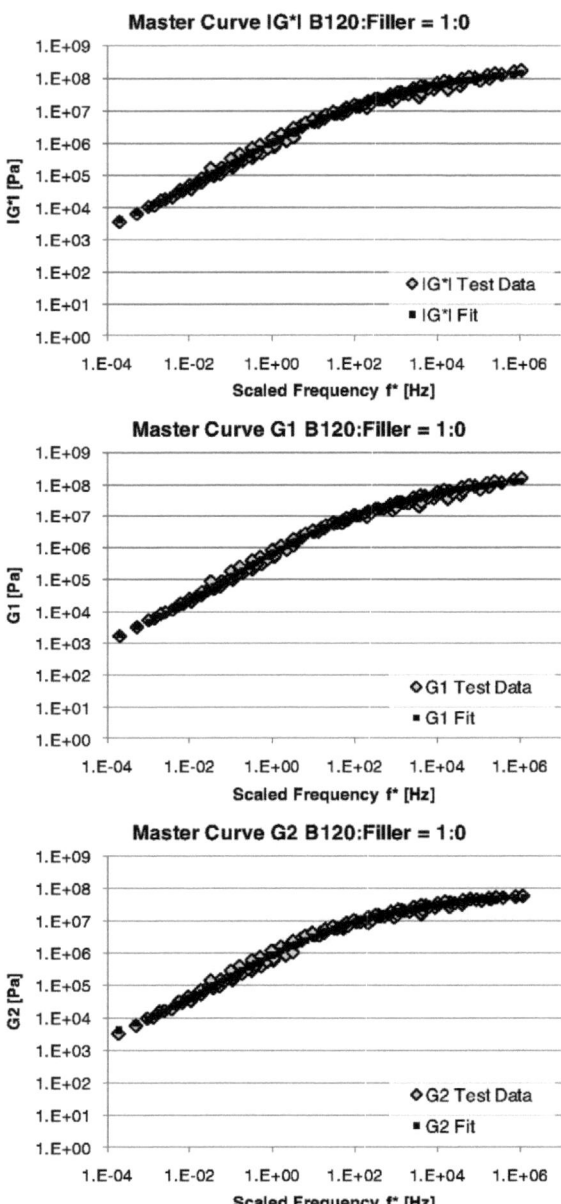

Figure B-1. Master curves from test data and regression analysis for PmB 25/55-65 ($|G^*|$ - top, G_1 – middle and G_2 – bottom).

Master Curve → Stiffness Ratio (Functional Relation)

To derive up-scaling relations between the bitumen level and the mastic, the stiffness ratio between mastic and bitumen $S_{mas,j}/S_{bit}$ was analyzed separately for each filler ratio j. In Figure B-2 three diagrams show this analysis for the dynamic shear modulus $|G^*|$, G_1 and G_2 again for the PmB and its mastics. A value of 1.0 in this diagram would indicate that the mastic shows the same stiffness as the binder at this point.

Looking at the different filler contents it is obvious and logic that a larger filler content leads to a higher stiffness in all cases, although a ratio r of 0.5 does not indicate a significant increase in stiffness. More interesting is the evolution of the stiffness in temperature/frequency domain. At high and low temperatures or frequencies the difference between binder and mastic is large (an r of 1.9 leads to a 5 times higher $|G^*|$). At intermediate temperatures (e.g. for 30°C or scaled frequencies ranging from 1 Hz to about 100 Hz) the increase in stiffness due to filler drops to a minimum. An r of 1.9 only produces about 2.5 times the $|G^*|$ of the pure binder at this point. Also the difference between various filler contents decreases. At these conditions it hardly matters whether r is 1.0 or 1.7. It can therefore be concluded that filler stiffens a binder to a far larger extent at high and low temperatures/frequencies. In between those two boundaries there is an area where the amount of filler does not influence the stiffness to the same extent. The potential for increasing stiffness in this domain seems to be limited.

The picture for the elastic part of the complex shear modulus G_1 is similar to the dynamic shear modulus for high to intermediate temperatures. When the temperature drops clearly below 30°C or the scaled frequency is raised above about 1,000 Hz for the reference temperature of 30°C, the stiffness increase due to added filler is stronger than for $|G^*|$ especially for high filler contents. The dynamic shear modulus increases about 4 to 5 times for an r of 1.7 to 1.9 at high frequencies whereas the elastic part increases about 6 to 7 times. This shows that adding filler especially accounts for a more elastic behavior at high frequencies or lower temperatures making it also more prone to low temperature cracking.

The diagram for the viscous part G_2 confirms the statements made above. Filler does not account for more elastic behavior at higher temperatures rather than at lower temperatures.

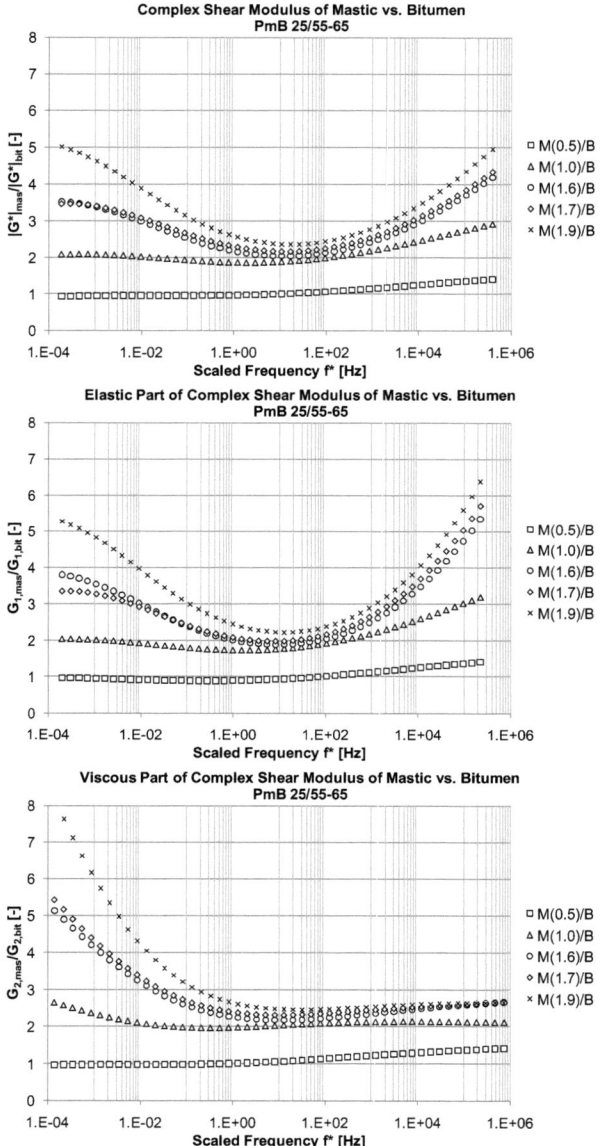

Figure B-2. Stiffness ratio of the mastic compared to the binder PmB 25/55-65 for |G*| (top), G_1 (middle), G_2 (bottom).

Figure B-3 contains the same analysis for the unmodified binder 70/100 and mastic with filler ratios r from 0.5 to 1.6. The top diagram shows the dynamic shear modulus $|G^*|$. Again, a higher filler content results in a stiffer behavior throughout the frequency range. But different from the modified binder, in this case a filler ratio of 0.5, doubles the stiffness. This filler content does not lead to significant stiffening for the modified binder. The amount of filler added to the bitumen does not affect the stiffening process to a high extent above a certain threshold level. Whether the filler ratio is 1.3, 1.4 or 1.6, the mastic is around 4 to 5 times stiffer at low frequencies or high temperatures and between 3 to 4 times stiffer at intermediate and higher frequencies. Taking into consideration the frequency domain, there is again a maximum of stiffening at low frequencies around 0.01 Hz but not like for the modified binder at the lowest measured frequency. There is also no dominant minimum of stiffening in the frequency domain. The stiffening drops down to a minimum of around 3 times the stiffness of the binder for the three higher filler ratios and slightly increases at the end of the curve. The frequency dependency seems to be more dominant the higher the filler content gets. In terms of $|G^*|$, the stiffening for the mastic is most effective at lower frequencies up to 1 Hz.

The situation for the elastic part G_1 is depicted in the middle diagram. From this and also from the analysis of the viscous part G_2 (lower diagram), it can be stated that adding filler does not account for more elastic behavior at high temperatures (or low frequencies), only from about 1 Hz on, the elastic part gets more dominant. Thus, the mastic gets stiffer but also more viscous in the higher temperatures range and stiffer and more elastic in the lower temperature domain.

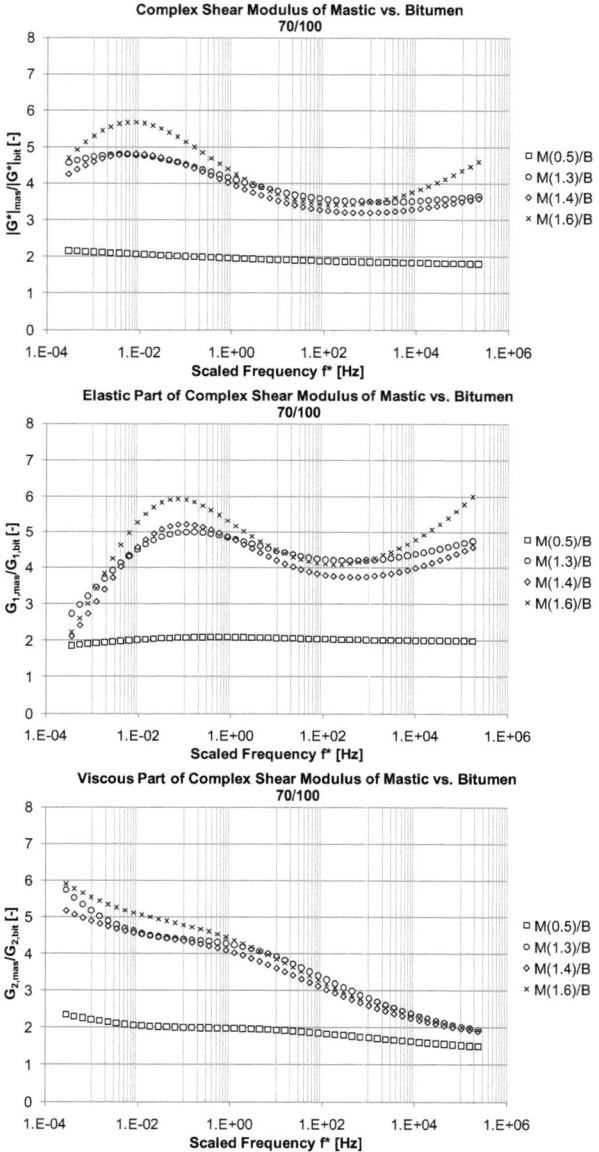

Figure B-3. Stiffness ratio of the mastic compared to the binder 70/100 for $|G^*|$ (top), G_1 (middle), G_2 (bottom).

The relation $S_{mas,j}/S_{bit}$ for each tested bitumen versus scaled frequency f^* can be described by a polynomial of 4th order in the following way:

$$\frac{S_{mas,j}}{S_{bit}}(f^*) = \sum_{i=0}^{4} a_{b-mj,i} \cdot \ln(f^*)^i \qquad (B.1)$$

$S_{mas,j}$......Stiffness parameter of mastic with ratio binder:mastic = 1:j, j>0
S_{bit}.........Stiffness parameter of binder
f^*...........Scaled frequency
$a_{b-mj,i}$.....Parameter of polynomial describing the relation between bitumen and mastic

For each mastic master curve, a set of five parameters $a_{b-mj,i}$ can be derived. Since the stiffness ratio $S_{mas,j}/S_{bit}$ is the relation between two analytical functions (i.e. the master curve of bitumen and mastic), the relation itself could be described and not only approximated by an analytical function. Still, it was found that the optimum in terms of fit quality vs. number of parameters can be reached if a polynomial of 4th order is used.

Since the approach taken in this annex was successful for $|G^*|$ and G_1, the analysis in the following is limited to those two parameters. If all three stiffness parameters were derived from the model, the received data would be over-determined since one of the stiffness parameters can always be calculated if the other two are given (see section V.1.2 for details). Thus, the derivation of only two stiffness values is perfectly sufficient to obtain all viscoelastic parameters. Table B-1 shows the parameters $a_{b-mj,i}$ for the PmB, Table B-2 gives the numbers for $a_{b-mj,i}$ for the standard bitumen 70/100. The coefficient of correlation R^2 of these approximations is always around or slightly below 1.0. By means of the functional relation shown in (B.1), the viscoelastic parameters of a mastic can be calculated for any arbitrary frequency f^* from the parameters of the respective binder. But this function has to be fitted for each filler content. The benefit of this relationship is therefore limited.

Table B-1. Parameters for function (B.1) describing the relation between a certain mastics M(j) with binder:mastic = 1:j and the respective binder PmB 25/55-65 for $|G^*|$ (top) and G_1 (bottom).

$|G^*|$	M(0)	M(0.5)	M(1.0)	M(1.6)	M(1.7)	M(1.9)
$a_{b-mj,0}$	1.00000E+00	9.70810E-01	1.86776E+00	2.19662E+00	2.31887E+00	2.62103E+00
$a_{b-mj,1}$	0.00000E+00	9.77778E-03	-1.20572E-02	-1.13927E-01	-1.14454E-01	-1.77708E-01
$a_{b-mj,2}$	0.00000E+00	2.16847E-03	7.06748E-03	1.74890E-02	1.66212E-02	2.56616E-02
$a_{b-mj,3}$	0.00000E+00	1.02989E-04	4.01088E-04	9.60738E-04	1.11481E-03	9.59038E-04
$a_{b-mj,4}$	0.00000E+00	-9.56145E-06	-2.99615E-05	-5.54356E-05	-6.13023E-05	-6.31240E-05
R^2	1.000	1.000	1.000	0.999	0.999	0.999

G_1	M(0)	M(0.5)	M(1.0)	M(1.6)	M(1.7)	M(1.9)
$a_{b-mj,0}$	1.00000E+00	8.97104E-01	1.73343E+00	2.03924E+00	2.11180E+00	2.49296E+00
$a_{b-mj,1}$	0.00000E+00	9.16996E-03	-1.04257E-02	-1.28625E-01	-1.12620E-01	-1.98578E-01
$a_{b-mj,2}$	0.00000E+00	3.43853E-03	9.53346E-03	2.23263E-02	1.96945E-02	3.04052E-02
$a_{b-mj,3}$	0.00000E+00	6.67774E-05	4.83740E-04	1.17555E-03	1.48330E-03	1.32726E-03
$a_{b-mj,4}$	0.00000E+00	-1.07177E-05	-3.30095E-05	-3.16744E-05	-3.51587E-05	-3.58928E-05
R^2	1.000	1.000	1.000	0.999	0.999	0.999

Table B-2. Parameters for function (B.1) describing the relation between a certain mastics M(j) with binder:mastic = 1:j and the respective binder 70/100 for |G*| (top) and G₁ (bottom).

|G*|	M(0)	M(0.5)	M(1.3)	M(1.4)	M(1.6)
$a_{b-mj,0}$	1.00000E+00	1.95810E+00	4.14424E+00	3.99602E+00	4.38255E+00
$a_{b-mj,1}$	0.00000E+00	-1.92155E-02	-1.62297E-01	-2.18474E-01	-3.28051E-01
$a_{b-mj,2}$	0.00000E+00	6.16856E-04	3.11819E-03	4.01232E-03	1.03074E-02
$a_{b-mj,3}$	0.00000E+00	-4.54251E-06	1.47288E-03	2.39693E-03	3.98604E-03
$a_{b-mj,4}$	0.00000E+00	-1.62961E-07	-7.51826E-05	-1.23033E-04	-2.09851E-04
R^2	1.000	1.000	1.000	0.999	0.998

G_1	M(0)	M(0.5)	M(1.3)	M(1.4)	M(1.6)
$a_{b-mj,0}$	1.00000E+00	2.08712E+00	4.79132E+00	4.78219E+00	5.25987E+00
$a_{b-mj,1}$	0.00000E+00	-1.91083E-03	-1.03114E-01	-1.96550E-01	-2.82260E-01
$a_{b-mj,2}$	0.00000E+00	-2.33323E-03	-1.54347E-02	-2.08436E-02	-1.81331E-02
$a_{b-mj,3}$	0.00000E+00	1.94672E-04	3.32636E-03	5.13731E-03	7.05078E-03
$a_{b-mj,4}$	0.00000E+00	-2.75339E-06	-1.10474E-04	-1.77993E-04	-2.62585E-04
R^2	1.000	0.995	0.983	0.975	0.977

An interesting question is, how well the stiffness parameters of the different tested mastics are approximated by the functional relationship in (B.1) with the fitted parameters from Table B-1 and Table B-2. To investigate this matter, $S_{mas,MC}$ from the mastic master curves (which were directly derived from the test data) were compared to $S_{mas,SR}$, which represent the calculated mastic stiffness parameters from the function (B.1) and the stiffness master curve of the binder $S_{bit,MC}$:

$$S_{mas_j,func}(f^*) = \left[\sum_{i=0}^{4} a_{b-m_j,i} \cdot \ln(f^*)^i\right] \cdot S_{bit,MC} \qquad (B.2)$$

In a next step the ratio $S_{mas,MC}/S_{mas,func}$ was calculated for the complete frequency range and the 2.5% and 97.5% quantiles as well as the median value were computed for this ratio. A value for the ratio of 1.0 would indicate that the derived stiffness from the functional relationship perfectly matches the stiffness from the respective mastic master curve. From the two quantile values the 95% confidence interval can be obtained.

Figure B-4 shows results of this analysis for the PmB 25/55-65 mastics. It becomes obvious that the deviation between the calculated values and the master curve values gets larger with larger amount of filler. But for both cases the dynamic shear modulus and its elastic part the 95% confidence interval is very small being largest for an r of 1.9 and G_1 ranging from 0.98 and 1.02. Thus in 95 out of 100 cases the difference between calculated values and the master curve is below ±2%.

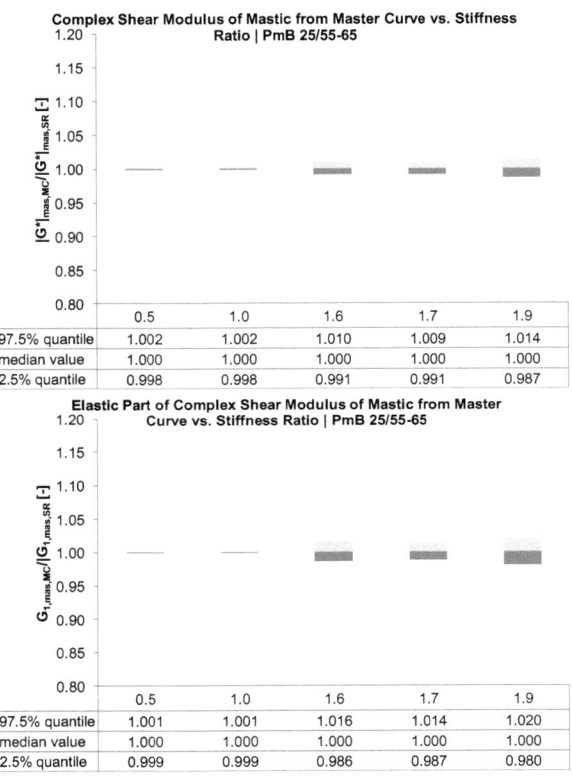

Figure B-4. Comparison of $|G^*|$ (top) and G_1 (bottom) of master curve and stiffness ratio according to (B.1) for PmB 25/55-65 mastic.

The same analysis for the standard bitumen 70/100 mastics is depicted in Figure B-5. Especially for G_1 the deviations are slightly higher. At a filler ratio r of 1.6, the 95% confidence interval ranges from 0.956 to 1.067, so the error runs between -4.4% and +6.7%. Still, these numbers promise a very good approximation.

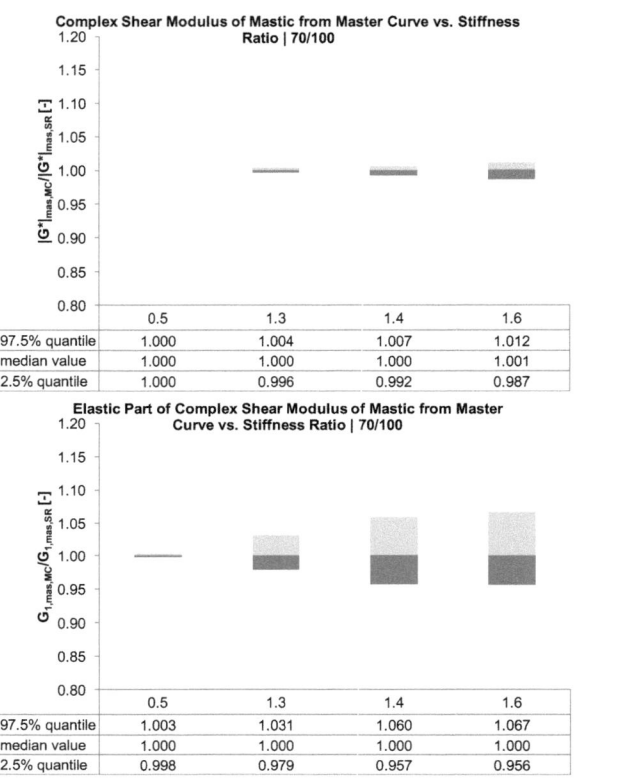

Figure B-5. Comparison of $|G^*|$ (top) and G_1 (bottom) of master curve and stiffness ratio according to (B.1) for 70/100 mastic.

Stiffness Ratio → B-M Model

To enhance the benefit of the data analysis, a further attempt was made to generalize the functional relationship stated in (B.1) to a model, which would allow the user to derive all viscoelastic parameters of a mastic with an arbitrary filler ratio for the complete frequency range directly from the viscoelastic parameters of a DSR-tested bitumen.

To manage this, possible connections between the five parameters $a_{b-mj,i}$ and the filler ratio r were taken into account. It was found, that each of the five $a_{b-mj,i}$ can be represented by the following function:

$$\hat{a}_{b-m,i}(r) = b_{b-m,i} \cdot r^{c_{b-m,i}} \tag{B.3}$$

Each parameter $a_{b-mj,i}$ is connected to the filler ratio r by a power function with two parameters $b_{b-m,i}$ and $c_{b-m,i}$ and can therefore be described by an analytical $\hat{a}_{b-m,i}$. The two parameters were derived for both mastic types and are summarized in Table B-3. The coefficient of correlation is clearly above 0.8 for all cases. To visualize what a lower R^2 and a high R^2 means graphically Figure B-6 shows two diagrams, the one on the top for

$\hat{a}_{b\text{-}m,1}$ for G_1 of the PmB 25/55-65 mastic with a low R^2 (0.821) and the lower diagram for $\hat{a}_{b\text{-}m,2}$ for $|G^*|$ of the PmB 25/55-65 mastic with a high R^2 (0.992).

Table B-3. Parameters for function (B.3) for PmB 25/55-65 mastic (left) and 70/100 mastic (right).

	PmB 25/55-65			70/100	
	$|G^*|$	G_1		$|G^*|$	G_1
$b_{b\text{-}m,0}$	1.65608E+00	1.53542E+00	$b_{b\text{-}m,0}$	3.22991E+00	3.69002E+00
$c_{b\text{-}m,0}$	6.99883E-01	7.06920E-01	$c_{b\text{-}m,0}$	7.10157E-01	8.12028E-01
R^2_0	0.967	0.963	R^2_0	0.987	0.994
$b_{b\text{-}m,1}$	-3.14342E-02	-3.06845E-02	$b_{b\text{-}m,1}$	-9.78046E-02	-3.82783E-02
$c_{b\text{-}m,1}$	2.30283E+00	2.43699E+00	$c_{b\text{-}m,1}$	2.37719E+00	4.33223E+00
R^2_1	0.841	0.821	R^2_1	0.995	0.997
$b_{b\text{-}m,2}$	7.31511E-03	9.96926E-03	$b_{b\text{-}m,2}$	2.45049E-03	-9.02684E-03
$c_{b\text{-}m,2}$	1.78762E+00	1.57271E+00	$c_{b\text{-}m,2}$	2.10601E+00	1.91073E+00
R^2_2	0.992	0.986	R^2_2	0.915	0.975
$b_{b\text{-}m,3}$	3.77595E-04	3.82169E-04	$b_{b\text{-}m,3}$	2.87655E-04	1.64611E-03
$c_{b\text{-}m,3}$	1.80100E+00	2.33424E+00	$c_{b\text{-}m,3}$	5.94607E+00	3.09145E+00
R^2_3	0.980	0.978	R^2_3	0.998	0.997
$b_{b\text{-}m,4}$	-2.74230E-05	-2.28782E-05	$b_{b\text{-}m,4}$	-1.31376E-05	-4.23858E-05
$c_{b\text{-}m,4}$	1.45037E+00	8.57102E-01	$c_{b\text{-}m,4}$	6.28865E+00	3.94652E+00
R^2_4	0.992	0.834	R^2_4	0.998	0.999

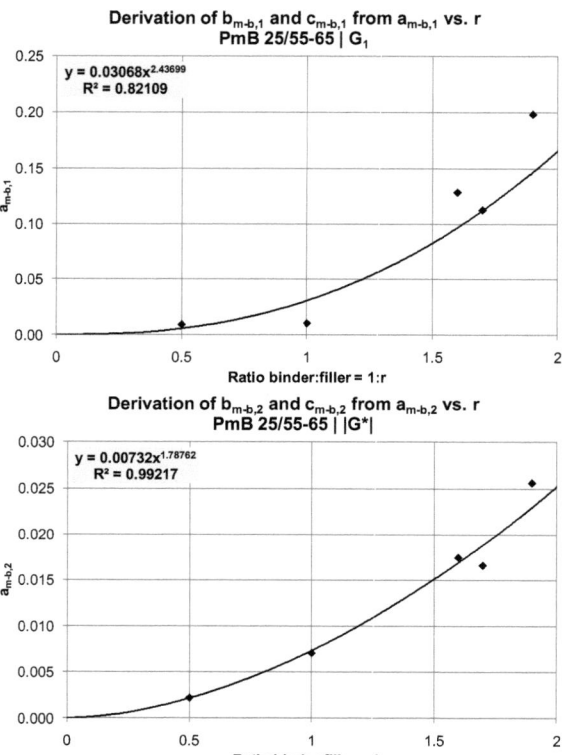

Figure B-6. Graphical representation of (B.3) for two cases: $a_{m\text{-}b,1}$ for G_1 of PmB 25/55-65 mastic (lower R^2) and $a_{m\text{-}b,2}$ for $|G^*|$ of PmB 25/55-65 mastic (high R^2).

Taking into consideration (B.3) and inserting it into (B.1), the *B-M Model* is achieved, which is presented in (B.4):

$$\frac{S_{mas}}{S_{bit}}(r, f^*) = \sum_{i=0}^{4} b_{b-m,i} \cdot r^{c_{b-m,i}} \cdot \ln(f^*)^i \qquad (B.4)$$

r .... Filler ratio (binder:filler = 1:r)
$b_{b\text{-}m,i}/c_{b\text{-}m,i}$ .... Model parameters (i=0 to 4)

With the *B-M Model*, the viscoelastic parameters of a mastic with an arbitrary filler ratio can be derived from the viscoelastic parameters of the respective binder for the complete frequency/temperature range. The ten model parameters are independent of filler ratio and frequency, but have to be calibrated for each binder type. One might argue that the disadvantage of this model is, that there is no physically explainable connection between the parameters and the material. However the major practical advantage is that the viscoelastic parameters can be derived for the complete range of frequencies or temperatures from down to -20°C up to +80°C.

Prediction Quality of B-M Model

It is important to know how well a model fits the reality, in this case how the predicted stiffness parameters from the model coincide with the stiffness parameters from the master curves $S_{mas,mod}$, which are directly related to the test data $S_{max,MC}$. The stiffness parameters of the model were derived from (B.4) using the master curve data of the binder:

$$S_{mas,mod}\left(r,f^*\right)=\left[\sum_{i=0}^{4}b_{b-m,i}\cdot r^{c_{b-m,i}}\cdot \ln\left(f^*\right)^i\right]\cdot S_{bit,MC} \qquad (B.5)$$

The ratio between the stiffness value $S_{mas,mod}/S_{mas,MC}$ was then calculated for the complete frequency range and the 2.5% and 97.5% quantiles, as well as median value were derived. This procedure was carried out for mastic from both binder types. For the mastic with the PmB, the results are presented in Figure B-7. The top diagram shows data for the dynamic shear modulus $/G^*/$, the lower diagram of the elastic part G_1. The bars in the diagrams represent the 95% confidence interval of the ratio. Again, the closer the value to 1.0, the better is the approximation of the model to the test data.

For example, for the mastic with a filler ratio of 0.5, the model predicts $/G^*/$ with a deviation to the master curve from test data between -11.5% and +18.2% for 95% of all cases. The diagrams show that the deviation lies within the ±20% range for all analyzed mastics. If it is taken into consideration that the measured stiffness values from the DSR tests run from some 10^3 Pa to around 10^8 Pa, therefore cover more than 4 decades, the deviations produced by the model can be seen as reasonable.

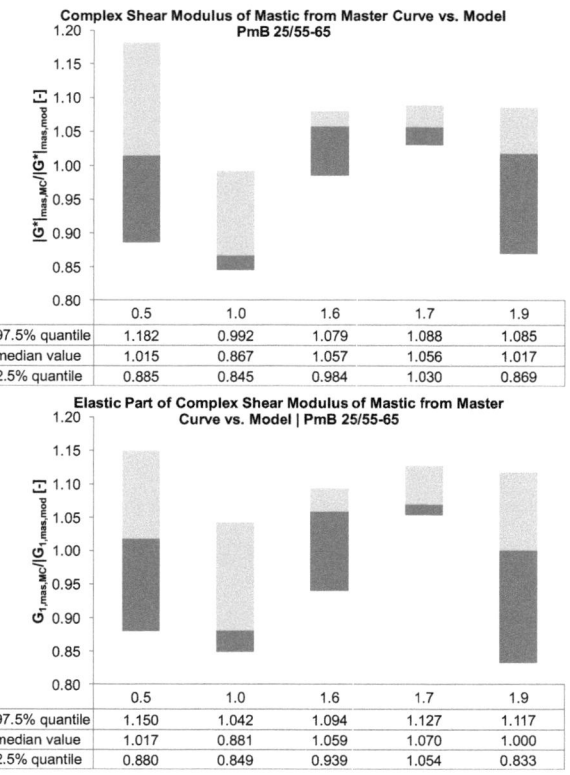

Figure B-7. Comparison of $|G^*|$ (top) and G_1 (bottom) from mastic master curve and *B-M Model* for PmB 25/55-65 mastic.

Figure B-8 shows analogue data for the 70/100 mastics. Interestingly enough the deviations of the model tend to be smaller for this bitumen type. The worst approximation can be found at a filler ratio of 1.6 with an error between -15.2% and +9.8% within the 95% confidence interval for $|G^*|$.

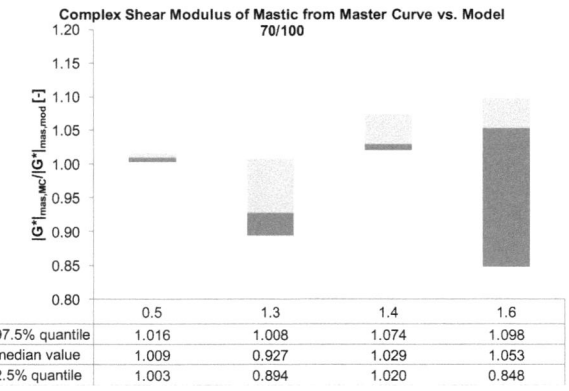

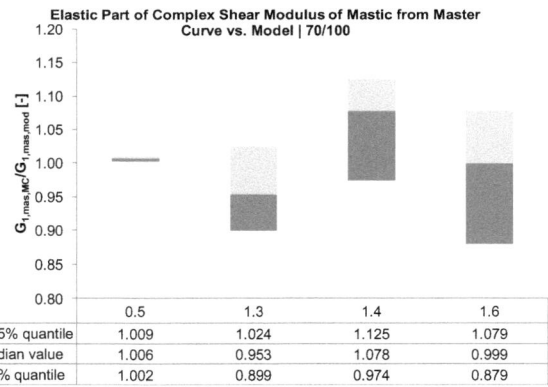

Figure B-8. Comparison of $|G^*|$ (top) and G_1 (bottom) from master curve and *B-M Model* for 70/100 mastic.

Yet another investigation was accomplished to bring the raw DSR test data into the picture. The top diagram in Figure B-9 compares the dynamic shear modulus $|G^*|$ of the PmB mastics directly from the DSR tests with the data from the master curve regression. Again the 95% confidence interval was derived. In 95 out of 100 cases the master curve values deviate from the actual tests data between around -15% to -20% and +20 to +30%.

In the lower diagram, the dynamic shear modulus of the DSR test data was compared to the modulus derived from the *B-M Model*. The deviations in the 95% confidence interval range between -30% to -10% and +18% to +40%. From this example it can be stated that the presented *B-M Model* predicts the viscoelastic parameters of the mastic only slightly worse than the analytically derived master curve. Since master curves have been used for the representation of material stiffness in numerous publications and can therefore be called state of the art and a benchmark, the *B-M Model* reaches this benchmark in terms of prediction quality.

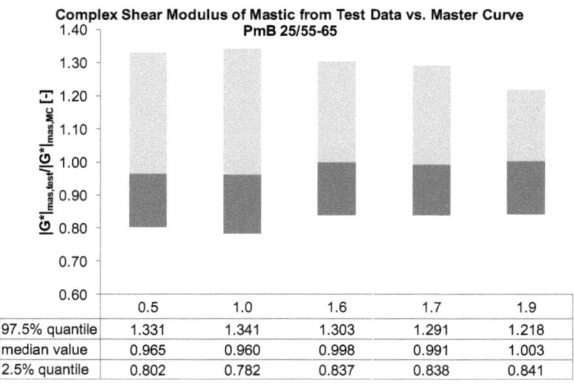

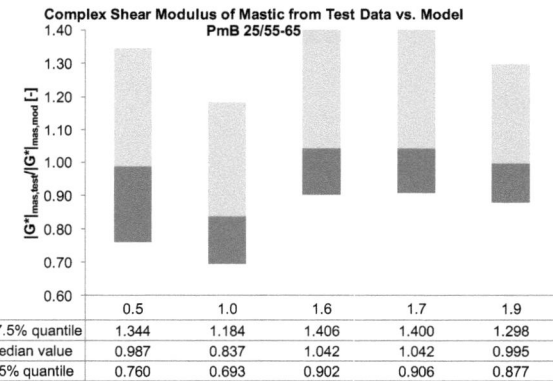

Figure B-9. Comparison of $|G^*|$ from test data and master curve (top) and test data and *B-M Model* (bottom) for PmB 25/55-65 mastic.

Sensitivity Analysis for B-M Model Parameters

As a last step, a sensitivity analysis for the *B-M Model* parameters was conducted. Therefore, the combination of $b_{b-m,i}$ and $c_{b-m,i}$ as shown in (B.3) (= $\hat{a}_{b-m,i}$) for $|G^*|$ of the PmB 25/55-65 mastic with a filler ratio of 1.9 was varied between 50% and 150% of the actual value $\hat{a}_{b-m,i}$. The ratio of the derived stiffness with the deviated model parameters $\hat{a}_{b-m,i,dev}$ to the stiffness with the actual model parameters was computed and the 95% confidence interval of this ratio was derived. A ratio of 1.0 would mean that the stiffness from deviated and actual model parameters coincides.

Figure B-10 shows the graphical presentation of the results for each of the five parameters sets $\hat{a}_{b-m,i}$ for *i* from 0 to 4. The grey filled area indicates the 95% confidence interval of the ratio of stiffness with deviated and actual model parameters. In all cases, the error caused by deviated model parameters is linear with the most severe impact for $\hat{a}_{b-m,0}$. For example, if this parameter is 10% below the actual parameter, then the derived stiffness would range between 89% and 95% of the actual stiffness value. In this case, the results would show a certain offset and scattering compared to the actual stiffness values. In another case, for example $\hat{a}_{b-m,1}$, if the parameters are 30% above the actual model parameters, the derived results would range between 90% and 108% of the actual model stiffness. In this case there would be hardly any offset but a scattering of results compared to the actual model parameters.

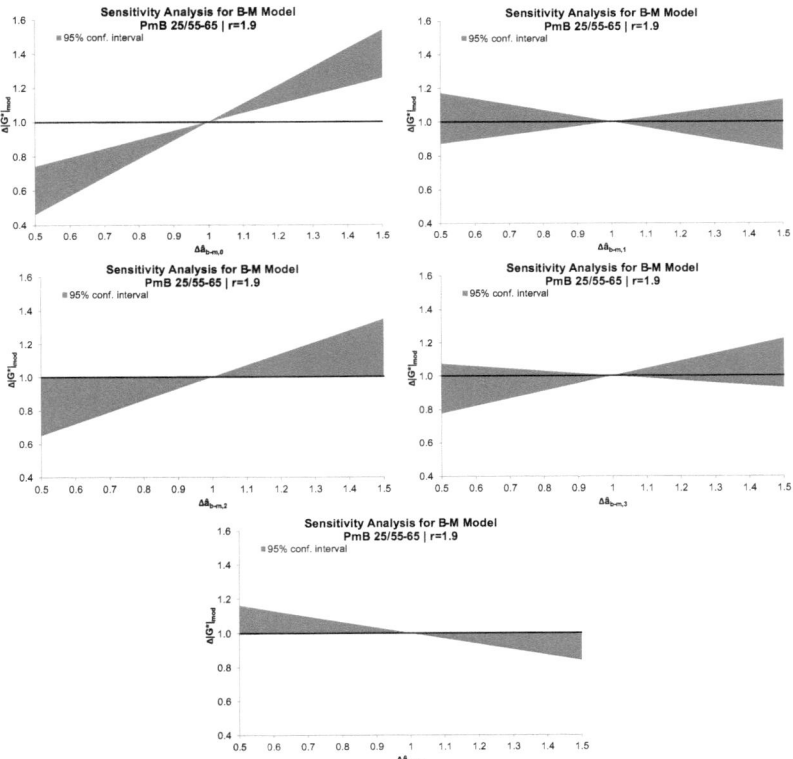

Figure B-10. Sensitivity analysis for the *B-M Model* parameters for $|G^*|$ of a PmB 25/55-65 mastic with a filler ratio of 1.9.

Summing up, it can be stated that deviations from $\hat{a}_{b-m,0}$ have a dominant effect on the model results, the same can be said about $\hat{a}_{b-m,2}$. Thus, these parameters must be fit with sufficient quality to ensure reliable modeling results. Deviations from $\hat{a}_{b-m,1}$, as well as from $\hat{a}_{b-m,3}$ show a lower effect, followed by $\hat{a}_{b-m,4}$.

Conclusions

From the findings of this annex, it can be stated that by following a three step approach an analytical model was derived successfully connecting the material behavior of the binder with the behavior of the mastic. The *B-M Model* consists of ten parameters that can easily be fitted from dynamic test data, if the data allows that master curves are obtained. To ensure this, the tests should cover a large range of frequencies and test temperatures. Model parameters were fitted for two binders and the respective mastics with different filler ratios, an unmodified bitumen 70/100 and a polymer modified binder PmB 25/55-65. For both materials the stiffness prediction by the model was successful and an in-depth study proved that the approximation quality of the model is on the same level as the fit quality of a standard master curve.

The model might not present physically based parameters but it achieves to predict the viscoelastic material behavior of the mastic with arbitrary filler contents over the complete frequency/temperature range with good quality in all cases investigated herein. A sensitivity analysis of the model parameters was also conducted. It shows that moderate deviations from the actual model parameters result only in moderate deviations of the derived stiffness. The linear behavior ensures that deviations from the actual model parameters do not result in progressive errors in the derived stiffness. The model can therefore be characterized as sound and stable.

As a next step in this research in the future, further DSR tests should cover an even larger temperature range in the high domain up to +80°C to reach the lower asymptote of the stiffness. This could enhance the model quality even further. If more bitumen types were tested in a next test program, correlations between the model parameters and other bitumen parameters (e.g. the penetration index,...) could be derived and the model could be generalized to a further extent. A distinction between unmodified and modified binders seems to be reasonable in a future analysis described above.

i want morebooks!

Buy your books fast and straightforward online - at one of world's fastest growing online book stores! Environmentally sound due to Print-on-Demand technologies.

Buy your books online at
www.get-morebooks.com

Kaufen Sie Ihre Bücher schnell und unkompliziert online – auf einer der am schnellsten wachsenden Buchhandelsplattformen weltweit! Dank Print-On-Demand umwelt- und ressourcenschonend produziert.

Bücher schneller online kaufen
www.morebooks.de

VDM Verlagsservicegesellschaft mbH
Heinrich-Böcking-Str. 6-8   Telefon: +49 681 3720 174   info@vdm-vsg.de
D - 66121 Saarbrücken   Telefax: +49 681 3720 1749   www.vdm-vsg.de

Printed by Books on Demand GmbH, Norderstedt / Germany